Cluster Analysis for Data Mining and System Identification

János Abonyi
Balázs Feil

Birkhäuser
Basel · Boston · Berlin

Authors:

János Abonyi
University of Pannonia
Department of Process Engineering
PO Box 158
8200 Veszprem
Hungary

Balázs Feil
University of Pannonia
Department of Process Engineering
PO Box 158
8200 Veszprem
Hungary

2000 Mathematical Subject Classification: Primary 62H30, 91C20; Secondary 62Pxx, 65C60

Library of Congress Control Number: 2007927685

Bibliographic information published by Die Deutsche Bibliothek:
Die Deutsche Bibliothek lists this publication in the Deutsche Nationalbibliographie; detailed bibliographic data is available in the internet at <http://dnb.ddb.de>

ISBN 978-3-7643-7987-2 Birkhäuser Verlag AG, Basel · Boston · Berlin

Basel · Boston · Berlin
P.O. Box 133, CH-4010 Basel, Switzerland
Part of Springer Science+Business Media
Printed on acid-free paper produced from chlorine-free pulp. TCF ∞
Cover design: Alexander Faust, Basel, Switzerland
Printed in Germany

ISBN 978-3-7643-7987-2

e-ISBN 978-3-7643-7988-9

9 8 7 6 5 4 3 2 1

www.birkhauser.ch

Contents

Preface

Data clustering is a common technique for statistical data analysis, which is used in many fields, including machine learning, data mining, pattern recognition, image analysis and bioinformatics. Clustering is the classification of similar objects into different groups, or more precisely, the partitioning of a data set into subsets (clusters), so that the data in each subset (ideally) share some common trait – often proximity according to some defined distance measure.

The aim of this book is to illustrate that advanced fuzzy clustering algorithms can be used not only for partitioning of the data, but it can be used for visualization, regression, classification and time-series analysis, hence fuzzy cluster analysis is a good approach to solve complex data mining and system identification problems.

Overview

In the last decade the amount of the stored data has rapidly increased related to almost all areas of life. The most recent survey was given by Berkeley University of California about the amount of data. According to that, data produced in 2002 and stored in pressed media, films and electronics devices only are about 5 exabytes. For comparison, if all the 17 million volumes of Library of Congress of the United States of America were digitalized, it would be about 136 terabytes. Hence, 5 exabytes is about 37,000 Library of Congress. If this data mass is projected into 6.3 billion inhabitants of the Earth, then it roughly means that each contemporary generates 800 megabytes of data every year. It is interesting to compare this amount with Shakespeare's life-work, which can be stored even in 5 megabytes. It is because the tools that make it possible have been developing in an impressive way, consider, e.g., the development of measuring tools and data collectors in production units, and their support information systems. This progress has been induced by the fact that systems are often been used in engineering or financial-business practice that we do not know in depth and we need more information about them. This lack of knowledge should be compensated by the mass of the stored data that is available nowadays. It can also be the case that the causality is reversed: the available data have induced the need to process and use them, e.g., web mining. The data reflect the behavior of the analyzed system, therefore there is at least the theoretical potential to obtain useful information and knowledge from data. On the ground of that need and potential a distinct science field grew up using many tools and results of other science fields: *data mining* or more general, *knowledge discovery in databases.*

Historically the notion of finding useful patterns in data has been given a variety of names including data mining, knowledge extraction, information discovery, and data pattern recognition. The term data mining has been mostly used by statisticians, data analysts, and the management information systems communities. The term knowledge discovery in databases (KDD) refers to the overall process of discovering knowledge from data, while data mining refers to a particular step of this process. Data mining is the application of specific algorithms for extracting patterns from data. The additional steps in the KDD process, such as data selection, data cleaning, incorporating appropriate prior knowledge, and proper interpretation of the results are essential to ensure that useful knowledge is derived form the data. Brachman and Anand give a practical view of the KDD process emphasizing the interactive nature of the process [51]. Here we broadly outline some of its basic steps depicted in Figure 1.

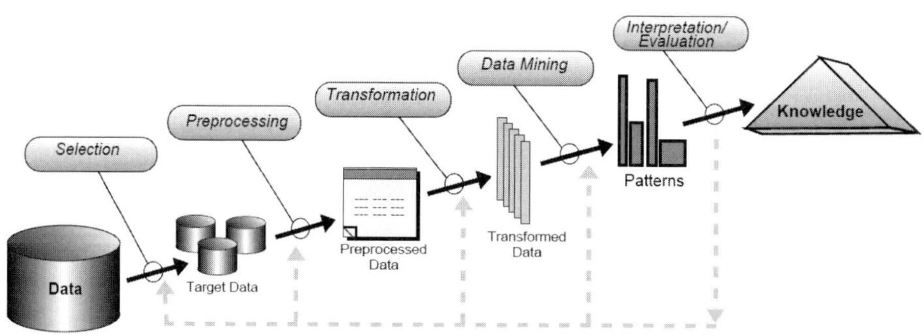

Figure 1: Steps of the knowledge discovery process.

1. *Developing and understanding of the application domain and the relevant prior knowledge, and identifying the goal of the KDD process.* This initial phase focuses on understanding the project objectives and requirements from a business perspective, then converting this knowledge into a data mining problem definition and a preliminary plan designed to achieve the objectives. The first objective of the data analyst is to thoroughly understand, from a business perspective, what the client really wants to accomplish. A business goal states objectives in business terminology. A data mining goal states project objectives in technical terms. For example, the business goal might be "Increase catalog sales to existing customers". A data mining goal might be "Predict how many widgets a customer will buy, given their purchases over

the past three years, demographic information (age, salary, city, etc.) and the price of the item." Hence, the prediction performance and the understanding of the hidden phenomenon are important as well. To understand a system, the system model should be as transparent as possible. The model transparency allows the user to effectively combine different types of information, namely linguistic knowledge, first-principle knowledge and information from data.

2. *Creating target data set.* This phase starts with an initial data collection and proceeds with activities in order to get familiar with the data, to identify data quality problems, to discover first insights into the data or to detect interesting subsets to form hypotheses for hidden information.

3. *Data cleaning and preprocessing.* The data preparation phase covers all activities to construct the final dataset (data that will be fed into the modelling tool(s)) from the initial raw data. Data preparation tasks are likely to be performed multiple times and not in any prescribed order. Tasks include table, record and attribute selection as well as transformation and cleaning of data for modelling tools. Basic operations such as the removal of noise, handling missing data fields.

4. *Data reduction and projection.* Finding useful features to represent the data depending on the goal of the task. Using dimensionality reduction or transformation methods to reduce the effective number of variables under consideration or to find invariant representation of data. Neural networks, cluster analysis, and neuro-fuzzy systems are often used for this purpose.

5. *Matching the goals of the KDD process to a particular* **data mining method**. Although the boundaries between prediction and description are not sharp, the distinction is useful for understanding the overall discovery goal. The goals of data mining are achieved via the following data mining tasks:

    - *Clustering*: Identification a finite set of categories or clusters to describe the data. Closely related to clustering is the method of probability density estimation. Clustering quantizes the available input-output data to get a set of prototypes and use the obtained prototypes (signatures, templates, etc.) as model parameters.

    - *Summation*: Finding a compact description for subset of data, e.g., the derivation of summary for association of rules and the use of multivariate visualization techniques.

    - *Dependency modelling*: finding a model which describes significant dependencies between variables (e.g., learning of belief networks).

    - *Regression*: Learning a function which maps a data item to a real-valued prediction variable based on the discovery of functional relationships between variables.

- *Classification*: learning a function that maps (classifies) a data item into one of several predefined classes (category variable).
- *Change and Deviation Detection:* Discovering the most significant changes in the data from previously measured or normative values.

6. *Choosing the data mining algorithm(s)*: Selecting algorithms for searching for patterns in the data. This includes deciding which model and parameters may be appropriate and matching a particular algorithm with the overall criteria of the KDD process (e.g., the end-user may be more interested in understanding the model than its predictive capabilities.) One can identify three primary components in any data mining algorithm: model representation, model evaluation, and search.

   - *Model representation*: The natural language is used to describe the discoverable patterns. If the representation is too limited, then no amount of training time or examples will produce an accurate model for the data. Note that more flexible representation of models increases the danger of overfitting the training data resulting in reduced prediction accuracy on unseen data. It is important that a data analysts fully comprehend the representational assumptions which may be inherent in a particular method.
     For instance, rule-based expert systems are often applied to classification problems in fault detection, biology, medicine etc. Among the wide range of computational intelligence techniques, fuzzy logic improves classification and decision support systems by allowing the use of overlapping class definitions and improves the interpretability of the results by providing more insight into the classifier structure and decision making process. Some of the computational intelligence models lend themselves to transform into other model structure that allows information transfer between different models (e.g., a decision tree mapped into a feedforward neural network or radial basis functions are functionally equivalent to fuzzy inference systems).
   - *Model evaluation criteria*: Qualitative statements or fit functions of how well a particular pattern (a model and its parameters) meet the goals of the KDD process. For example, predictive models can often be judged by the empirical prediction accuracy on some test set. Descriptive models can be evaluated along the dimensions of predictive accuracy, novelty, utility, and understandability of the fitted model. Traditionally, algorithms to obtain classifiers have focused either on accuracy or interpretability. Recently some approaches to combining these properties have been reported
   - *Search method*: Consists of two components: parameter search and model search. Once the model representation and the model evaluation criteria are fixed, then the data mining problem has been reduced

to purely an optimization task: find the parameters/models for the selected family which optimize the evaluation criteria given observed data and fixed model representation. Model search occurs as a loop over the parameter search method.
The automatic determination of model structure from data has been approached by several different techniques: neuro-fuzzy methods, genetic-algorithm and fuzzy clustering in combination with GA-optimization.

7. *Data mining*: Searching for patterns of interest in a particular representation form or a set of such representations: classification rules, trees or figures.

8. *Interpreting mined patterns*: Based on the results possibly return to any of steps 1–7 for further iteration. The data mining engineer interprets the models according to his domain knowledge, the data mining success criteria and the desired test design. This task interferes with the subsequent evaluation phase. Whereas the data mining engineer judges the success of the application of modelling and discovery techniques more technically, he/she contacts business analysts and domain experts later in order to discuss the data mining results in the business context. Moreover, this task only considers models whereas the evaluation phase also takes into account all other results that were produced in the course of the project. This step can also involve the visualization of the extracted patterns/models, or visualization of the data given the extracted models. By many data mining applications it is the user whose experience (e.g., in determining the parameters) is needed to obtain useful results. Although it is hard (and almost impossible or senseless) to develop totally automatical tools, our purpose in this book was to present as data-driven methods as possible, and to emphasize the transparency and interpretability of the results.

9. *Consolidating and using discovered knowledge*: At the evaluation stage in the project you have built a model (or models) that appears to have high quality from a data analysis perspective. Before proceeding to final deployment of the model, it is important to more thoroughly evaluate the model and review the steps executed to construct the model to be certain it properly achieves the business objectives. A key objective is to determine if there is some important business issue that has not been sufficiently considered. At the end of this phase, a decision on the use of the data mining results should be reached.

Creation of the model is generally not the end of the project. Even if the purpose of the model is to increase knowledge of the data, the knowledge gained will need to be organized and presented in a way that the customer can use it. It often involves applying "live" models within an organization's decision making processes, for example in real-time personalization of Web pages or repeated scoring of marketing databases. However, depending on the requirements, the deployment phase can be as simple as generating a report or as complex as implementing a repeatable data mining process across the enterprise. In many cases it is the customer, not the data analyst, who carries out the deployment steps. However, even if the analyst will not carry out the deployment effort it is important for the customer to understand up front what actions need to be carried out in order to actually make use of the created models.

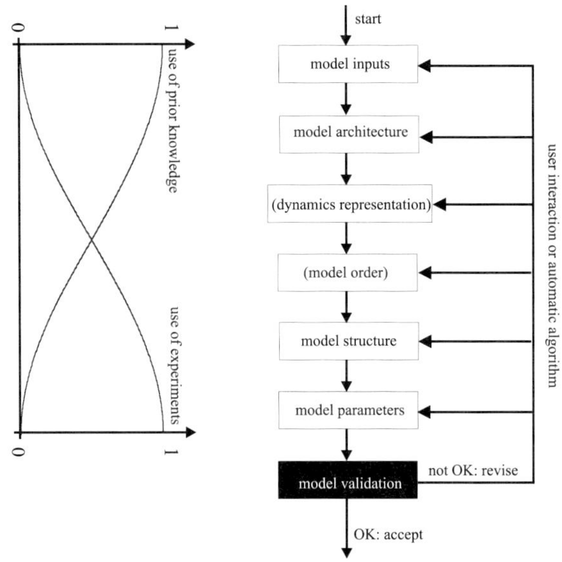

Figure 2: Steps of the knowledge discovery process.

Cross Industry Standard Process for Data Mining (www.crisp-dm.org) contains (roughly) these steps of the KDD process. However, the problems to be solved and their solution methods in KDD can be very similar to those occurred in system identification. The definition of system identification is the process of modelling from experimental data by Ljung [179]. The main steps of the system identification process are summarized well by Petrick and Wigdorowitz [216]:

1. Design an experiment to obtain the physical process input/output experimental data sets pertinent to the model application.
2. Examine the measured data. Remove trends and outliers. Apply filtering to remove measurement and process noise.
3. Construct a set of candidate models based on information from the experimental data sets. This step is the model structure identification.
4. Select a particular model from the set of candidate models in step 3 and estimate the model parameter values using the experimental data sets.

5. Evaluate how good the model is, using an objective function. If the model is not satisfactory then repeat step 4 until all the candidate models have been evaluated.

6. If a satisfactory model is still not obtained in step 5 then repeat the procedure either from step 1 or step 3, depending on the problem.

It can be seen also in Figure 2 from [204] that the system identification steps above may roughly cover the KDD phases. (The parentheses indicate steps that are necessary only when dealing with dynamic systems.) These steps may be complex and several other problem have to be solved during one single phase. Consider, e.g., the main aspects influencing the choice of a model structure:

- What type of model is needed, nonlinear or linear, static or dynamic, distributed or lamped?
- How large must the model set be? This question includes the issue of expected model orders and types of nonlinearities.
- How must the model be parameterized? This involves selecting a criterion to enable measuring the closeness of the model dynamic behavior to the physical process dynamic behavior as model parameters are varied.

To be successful the entire modelling process should be given as much information about the system as is practical. The utilization of prior knowledge and physical insight about the system are very important, but in nonlinear black-box situation no physical insight is available, we have 'only' observed inputs and outputs from the system.

When we attempt to solve real-world problems, like extracting knowledge from large amount of data, we realize that there are typically ill-defined systems to analyze, difficult to model and with large-scale solution spaces. In these cases, precise models are impractical, too expensive, or non-existent. Furthermore, the relevant available information is usually in the form of empirical prior knowledge and input-output data representing instances of the system's behavior. Therefore, we need an approximate reasoning systems capable of handling such imperfect information. computational intelligence (CI) and soft computing (SC) are recently coined terms describing the use of many emerging computing disciplines [2, 3, 13]. It has to be mentioned that KDD has evolved from the intersection of research fields such as machine learning, pattern recognition, databases, statistics, artificial intelligence, and more recently it gets new inspiration from computational intelligence. According to Zadeh (1994): "... in contrast to traditional, hard computing, soft computing is tolerant of imprecision, uncertainty, and partial truth." In this context Fuzzy Logic (FL), Probabilistic Reasoning (PR), Neural Networks (NNs), and Genetic Algorithms (GAs) are considered as main components of CI. Each of these technologies provide us with complementary *reasoning* and *searching* methods to solve complex, real-world problems. What is important to note is that soft computing is not a melange. Rather, it is a partnership in which each of the partners contributes a distinct methodology for addressing problems in its domain. In this perspective, the principal constituent methodologies in CI are complementary rather than competitive.

Because of the different data sources and user needs the purpose of data mining and computational intelligence methods, may be varied in a range field. The purpose of this book is not to overview all of them, many useful and detailed works have been written related to that. This book aims at presenting new methods rather than existing classical ones, while proving the variety of data mining tools and practical usefulness.

The aim of the book is to illustrate how effective data mining algorithms can be generated with the incorporation of fuzzy logic into classical cluster analysis models, and how these algorithms can be used not only for detecting useful knowledge from data by building transparent and accurate regression and classification models, but also for the identification of complex nonlinear dynamical systems. According to that, the new results presented in this book cover a wide range of topics, but they are similar in the applied method: fuzzy clustering algorithms were used for all of them. Clustering within data mining is such a huge topic that the whole overview exceeds the borders of this book as well. Instead of this, our aim was to enable the reader to take a tour in the field of data mining, while proving the flexibility and usefulness of (fuzzy) clustering methods. According to that, students and unprofessionals interested in this topic can also use this book mainly because of the Introduction and the overviews at the beginning of each chapter. However, this book is mainly written for electrical, process and chemical engineers who are interested in new results in clustering.

Organization

This book is organized as follows. The book is divided into six chapters. In Chapter 1, a deep introduction is given about clustering, emphasizing the methods and algorithms that are used in the remainder of the book. For the sake of completeness, a brief overview about other methods is also presented. This chapter gives a detailed description about fuzzy clustering with examples to illustrate the difference between them.

Chapter 2 is in direct connection with clustering: visualization of clustering results is dealt with. The presented methods enable the user to see the n-dimensional clusters, therefore to validate the results. The remainder chapters are in connection with different data mining fields, and the common is that the presented methods utilize the results of clustering.

Chapter 3 deals with fuzzy model identification and presents methods to solve them. Additional familiarity in regression and modelling is helpful but not required because there will be an overview about the basics of fuzzy modelling in the introduction.

Chapter 4 deals with identification of dynamical systems. Methods are presented with their help multiple input – multiple output systems can be modeled, *a priori* information can be built in the model to increase the flexibility and robustness, and the order of input-output models can be determined.

In Chapter 5, methods are presented that are able to use the label of data, therefore the basically unsupervised clustering will be able to solve classification problems. By the fuzzy models as well as classification methods transparency and interpretability are important points of view.

In Chapter 6, a method related to time-series analysis is given. The presented method is able to discover homogeneous segments in multivariate time-series, where the bounds of the segments are given by the change in the relationship between the variables.

Features

The book is abundantly illustrated by

- Figures (120);
- References (302) which give a good overview of the current state of fuzzy clustering and data mining topics concerned in this book;
- Examples (39) which contain simple synthetic data sets and also real-life case studies.

During writing this book, the authors developed a toolbox for MATLAB® called Clustering and Data Analysis Toolbox that can be downloaded from the File Exchange Web site of MathWorks. It can be used easily also by (post)graduate students and for educational purposes as well. This toolbox does not contain all of the programs used in this book, but most of them are available with the related publications (papers and transparencies) at the Web site: www.fmt.vein.hu/softcomp.

Acknowledgements

Many people have aided the production of this project and the authors are greatly indebted to all. These are several individuals and organizations whose support demands special mention and they are listed in the following.

The authors are grateful to the Process Engineering Department at the University of Veszprem, Hungary, where they have worked during the past years. In particular, we are indebted to *Prof. Ferenc Szeifert*, the former Head of the Department, for providing us the intellectual freedom and a stimulating and friendly working environment.

Balazs Feil is extremely grateful to his parents, sister and brother for their continuous financial, but most of all, mental and intellectual support. He is also indebted to all of his roommates during the past years he could (almost) always share his problems with.

Parts of this book are based on papers co-authored by *Dr. Peter Arva, Prof. Robert Babuska, Sandor Migaly, Dr. Sandor Nemeth, Peter Ferenc Pach, Dr. Hans Roubos*, and *Prof. Ferenc Szeifert*. We would like to thank them for their help and interesting discussions.

The financial support of the Hungarian Ministry of Culture and Education (FKFP-0073/2001) and Hungarian Research Founds (T049534) and the Janos Bolyai Research Fellowship of the Hungarian Academy of Sciences is gratefully acknowledged.

Chapter 1

Classical Fuzzy Cluster Analysis

1.1 Motivation

The goal of clustering is to determine the intrinsic grouping in a set of unlabeled data. Data can reveal clusters of different geometrical shapes, sizes and densities as demonstrated in Figure 1.1. Clusters can be spherical (a), elongated or "linear" (b), and also hollow (c) and (d). Their prototypes can be points (a), lines (b), spheres (c) or ellipses (d) or their higher-dimensional analogs. Clusters (b) to (d) can be characterized as linear and nonlinear subspaces of the data space ($\mathbb{R}^2$ in this case). Algorithms that can detect subspaces of the data space are of particular interest for identification. The performance of most clustering algorithms is influenced not only by the geometrical shapes and densities of the individual clusters but also by the spatial relations and distances among the clusters. Clusters can be well separated, continuously connected to each other, or overlapping each other. The separation of clusters is influenced by the scaling and normalization of the data (see Example 1.1, Example 1.2 and Example 1.3).

The goal of this section is to survey the core concepts and techniques in the large subset of cluster analysis, and to give detailed description about the fuzzy clustering methods applied in the remainder sections of this book.

Typical pattern clustering activity involves the following steps [128]:

1. **Pattern representation (optionally including feature extraction and/or selection) (Section** 1.2)

   Pattern representation refers to the number of classes, the number of available patterns, and the number, type, and scale of the features available to the clustering algorithm. Some of this information may not be controllable by the practitioner. **Feature selection** is the process of identifying the most effective subset of the original features to use in clustering. **Feature extraction** is the use of one or more transformations of the input features to produce new

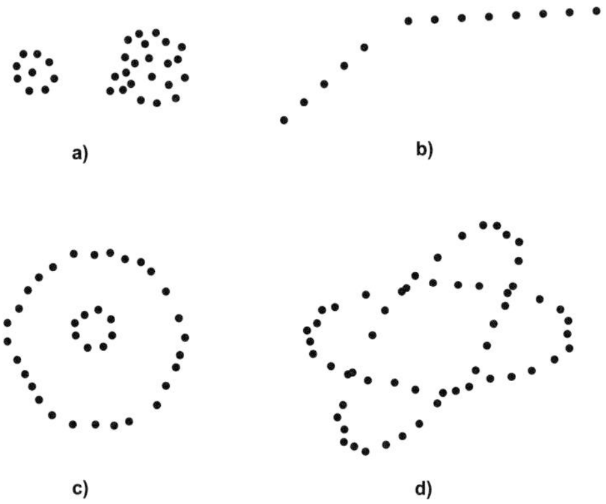

Figure 1.1: Clusters of different shapes in $\mathbb{R}^2$.

salient features. Either or both of these techniques can be used to obtain an appropriate set of features to use in clustering.

2. **Definition of a pattern proximity measure appropriate to the data domain (Section** 1.3**)**

   Dealing with clustering methods like in this book, 'What are clusters?' can be the most important question. Various definitions of a cluster can be formulated, depending on the objective of clustering. Generally, one may accept the view that a cluster is a group of objects that are more similar to one another than to members of other clusters. The term "similarity" should be understood as mathematical similarity, measured in some well-defined sense. In metric spaces, similarity is often defined by means of a *distance norm*. Distance can be measured among the data vectors themselves, or as a distance from a data vector to some prototypical object of the cluster. The prototypes are usually not known beforehand, and are sought by the clustering algorithms simultaneously with the partitioning of the data. The prototypes may be vectors of the same dimension as the data objects, but they can also be defined as "higher-level" geometrical objects, such as linear or nonlinear subspaces or functions. A variety of distance measures are in use in the various communities [21, 70, 128]. A simple distance measure like Euclidean distance can often be used to reflect dissimilarity between two patterns, whereas other similarity measures can be used to characterize the conceptual similarity between patterns [192] (see Section 1.3 for more details).

3. **Clustering or grouping (Section** 1.4**)**

   The grouping step can be performed in a number of ways. The output clustering (or clusterings) can be hard (a partition of the data into groups) or fuzzy (where each pattern has a variable degree of membership in each of the clusters). Hierarchical clustering algorithms produce a nested series of partitions based on a criterion for merging or splitting clusters based on similarity. Partitional clustering algorithms identify the partition that optimizes (usually locally) a clustering criterion. Additional techniques for the grouping operation include probabilistic [52] and graph-theoretic [299] clustering methods (see also Section 1.4).

4. **Data abstraction (if needed)**

   Data abstraction is the process of extracting a simple and compact representation of a data set. Here, simplicity is either from the perspective of automatic analysis (so that a machine can perform further processing efficiently) or it is human-oriented (so that the representation obtained is easy to comprehend and intuitively appealing). In the clustering context, a typical data abstraction is a compact description of each cluster, usually in terms of cluster prototypes or representative patterns such as the centroid [70]. A low-dimensional graphical representation of the clusters could also be very informative, because one can cluster by eye and qualitatively validate conclusions drawn from clustering algorithms. For more details see Chapter 2.

5. **Assessment of output (if needed) (Section** 1.7)

   How is the output of a clustering algorithm evaluated? What characterizes a 'good' clustering result and a 'poor' one? All clustering algorithms will, when presented with data, produce clusters – regardless of whether the data contain clusters or not. If the data does contain clusters, some clustering algorithms may obtain 'better' clusters than others. The assessment of a clustering procedure's output, then, has several facets. One is actually an assessment of the data domain rather than the clustering algorithm itself – data which do not contain clusters should not be processed by a clustering algorithm. The study of cluster tendency, wherein the input data are examined to see if there is any merit to a cluster analysis prior to one being performed, is a relatively inactive research area. The interested reader is referred to [63] and [76] for more information.

The goal of clustering is to determine the intrinsic grouping in a set of unlabeled data. But how to decide what constitutes a good clustering? It can be shown that there is no absolute 'best' criterion which would be independent of the final aim of the clustering. Consequently, it is the user which must supply this criterion, in such a way that the result of the clustering will suit their needs. In spite of that, a 'good' clustering algorithm must give acceptable results in many kinds of problems besides other requirements. In practice, the accuracy of a clustering algorithm is usually tested on well-known labeled data sets. It means that classes are known in the analyzed data set but certainly they are not used in the clustering. Hence, there is a benchmark to qualify the clustering method, and the accuracy can be represented by numbers (e.g., percentage of misclassified data).

Cluster validity analysis, by contrast, is the assessment of a clustering procedure's output. Often this analysis uses a specific criterion of optimality; however, these criteria are usually arrived at subjectively. Hence, little in the way of 'gold standards' exist in clustering except in well-prescribed subdomains. Validity assessments are objective [77] and are performed to determine whether the output is meaningful. A clustering structure is valid if it cannot reasonably have occurred by chance or as an artifact of a clustering algorithm. When statistical approaches to clustering are used, validation is accomplished by carefully applying statistical methods and testing hypotheses. There are three types of validation studies. An external assessment of validity compares the recovered structure to an a priori structure. An internal examination of validity tries to determine if the structure is intrinsically appropriate for the data. A relative test compares two structures and measures their relative merit. Indices used for this comparison are discussed in detail in [77] and [128], and in Section 1.7.

1.2 Types of Data

The expression 'data' has been mentioned several times previously. Being loyal to the traditional scientific conventionality, this expression needs to be explained.

Data can be 'relative' or 'absolute'. *'Relative data'* means that their values are not, but their pairwise distance are known. These distances can be arranged as a matrix called proximity matrix. It can also be viewed as a weighted graph. See also Section 1.4.1 where hierarchical clustering is described that uses this proximity matrix. In this book mainly *'absolute data'* is considered, so we want to give some more accurate expressions about this.

The types of absolute data can be arranged in four categories. Let x and x' be two values of the same attribute.

1. *Nominal type.* In this type of data, the only thing that can be said about two data is, whether they are the same or not: $x = x'$ or $x \neq x'$.

2. *Ordinal type.* The values can be arranged in a sequence. If $x \neq x'$, then it is also decidable that $x > x'$ or $x < x'$.

3. *Interval scale.* If the difference between two data items can be expressed as a number besides the above-mentioned terms.

4. *Ratio scale.* This type of data is interval scale but zero value exists as well. If $c = \frac{x}{x'}$, then it can be said that x is c times bigger x'.

In this book, the clustering of ratio scale data is considered. The data are typically observations of some phenomena. In these cases, not only one but n variables are measured simultaneously, therefore each observation consists of n measured variables, grouped into an n-dimensional column vector $\mathbf{x}_k = [x_{1,k}, x_{2,k}, \ldots, x_{n,k}]^T$, $\mathbf{x}_k \in \mathbb{R}^n$. These variables are usually not independent from each other, therefore *multivariate data analysis* is needed that is able to handle these observations. A set of N observations is denoted by $\mathbf{X} = \{\mathbf{x}_k | k = 1, 2, \ldots, N\}$, and is represented as an $N \times n$ matrix:

$$\mathbf{X} = \begin{bmatrix} x_{1,1} & x_{1,2} & \cdots & x_{1,n} \\ x_{2,1} & x_{2,2} & \cdots & x_{2,n} \\ \vdots & \vdots & \ddots & \vdots \\ x_{N,1} & x_{N,2} & \cdots & x_{N,n} \end{bmatrix}. \tag{1.1}$$

In pattern recognition terminology, the rows of $\mathbf{X}$ are called *patterns* or objects, the columns are called *features* or attributes, and $\mathbf{X}$ is called *pattern matrix*. In this book, $\mathbf{X}$ is often referred to simply as the *data matrix*. The meaning of the rows and columns of $\mathbf{X}$ with respect to reality depends on the context. In medical diagnosis, for instance, the rows of $\mathbf{X}$ may represent patients, and the columns are then symptoms, or laboratory measurements for the patients. When clustering is applied to the modelling and identification of dynamic systems, the rows of $\mathbf{X}$ contain samples of time signals, and the columns are, for instance, physical variables observed in the system (position, velocity, temperature, etc.).

In system identification, the purpose of clustering is to find relationships between independent system variables, called the regressors, and future values of dependent variables, called the regressands. One should, however, realize that the relations revealed by clustering are just acausal associations among the data vectors, and as such do not yet constitute a prediction model of the given system. To obtain such a model, additional steps are needed which will be presented in Section 4.3.

Data can be given in the form of a so-called dissimilarity matrix:

$$\begin{bmatrix} 0 & d(1,2) & d(1,3) & \cdots & d(1,N) \\ & 0 & d(2,3) & \cdots & d(2,N) \\ & & 0 & \ddots & \vdots \\ & & & & 0 \end{bmatrix} \tag{1.2}$$

where $d(i,j)$ means the measure of dissimilarity (distance) between object x_i and x_j. Because $d(i,i) = 0, \forall i$, zeros can be found in the main diagonal, and that matrix is symmetric because $d(i,j) = d(j,i)$. There are clustering algorithms that use that form of data (e.g., hierarchical methods). If data are given in the form of (1.1), the first step that has to be done is to transform data into dissimilarity matrix form.

1.3 Similarity Measures

Since similarity is fundamental to the definition of a cluster, a measure of the similarity between two patterns drawn from the same feature space is essential to most clustering procedures. Because of the variety of feature types and scales, the distance measure (or measures) must be chosen carefully. It is most common to calculate the dissimilarity between two patterns using a distance measure defined on the feature space. We will focus on the well-known distance measures used for patterns whose features are all continuous.

The most popular metric for continuous features is the **Euclidean distance**

$$d_2(\mathbf{x}_i, \mathbf{x}_j) = \left(\sum_{k=1}^{d} (x_{i,k} - x_{j,k})^2 \right)^{1/2} = \|\mathbf{x}_i - \mathbf{x}_j\|_2, \tag{1.3}$$

which is a special case ($p = 2$) of the **Minkowski metric**

$$d_p(\mathbf{x}_i, \mathbf{x}_j) = \left(\sum_{k=1}^{d} |x_{i,k} - x_{j,k}|^p \right)^{1/p} = \|\mathbf{x}_i - \mathbf{x}_j\|_p. \tag{1.4}$$

The Euclidean distance has an intuitive appeal as it is commonly used to evaluate the proximity of objects in two or three-dimensional space. It works well when a data set has "compact" or "isolated" clusters [186]. The drawback to direct use of the Minkowski metrics is the tendency of the largest-scaled feature to dominate the others. Solutions to this problem include normalization of the continuous features (to a common range or variance) or other weighting schemes. Linear correlation among features can also distort distance measures; this distortion can be alleviated by applying a whitening transformation to the data or by using the squared **Mahalanobis distance**

$$d_M(\mathbf{x}_i, \mathbf{x}_j) = (\mathbf{x}_i - \mathbf{x}_j)\mathbf{F}^{-1}(\mathbf{x}_i - \mathbf{x}_j)^T \tag{1.5}$$

where the patterns $\mathbf{x}_i$ and $\mathbf{x}_j$ are assumed to be row vectors, and $\mathbf{F}$ is the sample covariance matrix of the patterns or the known covariance matrix of the pattern generation process; $d_M(\cdot,\cdot)$ assigns different weights to different features based on their variances and pairwise linear correlations. Here, it is implicitly assumed that class conditional densities are unimodal and characterized by multidimensional spread, i.e., that the densities are multivariate Gaussian. The regularized Mahalanobis distance was used in [186] to extract hyperellipsoidal clusters. Recently, several researchers [78, 123] have used the Hausdorff distance in a point set matching context.

The norm metric influences the clustering criterion by changing the measure of dissimilarity. The Euclidean norm induces hyperspherical clusters, i.e., clusters whose surface of constant membership are hyperspheres. Both the diagonal and the Mahalanobis norm generate hyperellipsoidal clusters, the difference is that with the diagonal norm, the axes of the hyperellipsoids are parallel to the coordinate axes while with the Mahalanobis norm the orientation of the hyperellipsoids is arbitrary, as shown in Figure 1.2.

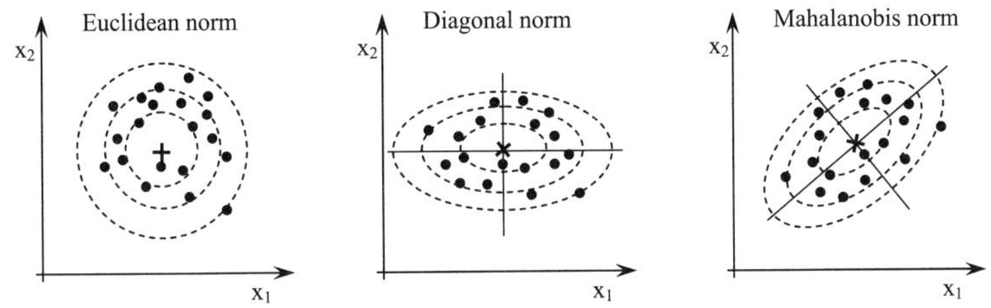

Figure 1.2: Different distance norms used in fuzzy clustering.

Some clustering algorithms work on a matrix of proximity values instead of on the original pattern set. It is useful in such situations to precompute all the $N(N-1)/2$ pairwise distance values for the N patterns and store them in a (symmetric) matrix (see Section 1.2).

Computation of distances between patterns with some or all features being non-continuous is problematic, since the different types of features are not comparable and (as an extreme example) the notion of proximity is effectively binary-valued for nominal-scaled features. Nonetheless, practitioners (especially those in machine learning, where mixed-type patterns are common) have developed proximity measures for heterogeneous type patterns. A recent example is [283], which proposes a combination of a modified Minkowski metric for continuous features and a distance based on counts (population) for nominal attributes. A variety of other metrics have been reported in [70] and [124] for computing the similarity between patterns represented using quantitative as well as qualitative features.

Patterns can also be represented using string or tree structures [155]. Strings are used in syntactic clustering [90]. Several measures of similarity between strings are described in [34]. A good summary of similarity measures between trees is given by Zhang [301]. A comparison of syntactic and statistical approaches for pattern recognition using several criteria was presented in [259] and the conclusion was that syntactic methods are inferior in every aspect. Therefore, we do not consider syntactic methods further.

There are some distance measures reported in the literature [100, 134] that take into account the effect of surrounding or neighboring points. These surrounding points are called context in [192]. The similarity between two points $\mathbf{x}_i$ and $\mathbf{x}_j$, given this context, is given by

$$s(\mathbf{x}_i, \mathbf{x}_j) = f(\mathbf{x}_i, \mathbf{x}_j, \mathcal{E}), \tag{1.6}$$

where $\mathcal{E}$ is the context (the set of surrounding points). One metric defined using context is the **mutual neighbor distance** (MND), proposed in [100], which is given by

$$MND(\mathbf{x}_i, \mathbf{x}_j) = NN(\mathbf{x}_i, \mathbf{x}_j) + NN(\mathbf{x}_j, \mathbf{x}_i), \tag{1.7}$$

where $NN(\mathbf{x}_i, \mathbf{x}_j)$ is the neighbor number of $\mathbf{x}_j$ with respect to $\mathbf{x}_i$. The MND is not a metric (it does not satisfy the triangle inequality [301]). In spite of this,

MND has been successfully applied in several clustering applications [99]. This observation supports the viewpoint that the dissimilarity does not need to be a metric. Watanabe's theorem of the ugly duckling [282] states:

"Insofar as we use a finite set of predicates that are capable of distinguishing any two objects considered, the number of predicates shared by any two such objects is constant, independent of the choice of objects."

This implies that it is possible to make any two arbitrary patterns equally similar by encoding them with a sufficiently large number of features. As a consequence, any two arbitrary patterns are equally similar, unless we use some additional domain information. For example, in the case of conceptual clustering [192], the similarity between $\mathbf{x}_i$ and $\mathbf{x}_j$ is defined as

$$s(\mathbf{x}_i, \mathbf{x}_j) = f(\mathbf{x}_i, \mathbf{x}_j, \mathcal{C}, \mathcal{E}), \tag{1.8}$$

where $\mathcal{C}$ is a set of pre-defined concepts.

So far, only continuous variables have been dealt with. There are a lot of similarity measures for binary variables (see, e.g., in [110]), but only continuous variables are considered in this book because continuous variables occur in system identification.

1.4 Clustering Techniques

Different approaches to clustering data can be described with the help of the hierarchy shown in Figure 1.3 (other taxonometric representations of clustering

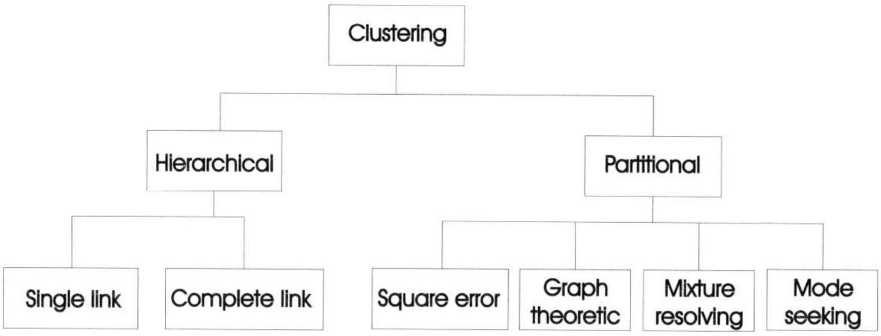

Figure 1.3: A taxonomy of clustering approaches.

methodology are possible; ours is based on the discussion in [128]). At the top level, there is a distinction between hierarchical and partitional approaches (hierarchical methods produce a nested series of partitions, while partitional methods produce only one).

1.4.1 Hierarchical Clustering Algorithms

A hierarchical algorithm yields a dendrogram representing the nested grouping of patterns and similarity levels at which groupings change. The dendrogram can be broken at different levels to yield different clusterings of the data. An example can be seen in Figure 1.4. On the left side, the interpattern distances can be seen in

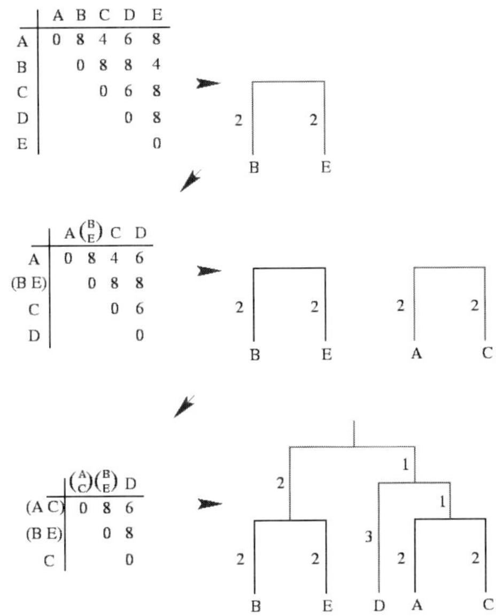

Figure 1.4: Dendrogram building [110].

a form of dissimilarity matrix (1.2). In this initial state every point forms a single cluster. The **first step** is to find the most similar two clusters (the nearest two data points). In this example, there are two pairs with the same distance, choose one of them arbitrarily (B and E here). Write down the signs of the points, and connect them according to the figure, where the length of the vertical line is equal to the half of the distance. In the **second step** the dissimilarity matrix should be refreshed because the connected points form a single cluster, and the distances between this new cluster and the former ones should be computed. These steps should be iterated until only one cluster remains or the predetermined number of clusters is reached.

Most hierarchical clustering algorithms are variants of the single-link [250], complete-link [153], and minimum-variance [141, 198] algorithms. Of these, the single-link and complete-link algorithms are most popular.

A simple example can be seen in Figure 1.5. On the left side, the small dots depict the original data. It can be seen that there are two well-separated clusters. The results of the single-linkage algorithm can be found on the right side. It can

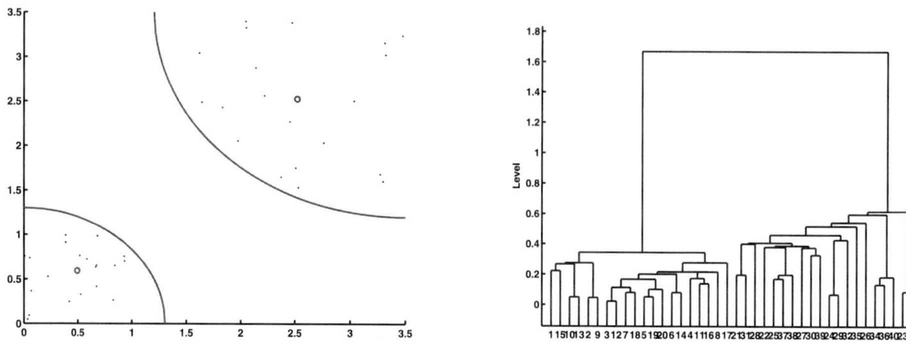

Figure 1.5: Partitional and hierarchical clustering results.

be determined that distances between data in the right cluster are greater than in the left one, but the two clusters can be separated well.

These two algorithms differ in the way they characterize the similarity between a pair of clusters. In the single-link method, the distance between two clusters is the minimum of the distances between all pairs of patterns drawn from the two clusters (one pattern from the first cluster, the other from the second). In the complete-link algorithm, the distance between two clusters is the maximum of all pairwise distances between patterns in the two clusters. In either case, two clusters are merged to form a larger cluster based on minimum distance criteria. The complete-link algorithm produces tightly bound or compact clusters [34]. The single-link algorithm, by contrast, suffers from a chaining effect [200]. It has a tendency to produce clusters that are straggly or elongated. The clusters obtained by the complete-link algorithm are more compact than those obtained by the single-link algorithm. The single-link algorithm is more versatile than the complete-link algorithm, otherwise. However, from a pragmatic viewpoint, it has been observed that the complete-link algorithm produces more useful hierarchies in many applications than the single-link algorithm [128].

1.4.2 Partitional Algorithms

A partitional clustering algorithm obtains a single partition of the data instead of a clustering structure, such as the dendrogram produced by a hierarchical technique. The difference of the two mentioned methods can be seen in Figure 1.5. Partitional methods have advantages in applications involving large data sets for which the construction of a dendrogram is computationally prohibitive.

Squared Error Algorithms

The most intuitive and frequently used criterion function in partitional clustering techniques is the squared error criterion, which tends to work well with isolated and compact clusters. The squared error for a clustering $\mathbf{V} = \{\mathbf{v}_i | i = 1, \ldots, c\}$ of a pattern set $\mathbf{X}$ (containing c clusters) is

$$J(\mathbf{X}; \mathbf{V}) = \sum_{i=1}^{c} \sum_{k \in i} \|\mathbf{x}_k^{(i)} - \mathbf{v}_i\|^2, \tag{1.9}$$

where $\mathbf{x}_k^{(i)}$ is the kth pattern belonging to the ith cluster and $\mathbf{v}_i$ is the centroid of the ith cluster (see Algorithm 1.4.1).

Algorithm 1.4.1 (Squared Error Clustering Method).

1. *Select an initial partition of the patterns with a fixed number of clusters and cluster centers.*
2. *Assign each pattern to its closest cluster center and compute the new cluster centers as the centroids of the clusters. Repeat this step until convergence is achieved, i.e., until the cluster membership is stable.*
3. *Merge and split clusters based on some heuristic information, optionally repeating step 2.*

The k-means is the simplest and most commonly used algorithm employing a squared error criterion [190]. It starts with a random initial partition and keeps reassigning the patterns to clusters based on the similarity between the pattern and the cluster centers until a convergence criterion is met (e.g., there is no reassignment of any pattern from one cluster to another, or the squared error ceases to decrease significantly after some number of iterations). The k-means algorithm is popular because it is easy to implement, and its time complexity is $O(N)$, where N is the number of patterns. A major problem with this algorithm is that it is sensitive to the selection of the initial partition and may converge to a local minimum of the criterion function value if the initial partition is not properly chosen. The whole procedure can be found in Algorithm 1.4.2.

Algorithm 1.4.2 (k-Means Clustering).

1. *Choose k cluster centers to coincide with k randomly-chosen patterns or k randomly defined points inside the hypervolume containing the pattern set.*
2. *Assign each pattern to the closest cluster center.*
3. *Recompute the cluster centers using the current cluster memberships.*
4. *If a convergence criterion is not met, go to step 2. Typical convergence criteria are: no (or minimal) reassignment of patterns to new cluster centers, or minimal decrease in squared error.*

Several variants of the k-means algorithm have been reported in the literature [21]. Some of them attempt to select a good initial partition so that the algorithm is more likely to find the global minimum value.

A problem accompanying the use of a partitional algorithm is the choice of the number of desired output clusters. A seminal paper [76] provides guidance on this key design decision. The partitional techniques usually produce clusters by optimizing a criterion function defined either locally (on a subset of the patterns) or globally (defined over all of the patterns). Combinatorial search of the set of possible labelings for an optimum value of a criterion is clearly computationally prohibitive. In practice, therefore, the algorithm is typically run multiple times with different starting states, and the best configuration obtained from all of the runs is used as the output clustering.

Another variation is to permit splitting and merging of the resulting clusters. Typically, a cluster is split when its variance is above a pre-specified threshold, and two clusters are merged when the distance between their centroids is below another pre-specified threshold. Using this variant, it is possible to obtain the optimal partition starting from any arbitrary initial partition, provided proper threshold values are specified. The well-known ISODATA algorithm employs this technique of merging and splitting clusters [36].

Another variation of the k-means algorithm involves selecting a different criterion function altogether. The dynamic clustering algorithm (which permits representations other than the centroid for each cluster) was proposed in [70, 256] and describes a dynamic clustering approach obtained by formulating the clustering problem in the framework of maximum-likelihood estimation. The regularized Mahalanobis distance was used in [186] to obtain hyperellipsoidal clusters.

The taxonomy shown in Figure 1.3 must be supplemented by a discussion of cross-cutting issues that may (in principle) affect all of the different approaches regardless of their placement in the taxonomy.

- *Hard vs. fuzzy.* A hard clustering algorithm allocates each pattern to a single cluster during its operation and in its output. A fuzzy clustering method assigns degrees of membership in several clusters to each input pattern. A fuzzy clustering can be converted to a hard clustering by assigning each pattern to the cluster with the largest measure of membership.

- *Agglomerative vs. divisive.* This aspect relates to algorithmic structure and operation (mostly in hierarchical clustering, see Section 1.4.1). An agglomerative approach begins with each pattern in a distinct (singleton) cluster, and successively merges clusters together until a stopping criterion is satisfied. A divisive method begins with all patterns in a single cluster and performs splitting until a stopping criterion is met.

- *Monothetic vs. polythetic.* This aspect relates to the sequential or simultaneous use of features in the clustering process. Most algorithms are polythetic; that is, all features enter into the computation of distances between patterns, and decisions are based on those distances. A simple monothetic algorithm reported in [21] considers features sequentially to divide the given collection of patterns. The major problem with this algorithm is that it generates $2n$ clusters where n is the dimensionality of the patterns. For large values of n ($n > 100$ is typical in information retrieval applications [233]), the number of clusters generated by this algorithm is so large that the data set is divided into uninterestingly small and fragmented clusters.

- *Deterministic vs. stochastic.* This issue is most relevant to partitional approaches designed to optimize a squared error function. This optimization can be accomplished using traditional techniques or through a random search of the state-space consisting of all possible labelings.
- *Incremental vs. non-incremental.* This issue arises when the pattern set to be clustered is large, and constraints on execution time or memory space affect the architecture of the algorithm. The early history of clustering methodology does not contain many examples of clustering algorithms designed to work with large data sets, but the advent of data mining has fostered the development of clustering algorithms that minimize the number of scans through the pattern set, reduce the number of patterns examined during execution, or reduce the size of data structures used in the algorithm's operations.

A cogent observation in [128] is that the specification of an algorithm for clustering usually leaves considerable flexibility in implementation. In the following, we briefly discuss other clustering techniques as well, but a separate section (Section 1.5) and deep description are devoted to fuzzy clustering methods which are the most important in this book. Note that methods described in the next chapters are based on mixture of models as it is described in the following as well.

Mixture-Resolving and Mode-Seeking Algorithms

The mixture resolving approach to cluster analysis has been addressed in a number of ways. The underlying assumption is that the patterns to be clustered are drawn from one of several distributions, and the goal is to identify the parameters of each and (perhaps) their number. Most of the work in this area has assumed that the individual components of the mixture density are Gaussian, and in this case the parameters of the individual Gaussians are to be estimated by the procedure. Traditional approaches to this problem involve obtaining (iteratively) a maximum likelihood estimate of the parameter vectors of the component densities [128]. More recently, the Expectation Maximization (EM) algorithm (a general purpose maximum likelihood algorithm [69] for missing-data problems) has been applied to the problem of parameter estimation. A recent book [194] provides an accessible description of the technique. In the EM framework, the parameters of the component densities are unknown, as are the mixing parameters, and these are estimated from the patterns. The EM procedure begins with an initial estimate of the parameter vector and iteratively rescores the patterns against the mixture density produced by the parameter vector. The rescored patterns are then used to update the parameter estimates. In a clustering context, the scores of the patterns (which essentially measure their likelihood of being drawn from particular components of the mixture) can be viewed as hints at the class of the pattern. Those patterns, placed (by their scores) in a particular component, would therefore be viewed as belonging to the same cluster. Nonparametric techniques for density-based clustering have also been developed [128]. Inspired by the Parzen window approach to nonparametric density estimation, the corresponding clustering procedure searches for bins with large counts in a multidimensional histogram of the input pattern set. Other approaches include the application of another partitional or hierarchical clustering algorithm using a distance measure based on a nonparametric density estimate.

Nearest Neighbor Clustering

Since proximity plays a key role in our intuitive notion of a cluster, nearest neighbor distances can serve as the basis of clustering procedures. An iterative procedure was proposed in [181]; it assigns each unlabeled pattern to the cluster of its nearest labeled neighbor pattern, provided the distance to that labeled neighbor is below a threshold. The process continues until all patterns are labeled or no additional labelings occur. The mutual neighborhood value (described earlier in the context of distance computation) can also be used to grow clusters from near neighbors.

Graph-Theoretic Clustering

The best-known graph-theoretic divisive clustering algorithm is based on construction of the minimal spanning tree (MST) of the data [299], and then deleting the MST edges with the largest lengths to generate clusters.

The hierarchical approaches are also related to graph-theoretic clustering. Single-link clusters are subgraphs of the minimum spanning tree of the data [101] which are also the connected components [98]. Complete-link clusters are maximal complete subgraphs, and are related to the node colorability of graphs [33]. The maximal complete subgraph was considered the strictest definition of a cluster in [25, 226]. A graph-oriented approach for non-hierarchical structures and overlapping clusters is presented in [207]. The Delaunay graph (DG) is obtained by connecting all the pairs of points that are Voronoi neighbors. The DG contains all the neighborhood information contained in the MST and the relative neighborhood graph (RNG) [266].

Figure 1.6 depicts the minimal spanning tree obtained from 75 two-dimensional points distributed into three clusters. The objects belonging to different clusters are marked with different dot notations. Clustering methods using a minimal spanning tree take advantages of the MST. For example building the minimal spanning tree of a dataset does not need any *a priori* information about the underlying data.

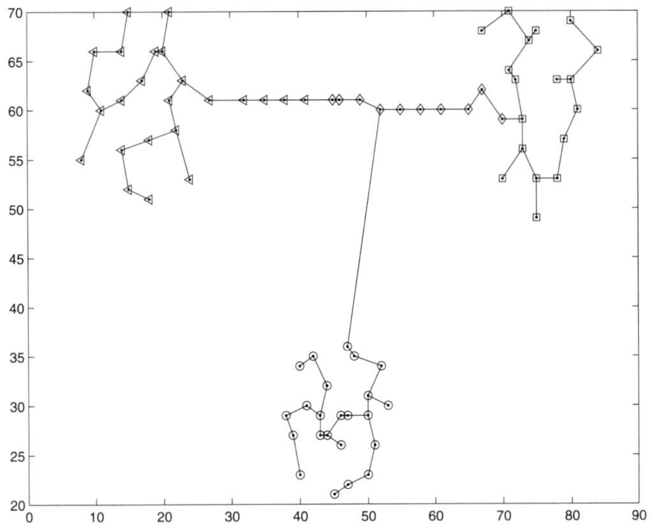

Figure 1.6: Example of a minimal spanning tree.

Moreover, as the MST ignores many possible connections between the data patterns, the cost of clustering can be decreased.

Using a minimal spanning tree for clustering was initially proposed by Zahn [299]. A minimal spanning tree is a weighted connected graph, where the sum of the weights is minimal. A graph G is a pair (V, E), where V is a finite set of the elements (samples by clustering), called vertices, and E is a collection of unordered pairs of V. An element of E, called edge, is $e_{i,j} = (v_i, v_j)$, where $v_i, v_j \in V$. In a weighted graph a weight function w is defined, which function determines a weight $w_{i,j}$ for each edge $e_{i,j}$. The complete graph K_N on a set of N vertices is the graph that has all the $\binom{N}{2}$ possible edges. Creating the minimal spanning tree means that we are searching the $G' = (V, E')$, the connected subgraph of G, where $E' \subset E$ and the cost is minimum. The cost is computed in the following way:

$$\sum_{e \in E'} w(e) \tag{1.10}$$

where $w(e)$ denotes the weight of the edge $e \in E$. In a graph G, where the number of the vertices is N, MST has exactly $N - 1$ edges.

A minimal spanning tree can be efficiently computed in $O(N^2)$ time using either Prim's [221] or Kruskal's [162] algorithm. Prim's algorithm starts with an arbitrary vertex as the root of a partial tree. In each step of the algorithm the partial tree grows by iteratively adding an unconnected vertex to it using the lowest cost edge, until no unconnected vertex remains. Kruskal's algorithm begins with the connection of the two nearest objects. In each step the nearest objects placed in different trees are connected. So the Kruskal's algorithm iteratively merges two trees (or a tree with a single object) in the current forest into a new tree. The algorithm continues until a single tree remains only, connecting all points.

However the use of minimal spanning trees in clustering algorithms also raises some interesting questions. How can we determine the edges at which the best cluster separations might be made? For finding the best clusters, when should we stop our algorithm? These questions cannot be answered in a trivial way. There are well-known criteria for that purpose but there are new results in this field, see, e.g., [273] where a synergistic combination of graph-theoretic and partitional algorithms was presented to avoid some drawbacks of these algorithms.

Criterion-1: The simplest way to delete edges from the minimal spanning tree is based on the distance between the vertices. By deleting the longest edge in each iteration step we get a nested sequence of subgraphs. As other hierarchical methods, this approach also requires a terminating condition. Several ways are known to stop the algorithms, for example the user can define the number of clusters, or we can give a threshold value on the length as well.

Similarly to Zahn [299] we suggest a global threshold value δ, which considers the distribution of the data in the feature space. In [299] this threshold (δ) is based on the average weight (distances) of the MST:

$$\delta = \lambda \frac{1}{N-1} \sum_{e \in E'} w(e) \tag{1.11}$$

where λ is a user defined parameter.

Criterion-2: Zahn [299] proposed also an idea to detect the hidden separations in the data. Zahn's suggestion is based on the distance of the separated subtrees. He suggested, that an edge is inconsistent if its length is at least f times as long as the average of the length of nearby edges. The input parameter f must be adjusted by the user. To determine which edges are 'nearby' is another question. It can be determined by the user, or we can say, that point $\mathbf{x}_i$ is nearby point of $\mathbf{x}_j$ if point $\mathbf{x}_i$ is connected to the point $\mathbf{x}_j$ by a path in a minimal spanning tree containing k or fewer edges. This method has the advantage of determining clusters which have different distances separating one another. Another use of the MST based clustering based on this criterion is to find dense clusters embedded in a sparse set of points. All that has to be done is to remove all edges longer than some predetermined length in order to extract clusters which are closer than the specified length to each other. If the length is chosen accordingly, the dense clusters are extracted from a sparse set of points easily. The drawback of this method is that the influence of the user is significant at the selection of the f and k parameters.

Several clustering methods based on linkage approach suffer from some discrepancies. In these cases the clusters are provided by merging or splitting of the objects or clusters using a distance defined between them. Occurrence of a data chain between two clusters can cause that these methods can not separate these clusters. This also happens with the basic MST clustering algorithm. To solve the chaining problem we suggest a new complementary condition for cutting the minimal spanning tree.

1.5 Fuzzy Clustering

Since clusters can formally be seen as subsets of the data set, one possible classification of clustering methods can be according to whether the subsets are fuzzy or crisp (hard). Hard clustering methods are based on classical set theory, and require that an object either does or does not belong to a cluster. Hard clustering in a data set $\mathbf{X}$ means partitioning the data into a specified number of mutually exclusive subsets of $\mathbf{X}$. The number of subsets (clusters) is denoted by c. Fuzzy clustering methods allow objects to belong to several clusters simultaneously, with different degrees of membership. The data set $\mathbf{X}$ is thus partitioned into c fuzzy subsets. In many real situations, fuzzy clustering is more natural than hard clustering, as objects on the boundaries between several classes are not forced to fully belong to one of the classes, but rather are assigned membership degrees between 0 and 1 indicating their partial memberships. The discrete nature of hard partitioning also causes analytical and algorithmic intractability of algorithms based on analytic functionals, since these functionals are not differentiable.

The remainder of this book focuses on fuzzy clustering with objective function and its applications. First let us define more precisely the concept of fuzzy partitions.

1.5.1 Fuzzy partition

The objective of clustering is to partition the data set $\mathbf{X}$ into c clusters. For the time being, assume that c is known, based on prior knowledge, for instance (for more details see Section 1.7). Fuzzy and possibilistic partitions can be seen as a generalization of hard partition.

A fuzzy partition of the data set $\mathbf{X}$ can be represented by a $c \times N$ matrix $\mathbf{U} = [\mu_{i,k}]$, where $\mu_{i,k}$ denotes the degree of membership that the kth observation belongs to the cth cluster ($1 \leq k \leq N$, $1 \leq i \leq c$). Therefore, the ith row of $\mathbf{U}$ contains values of the *membership function* of the ith fuzzy subset of $\mathbf{X}$. The matrix $\mathbf{U}$ is called the fuzzy partition matrix. Conditions for a fuzzy partition matrix are given by:

$$\mu_{i,k} \in [0,1], \; 1 \leq i \leq c, \; 1 \leq k \leq N, \tag{1.12}$$

$$\sum_{i=1}^{c} \mu_{i,k} = 1, \; 1 \leq k \leq N, \tag{1.13}$$

$$0 < \sum_{k=1}^{N} \mu_{i,k} < N, \; 1 \leq i \leq c. \tag{1.14}$$

Fuzzy partitioning space *Let* $\mathbf{X} = [\mathbf{x}_1, \mathbf{x}_2, \ldots, \mathbf{x}_N]$ *be a finite set and let* $2 \leq c < N$ *be an integer. The fuzzy partitioning space for* $\mathbf{X}$ *is the set*

$$M_{fc} = \left\{ \mathbf{U} \in \mathbb{R}^{c \times N} | \mu_{i,k} \in [0,1], \forall i,k; \sum_{i=1}^{c} \mu_{i,k} = 1, \forall k; 0 < \sum_{k=1}^{N} \mu_{i,k} < N, \forall i \right\}. \tag{1.15}$$

(1.13) constrains the sum of each column to 1, and thus the total membership of each $\mathbf{x}_k$ in $\mathbf{X}$ equals one. The distribution of memberships among the c fuzzy subsets is not constrained.

1.5.2 The Fuzzy c-Means Functional

A large family of fuzzy clustering algorithms is based on minimization of the *fuzzy c-means* objective function formulated as:

$$J(\mathbf{X}; U, V) = \sum_{i=1}^{c} \sum_{k=1}^{N} (\mu_{i,k})^m \|\mathbf{x}_k - \mathbf{v}_i\|_{\mathbf{A}}^2 \tag{1.16}$$

where

$$\mathbf{U} = [\mu_{i,k}] \tag{1.17}$$

is a fuzzy partition matrix of $\mathbf{X}$,

$$\mathbf{V} = [\mathbf{v}_1, \mathbf{v}_2, \ldots, \mathbf{v}_c], \ \mathbf{v}_i \in \mathbb{R}^n \tag{1.18}$$

is a matrix of *cluster prototypes* (centers), which have to be determined,

$$D_{i,k\mathbf{A}}^2 = \|\mathbf{x}_k - \mathbf{v}_i\|_{\mathbf{A}}^2 = (\mathbf{x}_k - \mathbf{v}_i)^T \mathbf{A} (\mathbf{x}_k - \mathbf{v}_i) \tag{1.19}$$

is a squared inner-product distance norm where $\mathbf{A}$ is the distance measure (see Section 1.3), and

$$m \in \langle 1, \infty) \tag{1.20}$$

is a weighting exponent which determines the fuzziness of the resulting clusters. The measure of dissimilarity in (1.16) is the squared distance between each data point $\mathbf{x}_k$ and the cluster prototype $\mathbf{v}_i$. This distance is weighted by the power of the membership degree of that point $(\mu_{i,k})^m$. The value of the cost function (1.16) is a measure of the total weighted within-group squared error incurred by the representation of the c clusters defined by their prototypes $\mathbf{v}_i$. Statistically, (1.16) can be seen as a measure of the total variance of $\mathbf{x}_k$ from $\mathbf{v}_i$.

1.5.3 Ways for Realizing Fuzzy Clustering

Having constructed the criterion function for clustering, this subsection will study how to optimize the objective function [286]. The existing ways were mainly classified into three classes: Neural Networks (NN), Evolutionary Computing (EC) and Alternative Optimization (AO). We briefly discuss the first two methods in this subsection, and the last one in the next subsections with more details.

- **Realization based on NN.** The application of neural networks in cluster analysis stems from the Kohonen's learning vector quantization (LVQ) [156], Self Organizing Mapping (SOM) [157] and Grossberg's adaptive resonance theory (ART) [59, 105, 106].

  Since NNs are of capability in parallel processing, people hope to implement clustering at high speed with network structure. However, the classical clustering NN can only implement spherical hard cluster analysis. So, people made much effort in the integrative research of fuzzy logic and NNs, which falls into two categories as follows. The first type of studies bases on the fuzzy competitive learning algorithm, in which the methods proposed by Pal et al [209], Xu [287] and Zhang [300] respectively are representatives of this type of clustering NN. These novel fuzzy clustering NNs have several advantages over the traditional ones. The second type of studies mainly focuses on the fuzzy logic operations, such as the fuzzy ART and fuzzy Min-Max NN.

- **Realization based on EC.** EC is a stochastic search strategy with the mechanism of natural selection and group inheritance, which is constructed on the basis of biological evolution. For its performance of parallel search, it can obtain the global optima with a high probability. In addition, EC has some advantages such as it is simple, universal and robust. To achieve clustering results quickly and correctly, evolutionary computing was introduced to fuzzy clustering with a series of novel clustering algorithms based on EC (see the review of Xinbo et al in [286]).

  This series of algorithms falls into three groups. The first group is simulated annealing based approach. Some of them can solve the fuzzy partition matrix by annealing, the others optimize the clustering prototype gradually. However, only when the temperature decreases slowly enough can the simulated annealing converge to the global optima. Hereby, the great CPU time limits its applications. The second group is the approach based on genetic algorithm and evolutionary strategy, whose studies are focused on such aspects as solution encoding, construction of fitness function, designing of genetic operators and choice of operation parameters. The third group, i.e., the approach based on Tabu search is only explored and tried by AL-Sultan, which is very initial and requires further research.

- **Realization based on Alternative Optimization.** The most popular technique is Alternative Optimization even today, maybe because of its simplicity [39, 79]. This technique will be presented in the following sections.

1.5.4 The Fuzzy c-Means Algorithm

The minimization of the c-means functional (1.16) represents a nonlinear optimization problem that can be solved by using a variety of available methods, ranging from grouped coordinate minimization, over simulated annealing to genetic algorithms. The most popular method, however, is a simple Picard iteration through the first-order conditions for stationary points of (1.16), known as the fuzzy c-means (FCM) algorithm.

The stationary points of the objective function (1.16) can be found by adjoining the constraint (1.13) to J by means of Lagrange multipliers:

$$\bar{J}(\mathbf{X}; U, V, \lambda) = \sum_{i=1}^{c}\sum_{k=1}^{N}(\mu_{i,k})^m D^2_{i,k\mathbf{A}} + \sum_{k=1}^{N}\lambda_k\left(\sum_{i=1}^{c}\mu_{i,k} - 1\right), \tag{1.21}$$

and by setting the gradients of $\bar{J}$ with respect to $\mathbf{U}, V$ and λ to zero. If $D^2_{i,k\mathbf{A}} > 0, \forall i, k$ and $m > 1$, then $(\mathbf{U}, V) \in M_{fc} \times \mathbb{R}^{n\times c}$ may minimize (1.16) only if

$$\mu_{i,k} = \frac{1}{\sum_{j=1}^{c}\left(D_{i,k\mathbf{A}}/D_{j,k\mathbf{A}}\right)^{2/(m-1)}}, \quad 1 \le i \le c,\ 1 \le k \le N, \tag{1.22}$$

and

$$\mathbf{v}_i = \frac{\sum\limits_{k=1}^{N}\mu_{i,k}^m \mathbf{x}_k}{\sum\limits_{k=1}^{N}\mu_{i,k}^m}, \quad 1 \le i \le c. \tag{1.23}$$

This solution also satisfies the remaining constraints (1.12) and (1.14). Note that equation (1.23) gives $\mathbf{v}_i$ as the weighted mean of the data items that belong to a cluster, where the weights are the membership degrees. That is why the algorithm is called "c-means". One can see that the FCM algorithm is a simple iteration through (1.22) and (1.23) (see Algorithm 1.5.1).

Example 1.1 (Demonstration for fuzzy c-means). *Consider a synthetic and a real data set in $\mathbb{R}^2$ (see Figure 1.7, Figure 1.8, Figure 1.9 and Figure 1.10). The dots represent the data points, the 'o' markers are the cluster centers. On the left side the membership values are also shown, on the right side the curves represent values that are inversely proportional to the distances. The figures shown in the further examples (Example 1.2 and Example 1.3) have the same structure.*

The synthetic data set in Figure 1.7 consist of three well-separated clusters of different shapes and size. The first cluster has ellipsoidal shape, the another two are round, but these are different in size. One can see that the FCM algorithm strictly imposes a circular shape, even though the clusters are rather elongated. Compare Figure 1.7 with Figure 1.8 it can be seen that in this case the normalization does not change the results because it has only a little effect on the distribution of data.

In Figure 1.9 a motorcycle data set can be seen: head acceleration of a human "post mortem test object" was plotted in time [295]. (This data set – among others – can be found on the webpage of this book www.fmt.vein.hu/softcomp/cluster.)

Algorithm 1.5.1 (Fuzzy c-Means).

Given the data set $\mathbf{X}$, *choose the number of clusters* $1 < c < N$, *the weighting exponent* $m > 1$, *the termination tolerance* $\epsilon > 0$ *and the norm-inducing matrix* $\mathbf{A}$. *Initialize the partition matrix randomly, such that* $\mathbf{U}^{(0)} \in M_{fc}$.

Repeat *for* $l = 1, 2, \ldots$

Step 1 *Compute the cluster prototypes (means):*

$$\mathbf{v}_i^{(l)} = \frac{\sum_{k=1}^{N} (\mu_{i,k}^{(l-1)})^m \mathbf{x}_k}{\sum_{k=1}^{N} (\mu_{i,k}^{(l-1)})^m}, \; 1 \leq i \leq c. \tag{1.24}$$

Step 2 *Compute the distances:*

$$D_{i,k\mathbf{A}}^2 = (\mathbf{x}_k - \mathbf{v}_i)^T \mathbf{A} (\mathbf{x}_k - \mathbf{v}_i), \quad 1 \leq i \leq c, \quad 1 \leq k \leq N. \tag{1.25}$$

Step 3 *Update the partition matrix:*

$$\mu_{i,k}^{(l)} = \frac{1}{\sum_{j=1}^{c} (D_{i,k\mathbf{A}}/D_{j,k\mathbf{A}})^{2/(m-1)}}. \tag{1.26}$$

until $||\mathbf{U}^{(l)} - \mathbf{U}^{(l-1)}|| < \epsilon$.

If the data are normalized (i.e., all the features have zero mean and unit variance), the clustering results will change as it can be seen in Figure 1.10. *The clusters have naturally circular shape but the cluster centers are different: these are rather located above each other by the original data (also with different initial states). Consequently the fuzzy c-means algorithm is sensitive to the scaling (normalization) of data.*

□

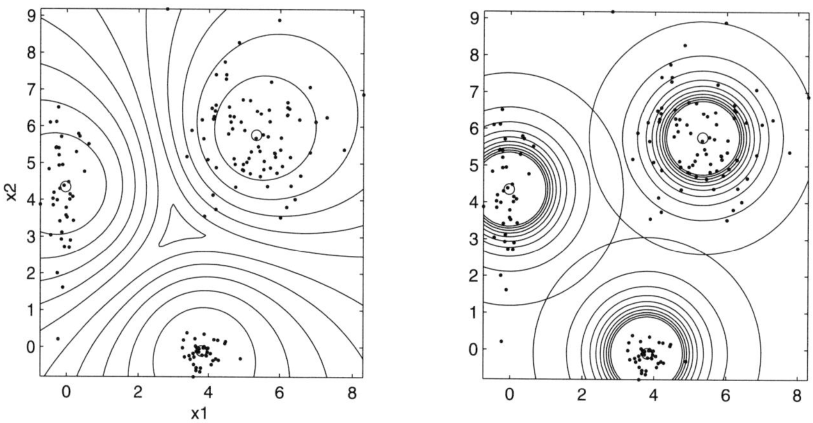

Figure 1.7: The results of the fuzzy c-means algorithm by the synthetic data set.

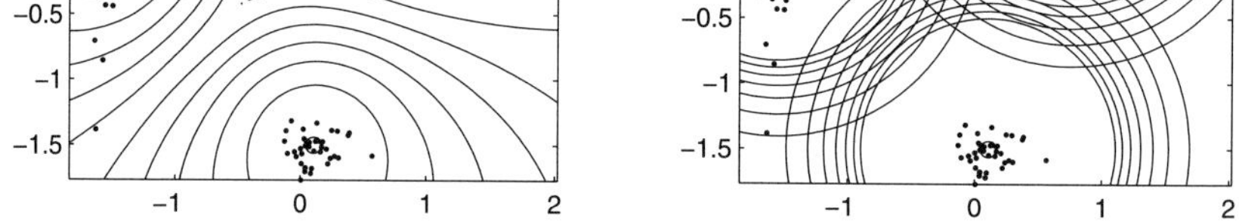

Figure 1.8: The results of the fuzzy c-means algorithm by the synthetic data set with normalization.

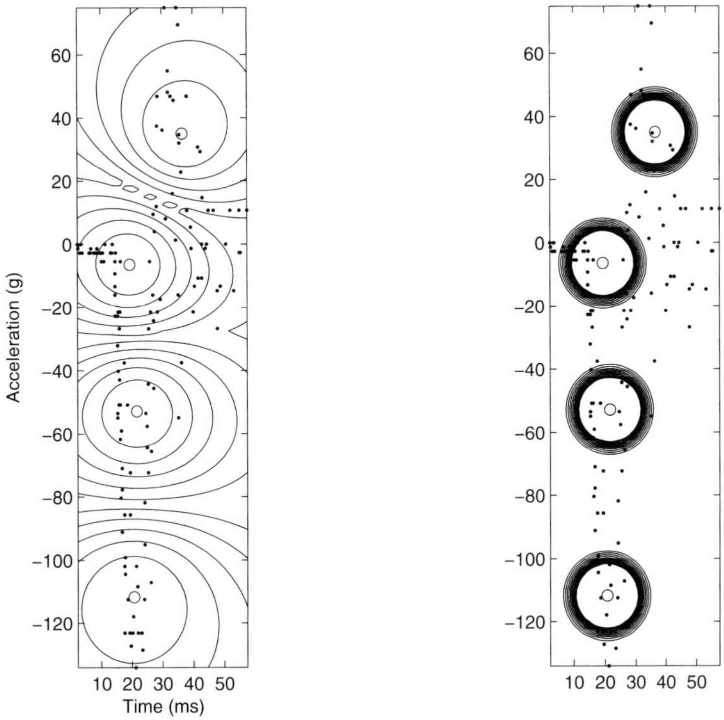

Figure 1.9: The results of the fuzzy c-means algorithm by the motorcycle data set.

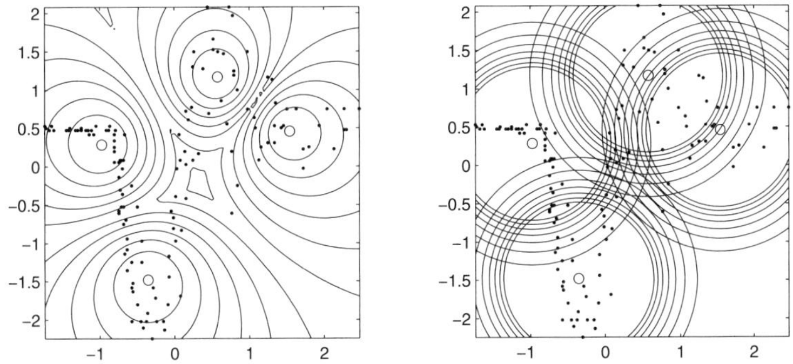

Figure 1.10: The results of the fuzzy c-means algorithm by the motorcycle data set with normalization.

1.5.5 Inner-Product Norms

The shape of the clusters is determined by the choice of the particular $\mathbf{A}$ in the distance measure (1.19). A common choice is $\mathbf{A} = \mathbf{I}$, which induces the standard Euclidean norm:

$$D^2_{i,k\mathbf{A}} = (\mathbf{x}_k - \mathbf{v}_i)^T(\mathbf{x}_k - \mathbf{v}_i). \tag{1.27}$$

The norm metric influences the clustering criterion by changing the measure of dissimilarity. The Euclidean norm induces hyperspherical clusters, i.e., clusters whose surface of constant membership are hyperspheres. Both the diagonal and the Mahalanobis norm generate hyperellipsoidal clusters, the difference is that with the diagonal norm, the axes of the hyperellipsoids are parallel to the coordinate axes while with the Mahalanobis norm the orientation of the hyperellipsoids is arbitrary, see also in Figure 1.2.

$\mathbf{A}$ can be chosen as an $n \times n$ diagonal matrix that accounts for different variances in the directions of the coordinate axes of $\mathbf{X}$:

$$\mathbf{A}_D = \begin{bmatrix} (1/\sigma_1)^2 & 0 & \dots & 0 \\ 0 & (1/\sigma_2)^2 & \dots & 0 \\ \vdots & \vdots & \ddots & \vdots \\ 0 & 0 & \dots & (1/\sigma_n)^2 \end{bmatrix}. \tag{1.28}$$

This matrix induces a diagonal norm on $\mathbb{R}^n$. Finally, $\mathbf{A}$ can be defined as the inverse of the $n \times n$ sample covariance matrix of $\mathbf{X}$: $\mathbf{A} = \mathbf{F}^{-1}$, with

$$\mathbf{F} = \frac{1}{N-1}\sum_{k=1}^{N}(\mathbf{x}_k - \overline{\mathbf{x}})(\mathbf{x}_k - \overline{\mathbf{x}})^T. \tag{1.29}$$

Here $\overline{\mathbf{x}}$ denotes the sample mean of the data. In this case, $\mathbf{A}$ induces the Mahalanobis norm on $\mathbb{R}^n$.

A common limitation of clustering algorithms based on a fixed distance norm is that such a norm induces a fixed topological structure on $\mathbb{R}^n$ and forces the objective function to prefer clusters of that shape even if they are not present. Generally, different matrices A_i are required for the different clusters, but there is no guideline as to how to choose them a priori. The norm-inducing matrix A can be adapted by using estimates of the data covariance, and can be used to estimate the statistical dependence of the data in each cluster. The Gustafson–Kessel algorithm (GK) and the fuzzy maximum likelihood estimation algorithm (Gath–Geva algorithm (GG)) are based on an adaptive distance measure.

1.5.6 Gustafson–Kessel Algorithm

Gustafson and Kessel extended the standard fuzzy c-means algorithm by employing an adaptive distance norm, in order to detect clusters of different geometrical shapes in one data set [108]. Each cluster has its own norm-inducing matrix A_i, which yields the following inner-product norm:

$$D^2_{i,k\mathbf{A}} = (\mathbf{x}_k - \mathbf{v}_i)^T \mathbf{A}_i (\mathbf{x}_k - \mathbf{v}_i), \quad 1 \leq i \leq c, \quad 1 \leq k \leq N. \tag{1.30}$$

The matrices $\mathbf{A}_i$ are used as optimization variables in the c-means functional, thus allowing each cluster to adapt the distance norm to the local topological structure of the data. Let $\mathbf{A}$ denote a c-tuple of the norm-inducing matrices: $\mathbf{A} = (\mathbf{A}_1, \mathbf{A}_2, \ldots, \mathbf{A}_c)$. The objective functional of the GK algorithm is defined by:

$$J(\mathbf{X}; \mathbf{U}, \mathbf{V}, \mathbf{A}) = \sum_{i=1}^{c} \sum_{k=1}^{N} (\mu_{i,k})^m D^2_{i,k\mathbf{A}_i}. \tag{1.31}$$

For a fixed $\mathbf{A}$, conditions (1.12), (1.13) and (1.14) can be directly applied. However, the objective function (1.31) cannot be directly minimized with respect to $\mathbf{A}_i$, since it is linear in $\mathbf{A}_i$. This means that J can be made as small as desired by simply making $\mathbf{A}_i$ less positive definite. To obtain a feasible solution, $\mathbf{A}_i$ must be constrained in some way. The usual way of accomplishing this is to constrain the determinant of $\mathbf{A}_i$. Allowing the matrix $\mathbf{A}_i$ to vary with its determinant fixed corresponds to optimizing the cluster's shape while its volume remains constant:

$$\det(\mathbf{A}_i) = \rho_i, \quad \rho > 0, \tag{1.32}$$

where ρ_i is fixed for each cluster. Using the Lagrange multiplier method, the following expression for $\mathbf{A}_i$ is obtained:

$$\mathbf{A}_i = [\rho_i \det(\mathbf{F}_i)]^{1/n} \mathbf{F}_i^{-1}, \tag{1.33}$$

where $\mathbf{F}_i$ is the *fuzzy covariance matrix* of the ith cluster defined by:

$$\mathbf{F}_i = \frac{\sum\limits_{k=1}^{N} (\mu_{i,k})^m (\mathbf{x}_k - \mathbf{v}_i)(\mathbf{x}_k - \mathbf{v}_i)^T}{\sum\limits_{k=1}^{N} (\mu_{i,k})^m}. \tag{1.34}$$

Note that the substitution of (1.33) and (1.34) into (1.30) gives a generalized squared Mahalanobis distance norm between $\mathbf{x}_k$ and the cluster mean $\mathbf{v}_i$, where the covariance is weighted by the membership degrees in $\mathbf{U}$. The formal description of GK clustering can be found in Algorithm 1.5.2.

Example 1.2 (Demonstration for Gustafson–Kessel algorithm). *The GK algorithm is applied to the data set from Example* 1.1. *Figure* 1.11 *shows the clustering results by the synthetic data set. From these results it can be seen that GK algorithm cannot reflect the different size of the clusters, however in Figure* 1.11 *the three*

Algorithm 1.5.2 (Gustafson–Kessel Algorithm).

Given the data set $\mathbf{X}$, *choose the number of clusters* $1 < c < N$, *the weighting exponent* $m > 1$, *the termination tolerance* $\epsilon > 0$ *and the norm-inducing matrix* $\mathbf{A}$. *Initialize the partition matrix randomly, such that* $\mathbf{U}^{(0)} \in M_{fc}$.

Repeat *for* $l = 1, 2, \ldots$

Step 1 *Calculate the cluster centers.*

$$\mathbf{v}_i^{(l)} = \frac{\sum_{k=1}^{N} (\mu_{i,k}^{(l-1)})^m \mathbf{x}_k}{\sum_{k=1}^{N} (\mu_{i,k}^{(l-1)})^m}, \ 1 \leq i \leq c. \tag{1.35}$$

Step 2 *Compute the cluster covariance matrices.*

$$\mathbf{F}_i^{(l)} = \frac{\sum_{k=1}^{N} (\mu_{i,k}^{(l-1)})^m \left(\mathbf{x}_k - \mathbf{v}_i^{(l)}\right) \left(\mathbf{x}_k - \mathbf{v}_i^{(l)}\right)^T}{\sum_{k=1}^{N} (\mu_{i,k}^{(l-1)})^m}, \ 1 \leq i \leq c. \tag{1.36}$$

Step 3 *Compute the distances.*

$$D_{i,k\mathbf{A}_i}^2(\mathbf{x}_k, \mathbf{v}_i) = \left(\mathbf{x}_k - \mathbf{v}_i^{(l)}\right)^T \left[(\rho_i \det(\mathbf{F}_i))^{1/n} F_i^{-1}\right] \left(\mathbf{x}_k - \mathbf{v}_i^{(l)}\right). \tag{1.37}$$

Step 4 *Update the partition matrix*

$$\mu_{i,k}^{(l)} = \frac{1}{\sum_{j=1}^{c} \left(D_{i,k\mathbf{A}_i}(\mathbf{x}_k, \mathbf{v}_i)/D_{j,k\mathbf{A}_j}(\mathbf{x}_k, \mathbf{v}_j)\right)^{2/(m-1)}}, \quad 1 \leq i \leq c, 1 \leq k \leq N. \tag{1.38}$$

until $||\mathbf{U}^{(l)} - \mathbf{U}^{(l-1)}|| < \epsilon$.

clusters by the synthetic data set are very different in size. (The normalization has a little effect as mentioned in Example 1.1, *so these results are not presented.)*

Figure 1.12 *shows that the GK algorithm can adapt the distance norm to the underlying distribution of the data. The shape of the clusters can be determined from the eigenstructure of the resulting covariance matrices* $\mathbf{F}_i$. *The GK algorithm is not so sensitive to the scaling (normalization) of data as the FCM algorithm, but the results can be significantly changed by different initial state.*

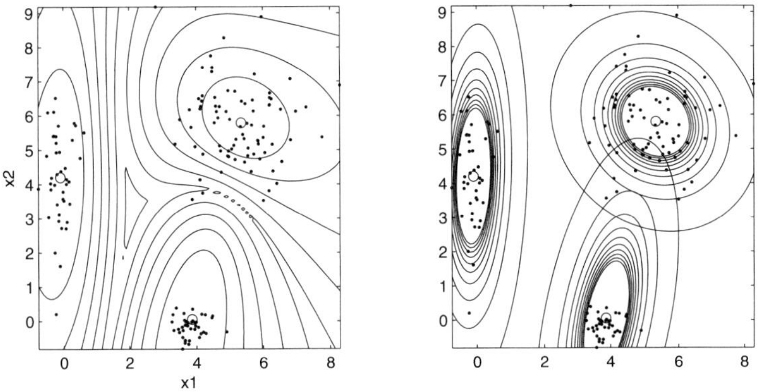

Figure 1.11: The results of the Gustafson–Kessel algorithm by the synthetic data set.

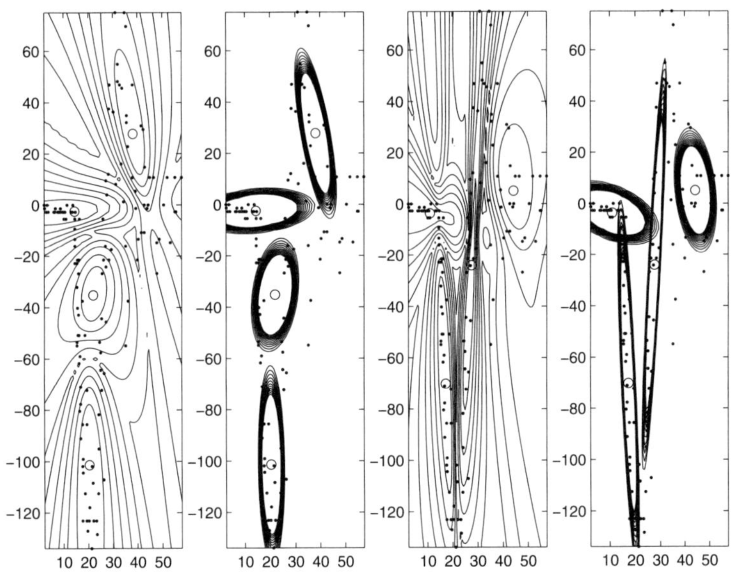

Figure 1.12: The results of the Gustafson–Kessel algorithm by the motorcycle data set.

The first two diagrams of Figure 1.12 *show the membership values and the inverse of the distances, respectively obtained by a randomly initialized partition matrix. This randomly initialized clustering does not give only this result, from another randomly selected initialization, we can obtain clusters that can be seen on the second two diagrams of Figure* 1.12. *If GK algorithm is initialized by the*

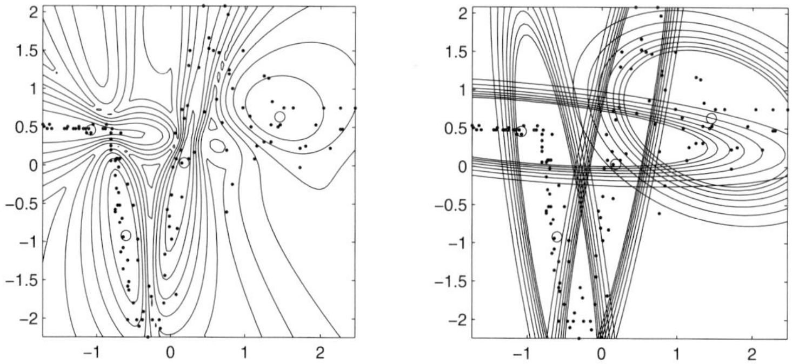

Figure 1.13: The results of the Gustafson–Kessel algorithm by the motorcycle data set with normalization.

results of FCM with normalized data set, then the results are always consistent and identical to the results plotted on the second two diagrams of Figure 1.12. *The objective function* (1.31) *is smaller in this second case.*

□

1.5.7 Gath–Geva Clustering Algorithm

The fuzzy maximum likelihood estimates clustering algorithm employs a distance norm based on the fuzzy maximum likelihood estimates, proposed by Bezdek and Dunn [40]:

$$D_{i,k}(\mathbf{x}_k, \mathbf{v}_i) = \frac{(2\pi)^{\left(\frac{n}{2}\right)}\sqrt{\det(\mathbf{F}_i)}}{\alpha_i} \exp\left(\frac{1}{2}\left(\mathbf{x}_k - \mathbf{v}_i\right)^T \mathbf{F}_i^{-1}\left(\mathbf{x}_k - \mathbf{v}_i\right)\right). \tag{1.39}$$

Note that, contrary to the GK algorithm, this distance norm involves an exponential term and thus decreases faster than the inner-product norm. $\mathbf{F}_i$ denotes the fuzzy covariance matrix of the ith cluster, similarly to GK algorithm, given by (1.34). The α_i is the prior probability of selecting cluster i, given by:

$$\alpha_i = \frac{1}{N}\sum_{k=1}^{N} \mu_{i,k}. \tag{1.40}$$

The membership degrees $\mu_{i,k}$ are interpreted as the posterior probabilities of selecting the ith cluster given the data point $\mathbf{x}_k$. Gath and Geva reported that the fuzzy maximum likelihood estimates clustering algorithm is able to detect clusters of varying shapes, sizes and densities. This is because the cluster covariance matrix is used in conjunction with an "exponential" distance, and the clusters are not constrained in volume. However, this algorithm is less robust in the sense that it needs a good initialization, since due to the exponential distance norm, it converges to a near local optimum. The minimum of (1.16) is sought by the alternating optimization (AO) method (Gath–Geva clustering algorithm) given in Algorithm 1.5.3:

Algorithm 1.5.3 (Gath–Geva Algorithm).

Given a set of data $\mathbf{X}$ *specify* c, *choose a weighting exponent* $m > 1$ *and a termination tolerance* $\epsilon > 0$. *Initialize the partition matrix such that* (1.12), (1.13) *and* (1.14) *holds.*

Repeat *for* $l = 1, 2, \ldots$

Step 1 *Calculate the cluster centers:* $\mathbf{v}_i^{(l)} = \frac{\sum_{k=1}^{N} (\mu_{i,k}^{(l-1)})^m \mathbf{x}_k}{\sum_{k=1}^{N} (\mu_{i,k}^{(l-1)})^m}$, $1 \le i \le c$.

Step 2 *Compute the distance measure* $D_{i,k}^2$.

The distance to the prototype is calculated based on the fuzzy covariance matrices of the cluster

$$\mathbf{F}_i^{(l)} = \frac{\sum_{k=1}^{N} (\mu_{i,k}^{(l-1)})^m \left(\mathbf{x}_k - \mathbf{v}_i^{(l)}\right)\left(\mathbf{x}_k - \mathbf{v}_i^{(l)}\right)^T}{\sum_{k=1}^{N} (\mu_{i,k}^{(l-1)})^m}, \; 1 \le i \le c. \tag{1.41}$$

The distance function is chosen as

$$D_{i,k}^2(\mathbf{x}_k, \mathbf{v}_i) = \frac{(2\pi)^{\left(\frac{n}{2}\right)} \sqrt{\det(\mathbf{F}_i)}}{\alpha_i} \exp\left(\frac{1}{2}\left(\mathbf{x}_k - \mathbf{v}_i^{(l)}\right)^T \mathbf{F}_i^{-1}\left(\mathbf{x}_k - \mathbf{v}_i^{(l)}\right)\right) \tag{1.42}$$

with the a priori probability $\alpha_i = \frac{1}{N}\sum_{k=1}^{N} \mu_{i,k}$.

Step 3 *Update the partition matrix*

$$\mu_{i,k}^{(l)} = \frac{1}{\sum_{j=1}^{c} \left(D_{i,k}(\mathbf{x}_k, \mathbf{v}_i)/D_{j,k}(\mathbf{x}_k, \mathbf{v}_j)\right)^{2/(m-1)}}, \quad 1 \le i \le c, 1 \le k \le N. \tag{1.43}$$

until $||\mathbf{U}^{(l)} - \mathbf{U}^{(l-1)}|| < \epsilon$.

Example 1.3 (Demonstration for Gath–Geva algorithm). *The GG algorithm is applied to the data set from Example* 1.1. *In Figure* 1.14 *the effect of the different cluster size can be seen. Since GK algorithm cannot detect the different cluster size (see Figure* 1.11*), GG algorithm can cluster the data perfectly.*

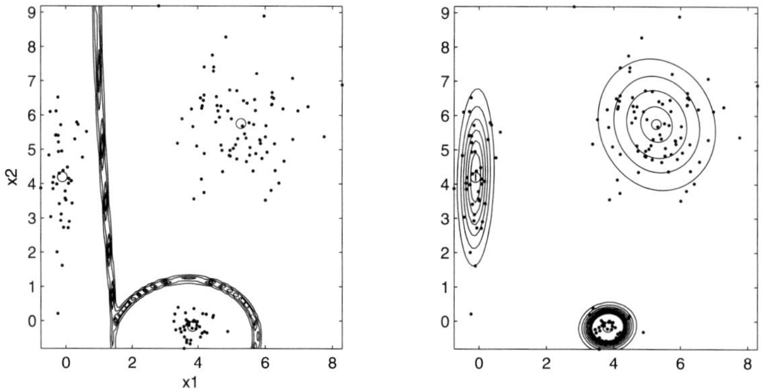

Figure 1.14: The results of the Gath–Geva algorithm by the synthetic data set.

Similarly to Example 1.2*, the results obtained by two different initializations are compared. On the left part of Figure* 1.15 *the membership values and the distances obtained by random initialization are shown, while on the remaining two subplots the results of the clustering initialized by fuzzy c-means are shown (i.e., the structure of Figure* 1.15 *is the same as in Figure* 1.12*). As the difference between these two plots shows, the GG algorithm is very sensitive to initialization, and it can adapt the distance norm to the underlying distribution of the data which is reflected in the different sizes of the clusters.*

The motorcycle data set can be considered a function with additional noise. Therefore the human analyzer finds better the results shown on the second two diagrams of Figure 1.15*. However, the value of the fuzzy objective function* (1.16) *is bigger in the second case than in the first case.*

□

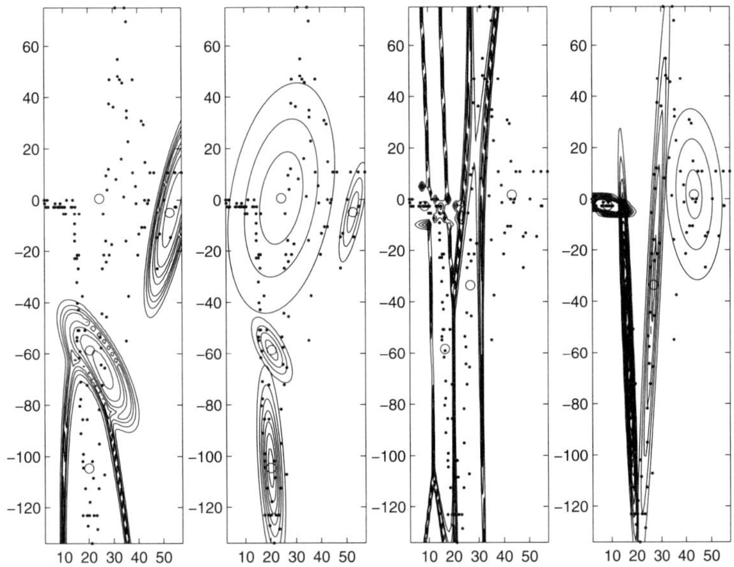

Figure 1.15: The results of the Gath–Geva algorithm by the motorcycle data set.

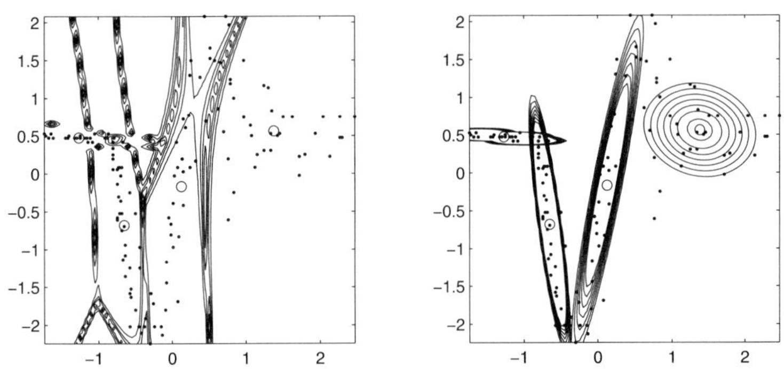

Figure 1.16: The results of the Gath–Geva algorithm by the motorcycle data set with normalization.

1.6 Cluster Analysis of Correlated Data

In practical situations, there is often functional relationship among variables. Therefore, there are independent (input) and dependent (output) variables. A sample or observation vector $\mathbf{z}_k$ contains independent and dependent variables as well: $\mathbf{z}_k = [\mathbf{x}_k^T \, y_k]^T$ in case of n input and one output variables. The set of the available input-output data do not fill the space but they are close to a surface defined by the relationship among them called regression surface. (Data may contain noise, that is why they do not form an ideal surface.) When the available input-output data set is clustered and the appropriate number of regressors are used, the clusters will be 'flat' and locally approximate the regression surface as it is depicted in Figure 1.17 in case of one input and one output variable. In this case the clusters can be approximately regarded as local linearizations of the modeled system, i.e., local (linear) models of the relationship within a particular region.

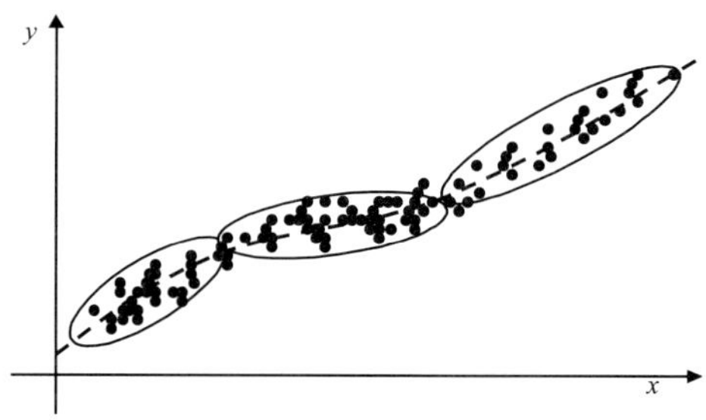

Figure 1.17: Example for clusters approximating the regression surface.

In other words, the clusters can be approximately regarded as local linear subspaces. This phenomenon is reflected by the smallest eigenvalues $\lambda_{i,n+1}$ of the cluster covariance matrices $\mathbf{F}_i$ that are typically in orders of magnitude smaller than the remaining eigenvalues [26] (see Figure 1.18).

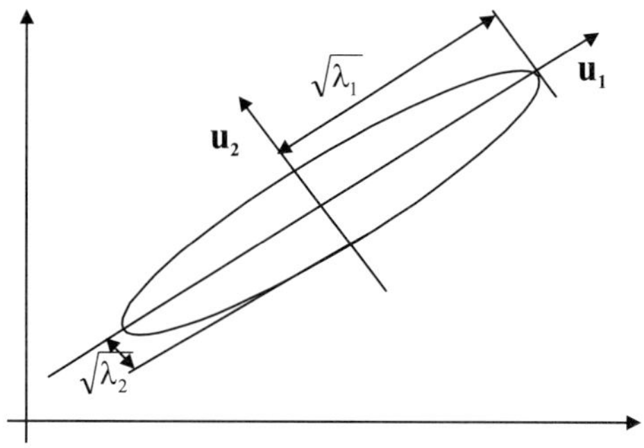

Figure 1.18: Eigenvalues of clusters obtained by GG clustering.

Based on the assumption that the clusters somehow represent the local linear approximation of the system, two methods can be presented for the estimation of the parameters of the local linear models.

1. **Ordinary least-squares estimation**

   The ordinary weighted least-squares method can be applied to estimate the parameters of the local linear models θ_i separately by minimizing the following criteria:

$$\min_{\boldsymbol{\theta}_i} \left(\mathbf{y} - \mathbf{X}_e\boldsymbol{\theta}_i\right)^T \boldsymbol{\Phi}_i \left(\mathbf{y} - \mathbf{X}_e\boldsymbol{\theta}_i\right) \tag{1.44}$$

   where $\mathbf{X}_e = [\mathbf{X}\,\mathbf{1}]$ is the regressor matrix extended by a unitary column and $\boldsymbol{\Phi}_i$ is a matrix having the membership degrees on its main diagonal:

$$\boldsymbol{\Phi}_i = \begin{bmatrix} \mu_{i,1} & 0 & \cdots & 0 \\ 0 & \mu_{i,2} & \cdots & 0 \\ \vdots & \vdots & \ddots & \vdots \\ 0 & 0 & \cdots & \mu_{i,N} \end{bmatrix}. \tag{1.45}$$

   The weighted least-squares estimate of the consequent parameters is given by

$$\boldsymbol{\theta}_i = \left(\mathbf{X}_e^T\boldsymbol{\Phi}_i\mathbf{X}_e\right)^{-1}\mathbf{X}_e^T\boldsymbol{\Phi}_i\mathbf{y}. \tag{1.46}$$

   Consider the case when there is one output variable. Similarly to the observation vector $\mathbf{z}_k = [\mathbf{x}_k^T\, y_k]^T$, the prototype vector is partitioned as $\mathbf{v}_i = \left[(\mathbf{v}_i^x)^T\; v_i^y\right]$ into a vector $\mathbf{v}_i^x$ corresponding to the regressor $\mathbf{x}$, and a scalar v_i^y corresponding to the output y. In this case, the output can be written in the following form:

$$y = \mathbf{a}_i^T\mathbf{x} + b_i \tag{1.47}$$

   where the parameters of the ith local linear model are $\boldsymbol{\theta}_i = [\mathbf{a}_i^T\; b_i]$. When $\mu_{i,k}$ is obtained by the Gath–Geva clustering algorithm, the covariance matrix can directly be used to obtain the estimate instead of (1.46):

$$\begin{aligned} \mathbf{a}_i &= \left(\mathbf{F}^{xx}\right)^{-1}\mathbf{F}^{xy} \\ b_i &= v_i^y - \mathbf{a}_i^T\mathbf{v}_i^x, \end{aligned} \tag{1.48}$$

   where $\mathbf{F}_i^{xx}$ is obtained by partitioning the covariance matrix $\mathbf{F}_i$ as follows

$$\mathbf{F}_i = \begin{bmatrix} \mathbf{F}_i^{xx} & \mathbf{F}_i^{xy} \\ \mathbf{F}_i^{yx} & \mathbf{F}_i^{yy} \end{bmatrix} \tag{1.49}$$

   where

   - $\mathbf{F}_i^{xx}$ is the $n \times n$ submatrix containing the first n rows and columns of $\mathbf{F}_i$,

- $\mathbf{F}_i^{xy}$ is an $n \times 1$ column vector containing the first n elements of last column of $\mathbf{F}_i$,
- $\mathbf{F}_i^{yx}$ is an $1 \times n$ row vector containing the first n elements of the last row of $\mathbf{F}_i$, and
- $\mathbf{F}_i^{yy}$ is the last element in the last row of $\mathbf{F}_i$.

This follows directly from the properties of least-squares estimation [74]. This method can easily be extended to more than one output variables.

2. **Total least-squares estimation**

The eigenvector corresponding to the smallest eigenvalue, $\mathbf{u}_{n+1}^i$, determines the normal vector to the hyperplane spanned by the remaining eigenvectors of that cluster (see Figure 1.18)

$$(\mathbf{u}_{n+1}^i)^T(\mathbf{z}_k - \mathbf{v}_i) = 0. \tag{1.50}$$

The smallest eigenvector is partitioned in the same way as the observation and prototype vectors, $\mathbf{u}_{n+1}^i = \left[\left(\mathbf{u}_{n+1}^{i,x}\right)^T \; u_{n+1}^{i,y}\right]^T$. By using this partitioned vectors (1.50) can be written as

$$\left[\left(\mathbf{u}_{n+1}^{i,x}\right)^T \; u_{n+1}^{i,y}\right]\left([\mathbf{x}_k^T \; y_k]^T - \left[(\mathbf{v}_i^x)^T \; v_i^y\right]^T\right) = 0 \tag{1.51}$$

from which the parameters of the hyperplane defined by the cluster can be obtained:

$$y_k = \underbrace{\frac{-1}{u_{n+1}^{i,y}}\left(\mathbf{u}_{n+1}^{i,x}\right)^T}_{\mathbf{a}_i^T}\mathbf{x}_k + \underbrace{\frac{1}{u_{n+1}^{i,y}}\left(\mathbf{u}_{n+1}^{i}\right)^T\mathbf{v}_i}_{b_i}. \tag{1.52}$$

Comparing to ordinary least-squares estimation, the TLS algorithm can be powerful when there are errors in the input variables. However, the TLS algorithm does not minimize the mean-square prediction error of the model. Furthermore, if the input variables of the model are locally greatly correlated the smallest eigenvector then does not define a hyperplane related to the regression problem, but it reflects the dependency of the input variables.

This section showed that hyperellipsoidal clusters can be effectively used to represent correlated data, and local linear models can be easily extracted from the parameters of these clusters. Of course, different cluster shapes can be obtained with different norms as suggested in the GK or GG algorithm, or with different kinds of prototypes, e.g., linear varieties (FCV), where the clusters are linear subspaces of the feature space. An r-dimensional linear variety is defined by the vector $\mathbf{v}_i$ and the directions $\mathbf{s}_{i,j}, j = 1, \ldots, r$. In this case, the distance between the data $\mathbf{x}_k$ and the ith cluster is:

$$D^2_{i,k} = \|\mathbf{x}_k - \mathbf{v}_i\|^2 - \sum_{j=1}^{r} \left((\mathbf{x}_k - \mathbf{v}_i)^T \mathbf{s}_{i,j}\right)^2 . \tag{1.53}$$

In Chapter 6 a similar distance norm based on Principal Component Analysis will be introduced and applied for the segmentation of multivariate time-series. In the following, a fuzzy clustering technique is introduced that is able to identify local models directly during the partitioning of the data.

Fuzzy c-Regression Models

Fuzzy c-regression models yield simultaneous estimates of parameters of c regression models together with a fuzzy c-partitioning of the data. The regression models take the following general form

$$y_k = f_i(\mathbf{x}_k, \boldsymbol{\theta}_i) \tag{1.54}$$

where $\mathbf{x}_k = [x_{k,1}, \ldots, x_{k,n}]$ denotes the kth data sample and the functions f_i are parameterized by $\boldsymbol{\theta}_i \in \mathbb{R}^{p_i}$. The membership degree $\mu_{i,k} \in \boldsymbol{U}$ is interpreted as a weight representing the extent to which the value predicted by the model $f_i(\mathbf{x}_k, \boldsymbol{\theta}_i)$ matches y_k. This prediction error is defined by:

$$E_{i,k} = \left(y_k - f_i(\mathbf{x}_k; \boldsymbol{\theta}_i)\right)^2 , \tag{1.55}$$

but other measures can be applied as well, provided they fulfill the minimizer property stated by Hathaway and Bezdek [113]. The family of objective functions for fuzzy c-regression models is defined by

$$E_m(\mathbf{U}, \{\boldsymbol{\theta}_i\}) = \sum_{i=1}^{c} \sum_{k=1}^{N} (\mu_{i,k})^m E_{i,k}(\boldsymbol{\theta}_i) , \tag{1.56}$$

where $m \in \langle 1, \infty)$ denotes a weighting exponent which determines the fuzziness of the resulting clusters. One possible approach to the minimization of the objective function (1.56) is the group coordinate minimization method that results in algorithm 1.6.1.

A specific situation arises when the regression functions f_i are linear in the parameters $\boldsymbol{\theta}_i$, $f_i(\mathbf{x}_k; \boldsymbol{\theta}_i) = \mathbf{x}^T_{i,k}\boldsymbol{\theta}_i$, where $\mathbf{x}_{i,k}$ is a known arbitrary function of $\mathbf{x}_k$. In this case, the parameters can be obtained as a solution of a set of weighted least-squares problem where the membership degrees of the fuzzy partition matrix $\mathbf{U}$ serve as the weights.

Algorithm 1.6.1 (Fuzzy c-Regression Models).

- **Initialization** *Given a set of data* $Z = \{(\mathbf{x}_1, y_1), \ldots, (\mathbf{x}_N, y_N)\}$ *specify* c, *the structure of the regression models* (1.55) *and the error measure* (1.56). *Choose a weighting exponent* $m > 1$ *and a termination tolerance* $\epsilon > 0$. *Initialize the partition matrix randomly.*

- **Repeat** *for* $l = 1, 2, \ldots$

  Step 1 *Calculate values for the model parameters* $\boldsymbol{\theta}_i$ *that minimize the cost function* $E_m(\mathbf{U}, \{\boldsymbol{\theta}_i\})$.

  Step 2 *Update the partition matrix*

$$\mu_{i,k}^{(l)} = \frac{1}{\sum_{j=1}^{c} (E_{i,k}/E_{j,k})^{2/(m-1)}}, \; 1 \le i \le c, \; 1 \le k \le N\,. \tag{1.57}$$

  until $||\mathbf{U}^{(l)} - \mathbf{U}^{(l-1)}|| < \epsilon$.

The N data pairs and the membership degrees are arranged in the following matrices.

$$\mathbf{X}_i = \begin{bmatrix} \mathbf{x}_{i,1}^T \\ \mathbf{x}_{i,2}^T \\ \vdots \\ \mathbf{x}_{i,N}^T \end{bmatrix}, \; \mathbf{y} = \begin{bmatrix} y_1 \\ y_2 \\ \vdots \\ y_N \end{bmatrix}, \; \boldsymbol{\Phi}_i = \begin{bmatrix} \mu_{i,1} & 0 & \cdots & 0 \\ 0 & \mu_{i,2} & \cdots & 0 \\ \vdots & \vdots & \ddots & \vdots \\ 0 & 0 & \cdots & \mu_{i,N} \end{bmatrix}. \tag{1.58}$$

The optimal parameters $\boldsymbol{\theta}_i$ are then computed by:

$$\boldsymbol{\theta}_i = [\mathbf{X}^T \boldsymbol{\Phi}_i \mathbf{X}]^{-1} \mathbf{X}^T \boldsymbol{\Phi}_i \mathbf{y}\,. \tag{1.59}$$

In the original paper that has introduced the fuzzy c-regression models [113], six issues were discussed for possible future research. Two of them were the following:

- Determining the number of clusters (the number of regression models).
- Avoiding local trap states.

The next section gives solutions to these problems by incorporating prior knowledge into the clustering procedure presented in this section.

Constrained Prototype based FCRM

This section deals with prototypes linear in the parameters (see also [1, 7]). Therefore, as it was shown, the parameters can be estimated by linear least-squares techniques. When linear equality and inequality constraints are defined on these prototypes, quadratic programming (QP) has to be used instead of the least-squares method. This optimization problem still can be solved effectively compared to other constrained nonlinear optimization algorithms.

The parameter constraints can be grouped into three categories:

- **Local constrains** are valid only for the parameters of a regression model, $\mathbf{\Lambda}_i \boldsymbol{\theta}_i \leq \boldsymbol{\omega}_i$.
- **Global constrains** are related to all of the regression models, $\mathbf{\Lambda}_{gl} \boldsymbol{\theta}_i \leq \boldsymbol{\omega}_{gl}$, $i = 1, \ldots, c$.
- **Relative constrains** define the relative magnitude of the parameters of two or more regression models,

$$\mathbf{\Lambda}_{rel,i,j} \begin{bmatrix} \boldsymbol{\theta}_i \\ \boldsymbol{\theta}_j \end{bmatrix} \leq \boldsymbol{\omega}_{rel,i,j}.$$

An example for these types of constrains are illustrated in Figure 1.19.

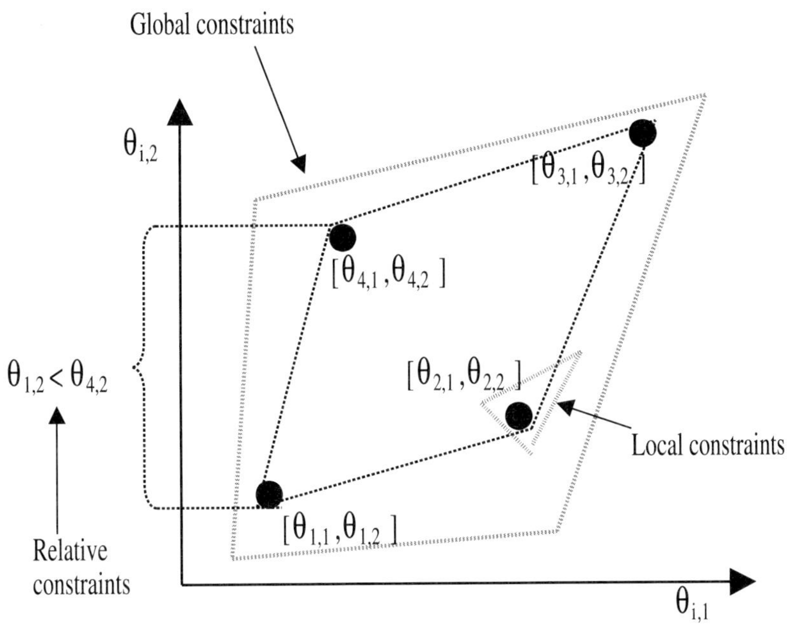

Figure 1.19: Example for local, global and relative constraints.

In order to handle relative constraints, the set of weighted optimization problems has to be solved simultaneously. Hence, the constrained optimization problem is formulated as follows:

$$\min_{\boldsymbol{\theta}} \left\{ \frac{1}{2} \boldsymbol{\theta}^T \mathbf{H} \boldsymbol{\theta} + \mathbf{c}^T \boldsymbol{\theta} \right\} \tag{1.60}$$

with $\mathbf{H} = 2\mathbf{X}^T\mathbf{\Phi}\mathbf{X}$, $\mathbf{c} = -2\mathbf{X}^T\mathbf{\Phi}\mathbf{y}'$, where

$$\mathbf{y}' = \begin{bmatrix} \boldsymbol{y} \\ \boldsymbol{y} \\ \vdots \\ \boldsymbol{y} \end{bmatrix}, \quad \boldsymbol{\theta} = \begin{bmatrix} \boldsymbol{\theta}_1 \\ \boldsymbol{\theta}_2 \\ \vdots \\ \boldsymbol{\theta}_c \end{bmatrix}, \tag{1.61}$$

$$\mathbf{X} = \begin{bmatrix} \boldsymbol{X}_1 & 0 & \cdots & 0 \\ 0 & \boldsymbol{X}_2 & \cdots & 0 \\ \vdots & \vdots & \ddots & \vdots \\ 0 & 0 & \cdots & \boldsymbol{X}_c \end{bmatrix}, \quad \mathbf{\Phi} = \begin{bmatrix} \mathbf{\Phi}_1 & 0 & \cdots & 0 \\ 0 & \mathbf{\Phi}_2 & \cdots & 0 \\ \vdots & \vdots & \ddots & \vdots \\ 0 & 0 & \cdots & \mathbf{\Phi}_c \end{bmatrix}. \tag{1.62}$$

and the constraints on $\boldsymbol{\theta}$:

$$\mathbf{\Lambda}\boldsymbol{\theta} \leq \boldsymbol{\omega} \tag{1.63}$$

with

$$\mathbf{\Lambda} = \begin{bmatrix} \mathbf{\Lambda}_1 & 0 & \cdots & 0 \\ 0 & \mathbf{\Lambda}_2 & \cdots & 0 \\ \vdots & \vdots & \ddots & \vdots \\ 0 & 0 & \cdots & \mathbf{\Lambda}_c \\ \mathbf{\Lambda}_{gl} & 0 & \cdots & 0 \\ 0 & \mathbf{\Lambda}_{gl} & \cdots & 0 \\ \vdots & \vdots & \ddots & \vdots \\ 0 & 0 & \cdots & \mathbf{\Lambda}_{gl} \\ \{\mathbf{\Lambda}_{rel}\} & & & \end{bmatrix}, \quad \boldsymbol{\omega} = \begin{bmatrix} \boldsymbol{\omega}_1 \\ \boldsymbol{\omega}_2 \\ \vdots \\ \boldsymbol{\omega}_c \\ \boldsymbol{\omega}_{gl} \\ \boldsymbol{\omega}_{gl} \\ \vdots \\ \boldsymbol{\omega}_{gl} \\ \{\boldsymbol{\omega}_{rel}\} \end{bmatrix}. \tag{1.64}$$

Example 1.4 (Identification of a Hammerstein system using constrained prototype based fuzzy c-regression model). *In this example, the presented method is used to approximate a Hammerstein system that consists of a series connection of a memoryless nonlinearity, f, and linear dynamics, G, as shown in Figure* 1.20*, where v represents the transformed input variable. For more details about dynamic systems and their modelling techniques, see Chapter* 3*, about Hammerstein and Wiener models, see Section* 4.1.

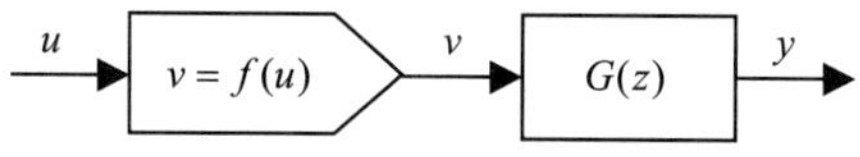

Figure 1.20: A series combination of a static nonlinearity and a linear dynamic system.

For transparent presentation, the Hammerstein system to be identified consists of a first-order linear part, $y(k+1) = 0.9y(k) + 0.1v(k)$, and a static nonlinearity *represented by a polynomial,* $v(k) = u(k)^2$*. The identification data consists of* 500 *input-output data. A FCRM with three regression models were identified,* $c = 3$*. The method has been included in the Fuzzy modelling and Identification Toolbox for MATLAB® and can be downloaded from the website of the book.*

As can be seen from Figure 1.21*, the fuzzy clusters obtained by standard FCRM do not result in convex membership functions.*

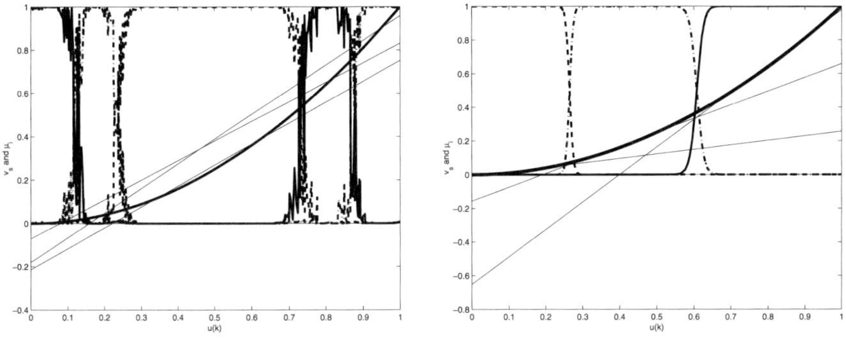

(a) *Unconstrained (standard) case.* (b) *FCRM with constrained prototypes.*

Figure 1.21: Projected membership degrees to the $u(k)$ domain obtained by fuzzy c-regression.

Two types of prior knowledge were used to define constraints on the cluster prototypes. Both of them resulted in relative constraints.

The first type of prior knowledge is that the process to be identified is a Hammerstein type. This means that the parameter at $y(k)$*, corresponding to the time constant of the system, is constant in all the rules. In the considered fuzzy model structure, each rule consists of a model. The output of a particular model is a linear combination of previous output(s) and input(s), and the output of the system is the weighted sum of the model outputs. The weight of the previous output and input with one sample time in the* i*th model is* $a_{i,1}$ *and* $b_{i,1}$*, respectively, and the bias is* c_i *(for more information see Section* 3.2*). With this notation, the statement above can be written as* $a_{i,1} = a_{j,1}, \forall i, j = 1, \ldots, c$ [30]*. In the unconstrained FCRM case, these poles of the local dynamic models are almost identical to each other and to the pole of the original system (see, Table* 1.1*), so this constraint has only little contribution to the resulted model. It can be seen in Table* 1.1*, that two of the regression models (*1 *and* 3*) have almost the same parameters. Even, the steady-state behavior of the resulted local LTI models do not relate to the quadratic nonlinearity of the system (see, Figure* 1.21*). This is because the clustering algorithm found a local minimum. To avoid these local minima, the second type of prior knowledge is that the regression models have to be different. As these models* *are forced to be different, the operating regions of these models will be also different, hence the fuzzy partition of the operating space will be more distinguishable. By using constraints, the resulted models have the same* $a_{i,1}$ *parameters that are identical to the parameters of the original system that generated the data,* 0.9, *(see, Table* 1.1*). Moreover, because of the gains of the local models are forced to be different from each other, the operating regions of the LTI models are clearly separated, Figure* 1.21.

Table 1.1: Parameters of the local models obtained by different identification method.

Unconstrained method

	1	2	3
$a_{i,1}$	0.9005	0.9020	0.9043
$b_{i,1}$	0.0962	0.1117	0.0864
c_i	-0.0213	-0.0177	-0.0068

Constrained method

	1	2	3
$a_{i,1}$	0.9000	0.9000	0.9000
$b_{i,1}$	0.0268	0.0816	0.1632
c_i	-0.0011	-0.0156	-0.0652

□

1.7 Validity Measures

Cluster validity refers to the problem whether a given fuzzy partition fits to the data at all. The clustering algorithm always tries to find the best fit for a fixed number of clusters and the parameterized cluster shapes. However this does not mean that even the best fit is meaningful at all. Either the number of clusters might be wrong or the cluster shapes might not correspond to the groups in the data, if the data can be grouped in a meaningful way at all. Two main approaches to determining the appropriate number of clusters in data can be distinguished:

- Starting with a sufficiently large number of clusters, and successively reducing this number by merging clusters that are similar (compatible) with respect to some predefined criteria. This approach is called *compatible cluster merging.*
- Clustering data for different values of c, and using *validity measures* to assess the goodness of the obtained partitions. This can be done in two ways:
  - The first approach is to define a validity function which evaluates a complete partition. An upper bound for the number of clusters must be estimated ($c_{\max}$), and the algorithms have to be run with each

$c \in \{2, 3, \ldots, c_{\max}\}$. For each partition, the validity function provides a value such that the results of the analysis can be compared indirectly.

– The second approach consists of the definition of a validity function that evaluates individual clusters of a cluster partition. Again, $c_{\max}$ has to be estimated and the cluster analysis has to be carried out for $c_{\max}$. The resulting clusters are compared to each other on the basis of the validity function. Similar clusters are collected in one cluster, very bad clusters are eliminated, so the number of clusters is reduced. The procedure can be repeated until there are *bad* clusters.

Different scalar validity measures have been proposed in the literature, none of them is perfect by itself, therefore it is advisable to use several indices simultaneously. The most important ones are described below:

1. **Partition Coefficient (PC):** measures the amount of "overlapping" between clusters. It is defined by Bezdek [39] as follows:

$$PC(c) = \frac{1}{N} \sum_{i=1}^{c} \sum_{k=1}^{N} (\mu_{i,k})^2 \tag{1.65}$$

where $\mu_{i,k}$ is the membership of data point k in cluster i. The disadvantage of PC is lack of direct connection to some property of the data themselves. The optimal number of clusters is at the maximum value.

2. **Classification Entropy (CE):** it measures the fuzzyness of the cluster partition only, which is similar to the Partition Coefficient.

$$CE(c) = -\frac{1}{N} \sum_{i=1}^{c} \sum_{k=1}^{N} \mu_{i,k} \ln(\mu_{i,k}) \,. \tag{1.66}$$

3. **Partition Index (SC):** is the ratio of the sum of compactness and separation of the clusters. It is a sum of individual cluster validity measures normalized through division by the fuzzy cardinality of each cluster [38].

$$SC(c) = \sum_{i=1}^{c} \frac{\sum_{k=1}^{N} (\mu_{i,k})^m ||\mathbf{x}_k - \mathbf{v}_i||^2}{\sum_{k=1}^{N} \mu_{i,k} \sum_{j=1}^{c} ||\mathbf{v}_j - \mathbf{v}_i||^2} \,. \tag{1.67}$$

SC is useful when comparing different partitions having equal number of clusters. A lower value of SC indicates a better partition.

4. **Separation Index (S):** on the contrary of partition index (SC), the separation index uses a minimum-distance separation for partition validity [38].

$$S(c) = \frac{\sum_{i=1}^{c} \sum_{k=1}^{N} (\mu_{i,k})^2 ||\mathbf{x}_k - \mathbf{v}_i||^2}{N \min_{i,j} ||\mathbf{v}_j - \mathbf{v}_i||^2} \,. \tag{1.68}$$

5. **Xie and Beni's Index (XB):** indirect indices like the partition coefficient suffer from three drawbacks. First, they are at best indirectly related to any real clusters in $\mathbf{X}$; second, they ignore additional parameters (such as $\mathbf{V}$); and third, they do not use $\mathbf{X}$ itself. Xie and Beni defined an index of fuzzy cluster validity that overcomes the second and third problems. It aims to quantify the ratio of the total variation within clusters and the separation of clusters [285].

$$XB(c) = \frac{\sum_{i=1}^{c}\sum_{k=1}^{N}(\mu_{i,k})^m ||\mathbf{x}_k - \mathbf{v}_i||^2}{N \min_{i,k} ||\mathbf{x}_k - \mathbf{v}_i||^2}. \tag{1.69}$$

The optimal number of clusters should minimize the value of the index.

6. **Dunn's Index (DI):** this index is originally proposed to use at the identification of "compact and well-separated clusters". So the result of the clustering has to be recalculated as it was a hard partition algorithm.

$$DI(c) = \min_{i \in c} \left\{ \min_{j \in c, i \neq j} \left\{ \frac{\min_{\mathbf{x} \in C_i, \mathbf{y} \in C_j} d(\mathbf{x}, \mathbf{y})}{\max_{k \in c}\{\max_{\mathbf{x},\mathbf{y} \in C} d(\mathbf{x}, \mathbf{y})\}} \right\} \right\}. \tag{1.70}$$

The main drawback of Dunn's index is the computational demand since calculating becomes computationally very expansive as c and N increase.

7. **Alternative Dunn Index (ADI):** the aim of modifying the original Dunn's index was that the calculation becomes more simple, when the dissimilarity function between two clusters ($\min_{\mathbf{x} \in C_i, \mathbf{y} \in C_j} d(\mathbf{x}, \mathbf{y})$) is rated in value from beneath by the triangle-nonequality:

$$d(\mathbf{x}, \mathbf{y}) \geq |d(\mathbf{y}, \mathbf{v}_j) - d(\mathbf{x}, \mathbf{v}_j)| \tag{1.71}$$

where $\mathbf{v}_j$ is the cluster center of the jth cluster.

$$ADI(c) = \min_{i \in c} \left\{ \min_{j \in c, i \neq j} \frac{\min_{\mathbf{x}_i \in C_i, \mathbf{x}_j \in C_j} |d(\mathbf{y}, \mathbf{v}_j) - d(\mathbf{x}_i, \mathbf{v}_j)|}{\max_{k \in c}\{\max_{\mathbf{x},\mathbf{y} \in C} d(\mathbf{x}, \mathbf{y})\}} \right\}. \tag{1.72}$$

8. **The fuzzy hyper volume:** this index is also widely applied and represents the volume of the clusters.

$$\mathcal{V} = \sum_{i=1}^{c} \det(\mathbf{F}_i). \tag{1.73}$$

Note, that the only difference of *SC*, *S* and *XB* is the approach of the separation of clusters. In the case of overlapped clusters the values of DI and ADI are not really reliable because of re-partitioning the results with the hard partition method.

Example 1.5 (Optimal number of clusters – validity indices). *In the course of every partitioning problem the number of clusters must be given by the user before the calculation, but it is rarely known a priori, in this case it must be searched also with using validity measures. In the following only a simple example is presented: the motorcycle data set is used to find out the optimal number of clusters. Validity measures mentioned above were used with Gustafson–Kessel algorithm to validate the partitioning of the motorcycle data with the current number of clusters.*

During the optimization parameters were fixed to the following values: $m = 2$ (1.31), $\rho = 1$ (1.33), $\epsilon = 0.001$ (1.38) *for each cluster,* $c \in [2, \ldots, 14]$. *The values of the validity measures depending from the number of cluster are plotted and embraced in Table* 1.2.

Table 1.2: **The numerical values of validity measures**

c	**2**	**3**	**4**	**5**	**6**	**7**
PC	0.8569	0.7743	0.7550	0.7422	0.7291	0.7139
CE	0.2531	0.4168	0.4588	0.5110	0.5679	0.6055
SC	1.6465	1.4591	1.0646	0.6593	0.6055	0.5126
S	0.0124	0.0160	0.0133	0.0067	0.0057	0.0069
XB	28.5271	4.8294	3.3545	4.4216	6.0929	5.1619
DI	0.1154	0.0317	0.0192	0.0230	0.0270	0.0183
ADI	0.0023	0.0017	0.0008	0.0003	0.0005	0.0004
c	**8**	**9**	**10**	**11**	**12**	**13**
PC	0.7075	0.6833	0.6957	0.6554	0.6682	0.6464
CE	0.6246	0.6882	0.6820	0.7590	0.7404	0.7689
SC	0.5181	0.5565	0.4868	0.5032	0.4354	0.4427
S	0.0064	0.0072	0.0062	0.0065	0.0055	0.0057
XB	5.1688	5.2679	3.4507	2.3316	5.0879	2.8510
DI	0.0203	0.0276	0.0196	0.0160	0.0144	0.0120
ADI	0.0000	0.0001	0.0003	0.0001	0.0001	0.0001

It should be mentioned again that no validation index is reliable only by itself, that is why all the programmed indexes are shown, and the optimum can be only detected with the comparison of all the results. We consider that partitions with less clusters are better when the differences between the values of a validation index are minor.

The main drawback of PC is the monotonic decreasing with c and the lack of direct connection to the data. CE has the same problems: monotonic increasing with c and hardly detectable connection to the data structure. On the score of Figure 1.22*, the number of clusters can be only rated to* 3.

In Figure 1.23 *more informative diagrams are shown: SC and S hardly decreases at the* $c = 5$ *point. The XB index reaches this local minimum at* $c = 4$. *Considering that SC and S are more useful, when comparing different clustering methods with the same c, we chose the optimal number of clusters to* 4*, which is confirmed by*

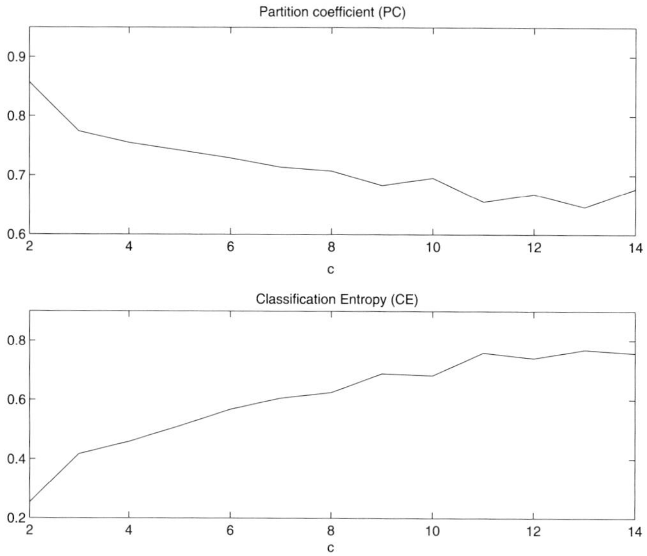

Figure 1.22: Values of Partition Coefficient and Classification Entropy.

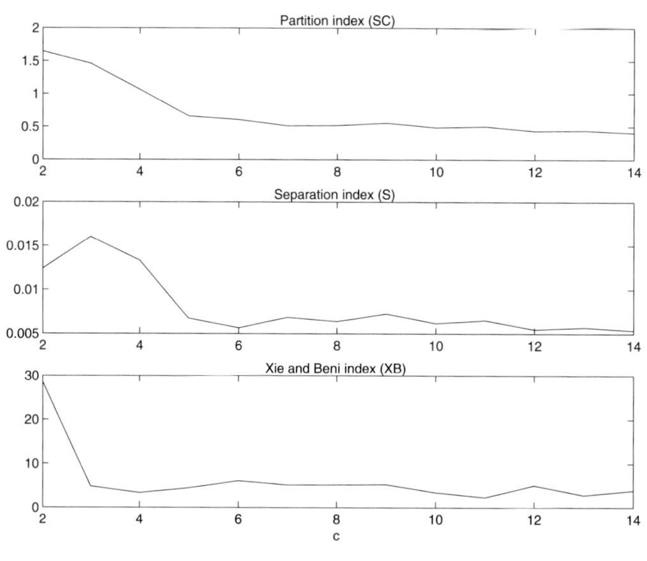

Figure 1.23: Values of Partition Index and Separation Index and Xie and Beni's Index.

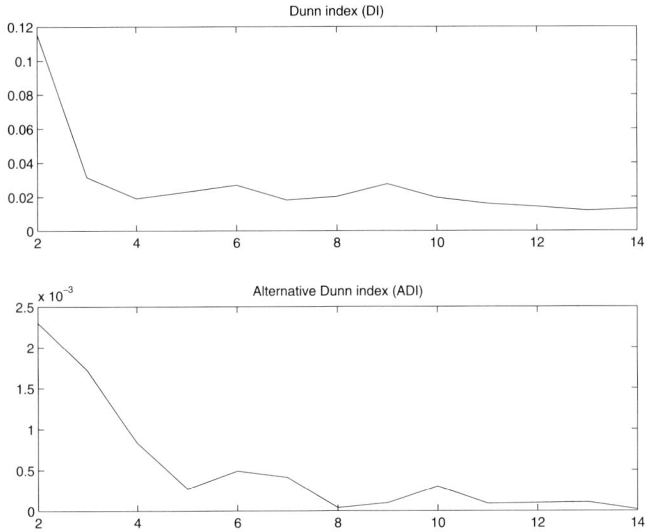

Figure 1.24: Values of Dunn's Index and the Alternative Dunn Index.

the Dunn's index too in Figure 1.24. (*The Alternative Dunn Index is not tested enough to know how reliable its results are.*)

□

Chapter 2

Visualization of the Clustering Results

Since in practical data mining problems high-dimensional data are clustered, the resulting clusters are high-dimensional geometrical objects which are difficult to analyze and interpret. Clustering always fits the clusters to the data, even if the cluster structure is not adequate for the problem. To analyze the adequateness of the cluster prototypes and the number of the clusters, cluster validity measures are used (see Section 1.7). However since validity measures reduce the overall evaluation to a single number, they cannot avoid a certain loss of information. A low-dimensional graphical representation of the clusters could be much more informative than such a single value of the cluster validity because one can cluster by eye and qualitatively validate conclusions drawn from clustering algorithms. This chapter introduces the reader to the visualization of high-dimensional data in general, and presents two new methods for the visualization of fuzzy clustering results.

2.1 Introduction: Motivation and Methods

The reduction of dimensionality of the feature space is also important because of the curse of dimensionality. In a nutshell, same number of examples fill more of the available space when the dimensionality is low, and its consequence is that exponential growth with dimensionality in the number of examples is required to accurately estimate a function.

Two general approaches for dimensionality reduction are:

(i) feature extraction: transforming the existing features into a lower-dimensional space, and

(ii) feature selection: selecting a subset of the existing features without a transformation.

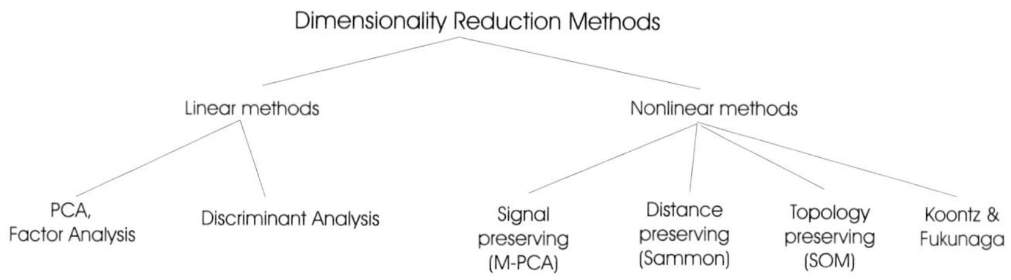

Figure 2.1: Taxonomy of dimensionality reduction methods.

Feature extraction means creating a subset of new features by combination of existing features. These methods can be grouped based on linearity (see Figure 2.1). A linear feature extraction or projection expresses the new features as linear combination of the original variables. The type of linear projection used in practice is influenced by the availability of category information about the patterns in the form of labels on the patterns. If no category information is available, the eigenvector projection (also called Principal Component Analysis (PCA)) is commonly used. Discriminant Analysis is a popular linear mapping technique when category labels are available. In many cases, linear projection cannot preserve the data structure because of its complexity. In these cases nonlinear projection methods should be used. Some of them will be described deeply later in this chapter (see Section 2.1.2 and Section 2.1.3).

2.1.1 Principal Component Analysis

PCA takes a data set $\mathbf{X} = [\mathbf{x}_1, \ldots, \mathbf{x}_N]^T$ where $\mathbf{x}_k = [x_{1,k}, \ldots, x_{n,k}]^T$ is the kth sample or data point in a given orthonormal basis in $\mathbf{R}^n$ and finds a new orthonormal basis, $\mathbf{U} = [\mathbf{u}_1, \ldots, \mathbf{u}_n], \mathbf{u}_i = [u_{1,i}, \ldots, u_{n,i}]^T$, with its axes ordered. This new basis is rotated in such a way that the first axis is oriented along the direction in which the data has its highest variance. The second axis is oriented along the direction of maximal variance in the data, orthogonal to the first axis. Similarly, subsequent axes are oriented so as to account for as much as possible of the variance in the data, subject to the constraint that they must be orthogonal to preceding axes. Consequently, these axes have associated decreasing 'indices', $\lambda_i, i = 1, \ldots, n$, corresponding to the variance of the data set when projected on the axes. The principal components are the new basis vectors, ordered by their corresponding variances. The vector with the largest variance corresponds to the first principal component.

By projecting the original data set on the q first principal components, with $q < n$, a new data set with lower dimensionality (mainly 2 or 3 for visualization purposes) can be obtained. If the principal components are first scaled by the corresponding inverse variances, the variables of the new data set will all have unit variance – a procedure known as whitening or sphering.

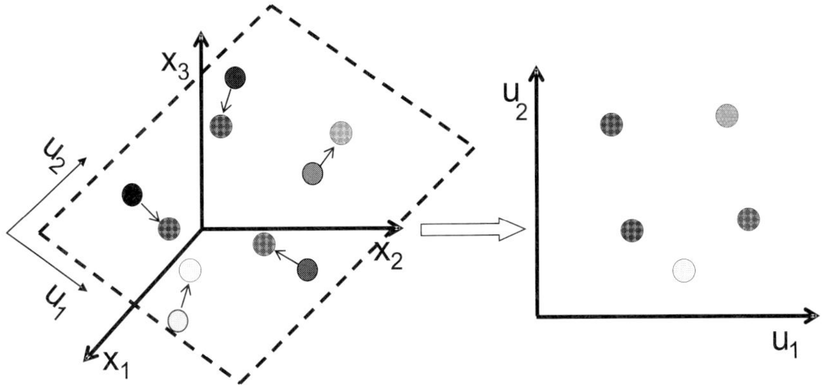

Figure 2.2: Illustration for PCA.

The traditional way of computing the principal components is to compute the sample covariance matrix of the data set,

$$\mathbf{F} = \frac{1}{N-1}\sum_{k=1}^{N}(\mathbf{x}_k - \overline{\mathbf{x}})(\mathbf{x}_k - \overline{\mathbf{x}})^T, \quad \overline{\mathbf{x}} = \frac{1}{N}\sum_{k=1}^{N}\mathbf{x}_k \tag{2.1}$$

and then find its eigenstructure

$$\mathbf{FU} = \mathbf{U}\Lambda. \tag{2.2}$$

$\mathbf{U}$ is an $n \times n$ matrix which has the unit lengths eigenvectors in its columns and Λ is diagonal matrix with the corresponding eigenvalues $\lambda_i, \ldots, \lambda_n$ along the diagonal. The variance of the data set is equal to

$$\sigma^2 = \sum_{i=1}^{n}\lambda_i, \tag{2.3}$$

hence, if only the first few (mostly 2) greatest eigenvalues are used to visualize the original multidimensional data, the sum of the remainder eigenvalues is lost. The eigenvectors are the principal components and the eigenvalues are the corresponding variances. In this case, $\mathbf{y}_k = \mathbf{U}^T\mathbf{x}_k$ is the representation of the kth sample in the new basis (with q vectors), and its approximation in the original space $\widehat{\mathbf{x}}_k = \mathbf{U}\mathbf{U}^T\mathbf{x}_k$.

An important property of the principal components is that they constitute the unique set of vectors that minimizes the reconstruction error,

$$Q = \sum_{k=1}^{N}(\mathbf{x}_k - \widehat{\mathbf{x}}_k)^T(\mathbf{x}_k - \widehat{\mathbf{x}}_k) = \sum_{k=1}^{N}||\mathbf{x}_k - \mathbf{U}\mathbf{U}^T\mathbf{x}_k||^2 = \sum_{k=1}^{N}\mathbf{x}_k^T(\mathbf{I} - \mathbf{U}\mathbf{U}^T)\mathbf{x}_k. \tag{2.4}$$

Q is the sum of the squared distances between the data points and their projections on the q principal components, summed over the data set. Thus, it is a decreasing function of q, equal to zero when $q = n$. Under this formulation, PCA is known as the Karhunen-Loeve transform, and it suggests an alternative way of finding the principal components, by minimizing (2.4). This approach has formed the basis for non-linear extensions.

The analysis of the distribution of the projected data is also informative. The Hotelling T^2 measure is often used to calculate the distance of the mapped data from the center of the linear subspace

$$T^2 = \sum_{k=1}^{N} \mathbf{y}_k^T \mathbf{y}_k. \tag{2.5}$$

Figure 2.3 illustrates these measures in case of two variables and one principal component.

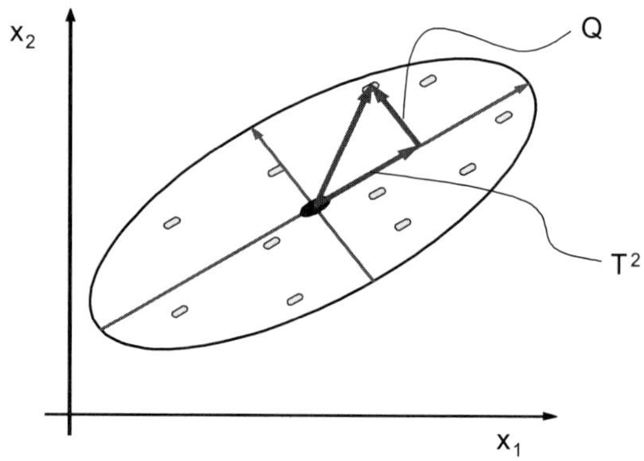

Figure 2.3: Distance measures based on the PCA model.

These T^2 and Q measures are often used for monitoring of multivariate systems and for exploration of the errors and the causes of the errors [203].

Example 2.1 (Visualization of Wine data based on Principal Component Analysis). *The Wine data, which is available from the University of California, Irvine, via anonymous ftp ftp.ics.uci.edu/pub/machine-learning-databases, contains the chemical analysis of* 178 *wines grown in the same region in Italy but derived from three different cultivars. The problem is to distinguish the three different types based on* 13 *continuous attributes derived from chemical analysis: Alcohol, Malic acid, Ash, Alcalinity of ash, Magnesium, Total phenols, Flavanoids, Nonflavanoids phenols, Proanthocyaninsm, Color intensity, Hue, OD*280/*OD*315 *of dilluted wines and Proline (Figure* 2.4*).*

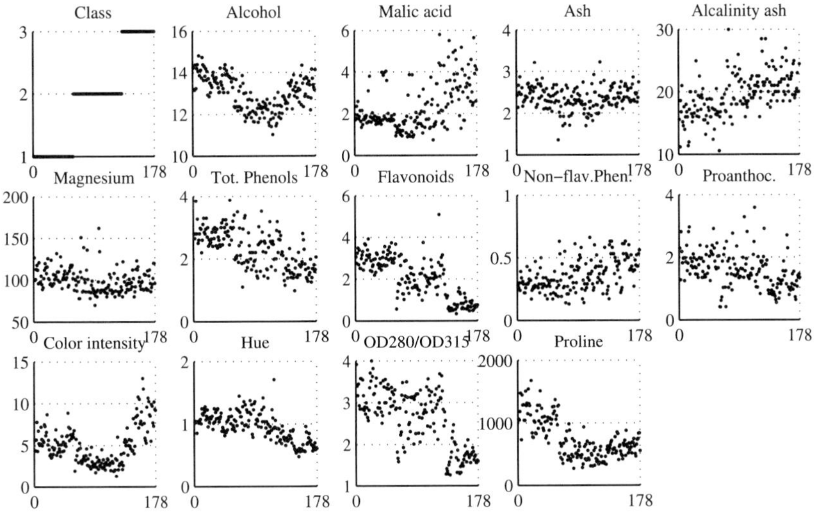

Figure 2.4: Wine data: 3 classes and 13 attributes.

PCA analysis can be used for visualization of the samples in two dimensions. The results can be seen in Figure 2.5*. The three different wine types locate in*

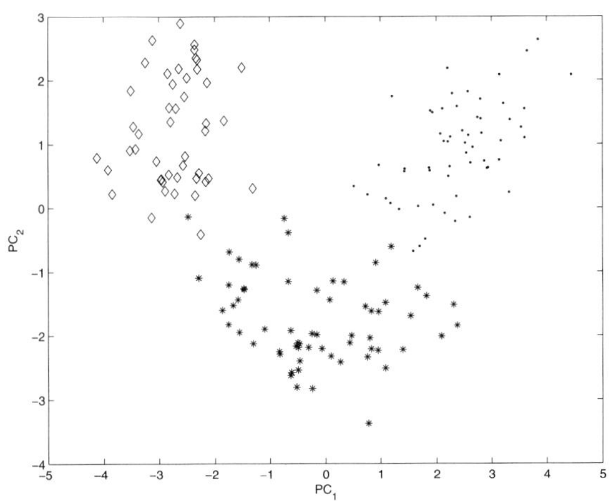

Figure 2.5: PCA analysis of wine data.

three well-separated regions of the space spanned by the first two PCs. However, it should be mentioned that the first two eigenvalues contain only the 55% *of the total variance, so PCA can result in false values in case of classification problem.*

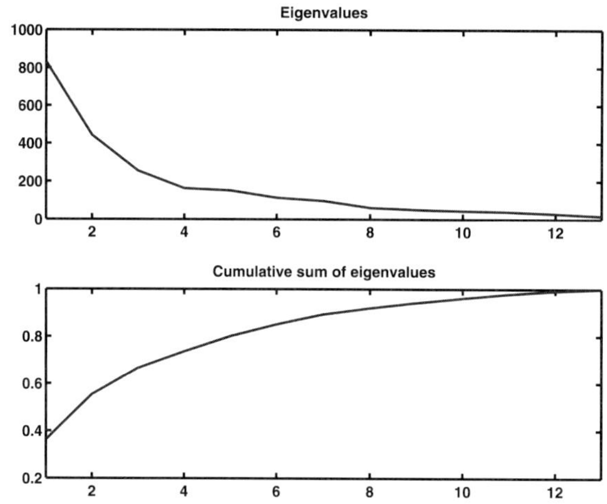

Figure 2.6: Screeplot of wine data.

The eigenvalues and their cumulative sum can be seen in Figure 2.6. *The top figure is the so-called screeplot that plots the ordered eigenvalues according to their contribution to the variance of data.*

□

2.1.2 Sammon Mapping

While PCA attempts to preserve the variance of the data during the mapping, Sammon's mapping tries to preserve the interpattern distances [187, 210]. For this purpose, Sammon defined the mean-square-error between the distances in the high-dimensional space and the distances in the projected low-dimensional space. This square-error formula is similar to the "stress" criterion from multidimensional scaling.

The Sammon mapping is a well-known procedure for mapping data from a high n-dimensional space onto a lower q-dimensional space by finding N points in the q-dimensional data space, such that the interpoint distances $d^*_{i,j} = d^*(\mathbf{y}_i, \mathbf{y}_j)$ in the q-dimensional space approximate the corresponding interpoint distances $d_{i,j} = d(\mathbf{x}_i, \mathbf{x}_j)$ in the n-dimensional space. See Figure 2.7.
This is achieved by minimizing an error criterion, called the Sammon's stress, E:

$$E = \frac{1}{\lambda} \sum_{i=1}^{N-1} \sum_{j=i+1}^{N} \frac{\left(d_{i,j} - d^*_{i,j}\right)^2}{d_{i,j}} \tag{2.6}$$

where $\lambda = \sum\limits_{i=1}^{N-1} \sum\limits_{j=i+1}^{N} d_{i,j}$.

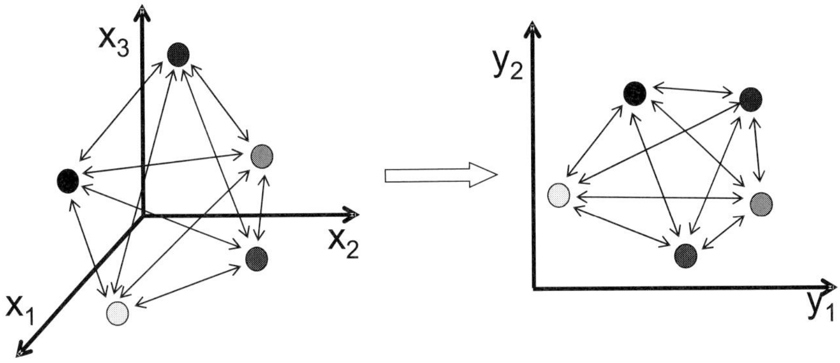

Figure 2.7: Illustration for Sammon mapping.

The minimization of E is an optimization problem in Nq variables $y_{i,l}$, $i = 1, 2, \ldots, N$, $l = 1, \ldots, q$, as $\mathbf{y}_i = [y_{i,1}, \ldots, y_{i,q}]^T$. Sammon applied the method of steepest decent to minimizing this function. Introduce the estimate of $y_{i,l}$ at the tth iteration

$$y_{i,l}(t+1) = y_{i,l}(t) - \alpha \left[\frac{\frac{\partial E(t)}{\partial y_{i,l}(t)}}{\frac{\partial^2 E(t)}{\partial^2 y_{i,l}(t)}} \right] \tag{2.7}$$

where α is a nonnegative scalar constant (recommended $\alpha \simeq 0.3 - 0.4$), i.e., the step size for gradient search in the direction of

$$\begin{aligned} \frac{\partial E(t)}{\partial y_{i,l}(t)} &= -\frac{2}{\lambda} \sum_{k=1,k\neq i}^{N} \left[\frac{d_{k,i} - d^*_{k,i}}{d_{k,i} d^*_{k,i}} \right] (y_{i,l} - y_{k,l}) \\ \frac{\partial^2 E(t)}{\partial^2 y_{i,l}(t)} &= -\frac{2}{\lambda} \sum_{k=1,k\neq i}^{N} \frac{1}{d_{k,i} d^*_{k,i}} \Bigg[(d_{k,i} - d^*_{k,i}) \\ &\quad - \left(\frac{(y_{i,l} - y_{k,l})^2}{d^*_{k,i}} \right) \left(1 + \frac{d_{k,i} - d^*_{k,i}}{d_{k,i}} \right) \Bigg] \end{aligned} \tag{2.8}$$

It is not necessary to maintain λ for a successful solution of the optimization problem, since the minimization of $\sum_{i=1}^{N-1} \sum_{j=i+1}^{N} \left(d_{i,j} - d^*_{i,j}\right)^2 / d_{i,j}$ gives the same result.

When the gradient-descent method is applied to search for the minimum of Sammon's stress, a local minimum in the error surface can be reached. Therefore a significant number of runs with different random initializations may be necessary. Nevertheless, the initialization of $\mathbf{y}$ can be based on information which is obtained from the data, such as the first and second norms of the feature vectors or the principal axes of the covariance matrix of the data [187].

Example 2.2 (Visualization of Wine data based on Sammon mapping). *In Figure 2.8 the results given by the classical Sammon mapping can be seen. Compared to the result by PCA (Figure 2.5), it can be determined that the classes are not as distinct as shown by PCA. For more details about the comparison see Example 2.4, Example 2.5 and Example 2.6.*

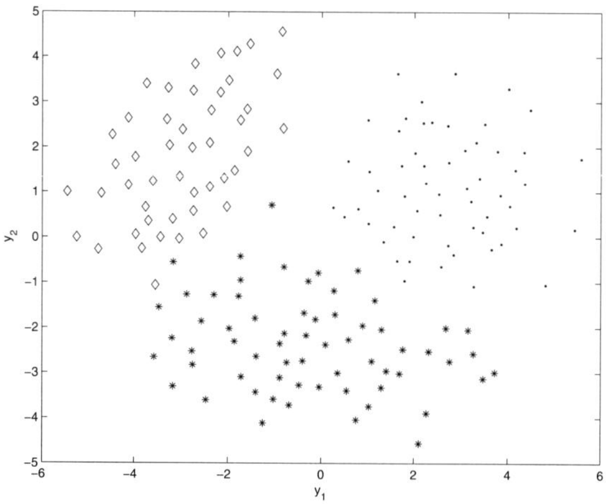

Figure 2.8: Wine data visualized by Sammon mapping.

□

2.1.3 Kohonen Self-Organizing Maps

The Self-Organizing Map (SOM) is a new, effective tool for the visualization of high-dimensional data. It implements an orderly mapping of a high-dimensional distribution onto a regular low-dimensional grid. Thereby it is able to convert complex, nonlinear statistical relationships between high-dimensional data items into simple geometric relationships on a low-dimensional display. As it compresses information while preserving the most important topological and metric relationships of the primary data items on the display, it may also be thought to produce some kind of abstractions. These two aspects, visualization and abstraction, can be utilized in a number of ways in complex tasks such as process analysis, machine perception, control, and communication [158].

SOM performs a topology preserving mapping from high-dimensional space onto map units so that relative distances between data points are preserved [157].

The map units, or neurons, form usually a two-dimensional regular lattice. Each neuron i of the SOM is represented by an n-dimensional weight, or model vector $\mathbf{v}_i = [v_{i,n}, \ldots, v_{i,n}]^T$. These weight vectors of the SOM form a codebook. The neurons of the map are connected to adjacent neurons by a neighborhood relation, which dictates the topology of the map. The number of the neurons determines the granularity of the mapping, which affects the accuracy and the generalization capability of the SOM.

SOM is a vector quantizer, where the weights play the role of the codebook vectors. This means, each weight vector represents a local neighborhood of the space, also called Voronoi cell. The response of a SOM to an input $\mathbf{x}$ is determined by the reference vector (weight) $\mathbf{v}_i^0$ which produces the best match of the input

$$i^0 = \arg\min_i \|\mathbf{v}_i - \mathbf{x}\| \tag{2.9}$$

where i^0 represents the index of the Best Matching Unit (BMU).

During the iterative training, the SOM forms an elastic net that folds onto "cloud" formed by the data. The net tends for approximate the probability density of the data: the codebook vectors tend to drift there where the data are dense, while there are only a few codebook vectors where the data are sparse. The training of SOM can be accomplished generally with a competitive learning rule as

$$\mathbf{v}_i^{(k+1)} = \mathbf{v}_i^{(k)} + \eta \Lambda_{i^0,i}(\mathbf{x} - \mathbf{v}_i^{(k)}) \tag{2.10}$$

where $\Lambda_{i^0,i}$ is a spatial neighborhood function and η is the learning rate. Usually, the neighborhood function is

$$\Lambda_{i^0,i} = \exp\left(\frac{\|\mathbf{r}_i - \mathbf{r}_i^0\|^2}{2\sigma^{2(k)}}\right) \tag{2.11}$$

where $\|\mathbf{r}_i - \mathbf{r}_i^0\|$ represents the Euclidean distance in the output space between the ith vector and the winner.

The whole procedure wants to be illustrated by Figure 2.9 and Figure 2.10.

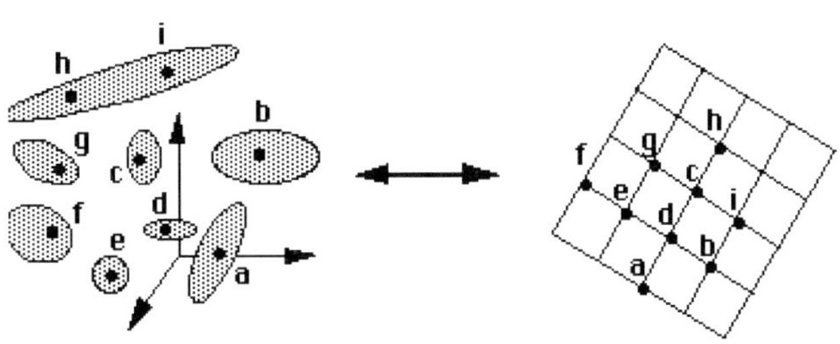

Figure 2.9: Illustration of the topology preserving property of SOM.

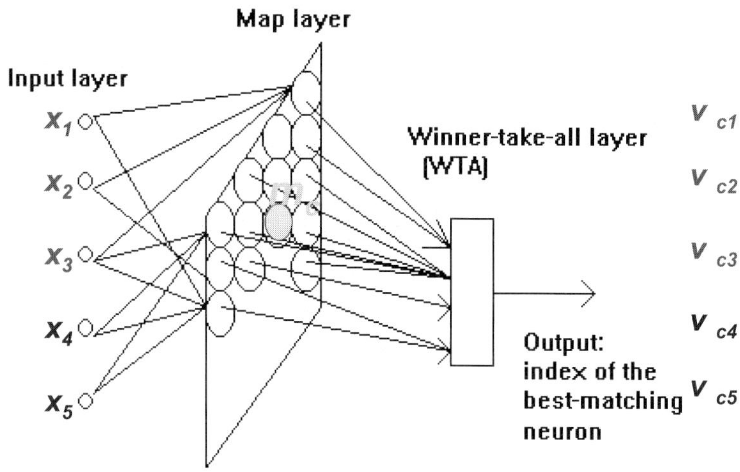

Figure 2.10: Illustration of the BMU computation.

Usually, with the use of this tool the cluster centers (the codebook of the SOM) are mapped into a two-dimensional space [274]. Recently, several related approaches have been proposed to increase the performance of SOM by the incorporation of fuzzy logic. In [275], a modified SOM algorithm was introduced. Neurons are replace by fuzzy rules which allows an efficient modelling of continuous valued functions. In [19] fuzzy clustering combined with SOM is used to project the data to lower dimensions. In [267], some aspects of the fuzzy c-means model are integrated into the classical Kohonen-type hard clustering framework. Another approach has been presented in [211], where a fuzzy self-organizing map is developed based on the modifications of the fuzzy c-means functional. Fuzzy c-means cluster analysis has also been combined with similar mappings and successfully applied to map the distribution of pollutants and to trace their sources to access potential environmental hazard on a soil database from Austria [111].

Example 2.3 (Visualization of Wine data based on Self-Organizing Map). *The SOM has been utilized to visualize the Wine data. SOM can be effectively used for correlation hunting, which procedure is useful for detecting the redundant features. It is interesting to note that rules can be given by the map of the variables given in Figure* 2.11.

For example, **if** *Alcohol is high* **and** *Flavonoids are high* **and** *Color intensity is medium,* **then** *the wine is in class 1. It is visible also in the data. This knowledge can be easily validated by analyzing the SOM of the data given in Figure* 2.11.

□

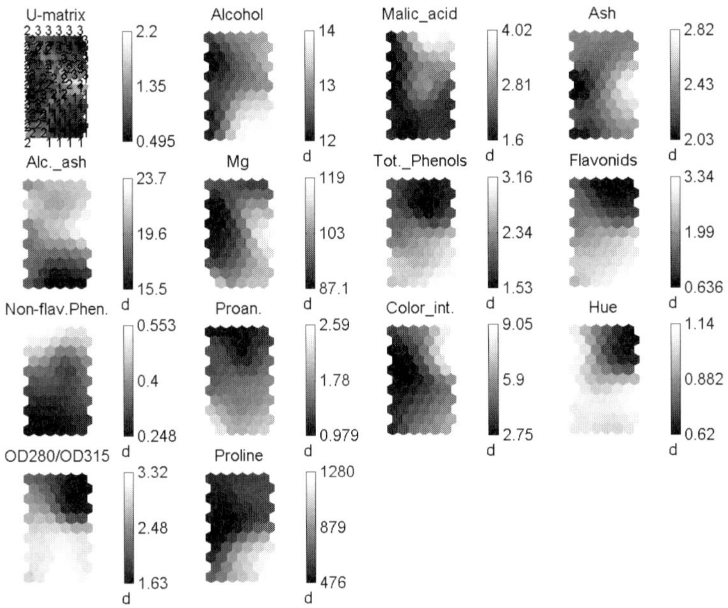

Figure 2.11: Self-Organizing Map of the Wine data.

There are several other methods beyond the ones described above.

- **Projection Pursuit.** Projection Pursuit (PP) is an unsupervised technique that searches interesting low-dimensional linear projections of a high-dimensional data by optimizing a certain objective function called Projection Index (PI). The projection index defines the intent of PP procedure. The notation of interesting obviously varies with the application. The goal of data mining (i.e., revealing data clustering tendency, an edge or jump of data density) should be translated into a numerical index, being a functional of the projected data distribution. This function should change continuously with the parameters defining the projection and have a large value when the projected distribution is defined to be interesting and small otherwise. Most projection indices are developed from the standpoint that normality represents the notion of "uninterestingness". They differ in the assumptions about the nature of deviation from normality and in their computational efficiency. Generally, they can be divided into two groups: parametric projection indices and the nonparametric ones. Parametric projection indices are designed to capture any departure of data distribution from a specified distribution, whereas nonparametric indices are more general and they are not focused on a particular distribution. There is a huge collection of proposed projection indices, e.g., Shannon entropy [244] and Yenyukov index [296].

- **Generative Topographic Maps.** Generative Topographic Mapping (GTM), introduced by Bishop et al. [45, 46], can be considered as a probabilistic reformulation of the self-organizing map (SOM) approach. The aim of the GTM procedure is to model the distribution of data in a high-dimensional space in terms of a smaller number of latent variables.

- **Auto-associative feed-forward networks.** Feed-forward network is usually used in supervised settings. Nevertheless, it can be applied as a nonlinear projection method. In this case the net is trained to map a vector to itself through a "bottleneck" layer, i.e., the layer with a smaller number of nodes than input (and output) layer [161]. The bottleneck layer of the network performs the dimension reduction, because the number of neurons in this layer is smaller than that in the input and output layers, so that the network is forced to develop compact representation of the input data. Figure 2.12 shows a schematic picture of an auto-associative network.

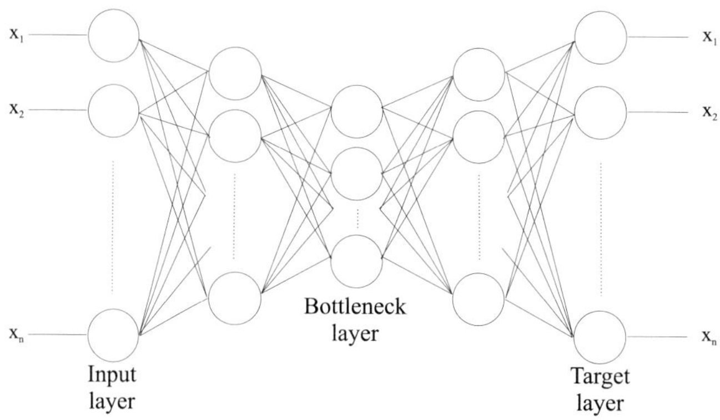

Figure 2.12: Auto-associative nonlinear PCA network.

If all the units in this network are taken to be linear, in which case any intermediary layers between inputs and targets and the bottleneck layer can be removed, and the network is trained using the sum-of-squares error function, this training corresponds to the minimization of the reconstruction error in (2.4). This will result in the network performing standard PCA. In fact, it can be shown that this will also be the case for a network with a single bottleneck layer of non-linear units.

- **Discriminant Analysis.** The discriminant analysis projection maximizes the between-group scatter while holding the within-group scatter constant [62]. This projection requires that all patterns have pattern class or category labels. Even when no extrinsic category labels are available, the patterns can be clustered and cluster labels be used as category information for projection purposes. See Figure 2.13.

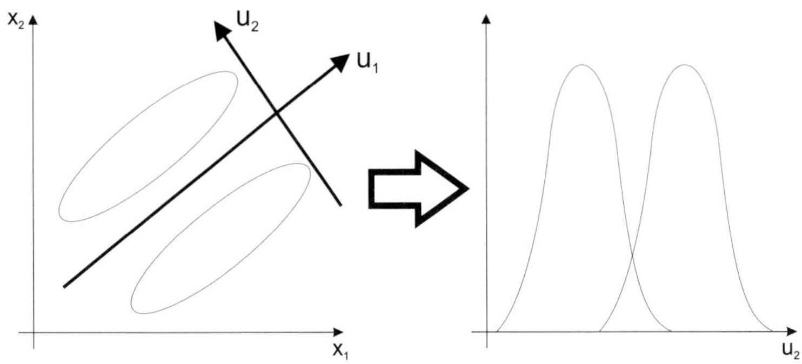

Figure 2.13: Scheme of Discriminant Analysis.

Multidimensional scaling is a generic name for a body of procedures and algorithms that start with an ordinal proximity matrix and generate configurations of points in one, two or three dimensions. Multidimensional scaling translates an ordinal scale to a set of ratio scales and is an example of ordination. MDSCAL developed by Kruskal [163, 164, 165] is one of the most popular techniques in this field. Since the objective of a multidimensional scaling method is to create a set of scales or dimensions that represent the data, natural links exist between multidimensional scaling, intrinsic dimensionality and nonlinear projection.

The techniques mentioned so far in this chapter are general visualization methods, which do not have a direct connection with clustering (except SOM). In the next two sections new methods will be presented for visualization of clustering results. The first method is based on the results of classical fuzzy clustering algorithms and an iterative projection method is applied to preserve the data structure in two dimensions in the sense that the distances between the data points and the cluster centers should be similar in the original high and in the projected two dimensions as well. The second approach is a modified fuzzy clustering method and the purpose of the modification is to increase the applicability of the classical fuzzy c-means algorithm by ordering the cluster centers (prototypes) in an easily visualizable low-dimensional space.

2.2 Visualization of Fuzzy Clustering Results by Modified Sammon Mapping

This section focuses on the application of Sammon mapping for the visualization of the results of clustering, as the mapping of the distances is much closer to the task of clustering than preserving the variances. This section is mainly based on a previous work of the authors, for more details see [160]. There are two main problems encountered in the application of Sammon mapping to the visualization of fuzzy clustering results:

- The prototypes of clustering algorithms may be vectors (centers) of the same dimension as the data objects, but they can also be defined as "higher-level" geometrical objects, such as linear or non-linear subspaces or functions. Hence, classical projection methods based on the variance of the data (PCA) or based on the preservation of the Euclidian interpoint distance of the data (Sammon mapping) are not applicable when the clustering algorithms do not use the Euclidian distance norm.

- As Sammon mapping attempts to preserve the structure of high n-dimensional data by finding N points in a much lower q-dimensional data space, where the interpoint distances measured in the q-dimensional space approximate the corresponding interpoint distances in the n-dimensional space, the algorithm involves a large number of computations as in every iteration step it requires the computation of $N(N-1)/2$ distances. Hence, the application of Sammon mapping becomes impractical for large N [68, 210].

By using the basic properties of fuzzy clustering algorithms a useful and easily applicable idea is to map the cluster centers and the data such that the distances between the clusters and the data-points will be preserved (see Figure 2.14). During the iterative mapping process, the algorithm uses the membership values of the data and minimizes an objective function that is similar to the objective function of the original clustering algorithm.

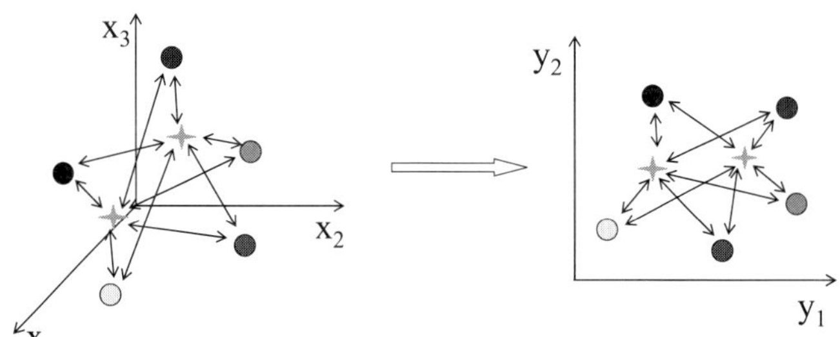

Figure 2.14: Illustration for fuzzy Sammon method.

2.2.1 Modified Sammon Mapping

To avoid the problem mentioned above, in the following we introduce some modifications in order to tailor Sammon mapping for the visualization of fuzzy clustering results. By using the basic properties of fuzzy clustering algorithms where only the distance between the data points and the cluster centers are considered to be important, the modified algorithm takes into account only $N \times c$ distances, where

c represents the number of clusters, weighted by the membership values:

$$E_{fuzz} = \sum_{i=1}^{c} \sum_{k=1}^{N} (\mu_{i,k})^m \left(d(\mathbf{x}_k, \eta_i) - d^*(\mathbf{y}_k, \mathbf{z}_i)\right)^2 \tag{2.12}$$

where $d(\mathbf{x}_k, \eta_i)$ represents the distance between the $\mathbf{x}_k$ datapoint and the η_i cluster center measured in the original n-dimensional space, while $d^*(\mathbf{y}_k, \mathbf{z}_i)$ represents the Euclidian distance between the projected cluster center $\mathbf{z}_i$ and the projected data $\mathbf{y}_k$. This means that in the projected space, every cluster is represented by a single point, regardless of the form of the original cluster prototype, η_i. The application of the simple Euclidian distance measure increases the interpretability of the resulted plots (typically in two dimensions, although three-dimensional plots can be used as well). If the type of cluster prototypes is properly selected, the projected data will fall close to the projected cluster center represented by a point resulting in an approximately spherically shaped cluster.

The resulting algorithm is similar to the original Sammon mapping, but in this case in every iteration after the adaptation of the projected data points, the projected cluster centers are recalculated based on the weighted mean formula of the fuzzy clustering algorithms (see Algorithm 2.2.1).

The resulted two-dimensional plot of the projected data and the cluster centers are easily interpretable since it is based on normal Euclidian distance measures between the cluster centers and the data points. Based on these mapped distances, the membership values of the projected data can be also plotted based on the classical formula of the calculation of the membership values:

$$\mu^*_{i,k} = 1 / \sum_{j=1}^{c} \left(\frac{d^*(\mathbf{x}_k, \eta_i)}{d^*(\mathbf{x}_k, \eta_j)} \right)^{\frac{2}{m-1}}. \tag{2.13}$$

Of course, the resulting 2D plot will only approximate the original high-dimensional clustering problem. The quality of the approximation can easily be evaluated based on the mean square error of the original and the recalculated membership values.

$$P = \|\mathbf{U} - \mathbf{U}^*\| \tag{2.14}$$

where $\mathbf{U}^* = [\mu^*_{i,k}]$ represents the matrix of the recalculated memberships.

Of course there are other tools to get information about the quality of the mapping of the clusters. For example, the comparison of the cluster validity measures calculated based on the original and mapped membership values can also be used for this purpose.

2.2.2 Application Examples

In this section several numerical experiments will be given to demonstrate the applicability of the presented visualization tool. For the sake of comparison, the

Algorithm 2.2.1 (Modified Sammon Mapping).

- **[Input]** *: Desired dimension of the projection, usually* $q = 2$*, the original data set,* $\mathbf{X}$*; and the results of fuzzy clustering: cluster prototypes,* η_i*, membership values,* $\mathbf{U} = [\mu_{i,k}]$*, and the distances* $D = [d_{k,i} = d(\mathbf{x}_k, \eta_i)]_{N\times c}$.
- **[Initialize]** *the* $\mathbf{y}_k$ *projected data points by PCA based projection of* $\mathbf{x}_k$*, and compute the projected cluster centers by*

$$\mathbf{z}_i = \frac{\sum_{k=1}^{N} (\mu_{i,k})^m \mathbf{y}_k}{\sum_{k=1}^{N} (\mu_{i,k})^m} \tag{2.15}$$

and compute the distances with the use of these projected points $D^* = \left[d^*_{k,i} = d(\mathbf{y}_k, \mathbf{z}_i)\right]_{N\times c}$*. Random initialization can also be used but PCA based projection is a better choice in many cases because it may reduce the number of iteration in the next phase.*

- **[While]** $(E_{fuzz} > \varepsilon)$ *and* $(t \leq maxstep)$

$\{$*for* $(i = 1 : i \leq c : i{+}{+}$

$\quad\{$*for* $(j = 1 : j \leq N : j{+}{+}$

$\quad\quad\{$*Compute* $\frac{\partial E(t)}{\partial y_{i,l}(t)}, \frac{\partial^2 E(t)}{\partial^2 y_{i,l}(t)},$

$$\Delta y_{i,l} = \Delta y_{i,l} + \left[\frac{\frac{\partial E(t)}{\partial y_{i,l}(t)}}{\frac{\partial^2 E(t)}{\partial^2 y_{i,l}(t)}}\right] \quad \}$$

$\quad\}$

$y_{i,l} = y_{i,l} + \Delta y_{i,l}, \forall i = 1, \ldots, N,\ l = 1, \ldots, q$

Compute $\mathbf{z}_i = \sum_{k=1}^{N} (\mu_{i,k})^m \mathbf{y}_k / \sum_{k=1}^{N} (\mu_{i,k})^m$

$D^* = \left[d^*_{k,i} = d(\mathbf{y}_k, \mathbf{z}_i)\right]_{N\times c}$

$\}$

Compute E_{fuzz} *by* (2.12).

where the derivatives are

$$\begin{aligned}
\frac{\partial E(t)}{\partial y_{i,l}(t)} &= -\frac{2}{\lambda} \sum_{k=1,k\neq i}^{N} \left[\frac{d_{k,i} - d^*_{k,i}}{d^*_{k,i}} (\mu_{i,k})^m\right] (y_{i,l} - y_{k,l}) \\
\frac{\partial^2 E(t)}{\partial^2 y_{i,l}(t)} &= -\frac{2}{\lambda} \sum_{k=1,k\neq i}^{N} \frac{(\mu_{i,k})^m}{d^*_{k,i}} \Bigg[(d_{k,i} - d^*_{k,i}) \\
&\quad - \frac{(y_{i,l} - y_{k,l})^2}{d^*_{k,i}} \left(1 + (d_{k,i} - d^*_{k,i})(\mu_{i,k})^m\right)\Bigg]
\end{aligned} \tag{2.16}$$

data and the cluster centers are also projected by Principal Component Analysis (PCA) and the standard Sammon projection. Beside the visual inspection of the results the mean square error of the recalculated membership values, P, see (2.14), the difference between the original F and the recalculated F^* partition coefficient (1.65) (one of the cluster validity measures described in Section 1.7), and the Sammon stress coefficient (2.6) will be analyzed.

Example 2.4 (Synthetic data to illustrate the distance mapping of the cluster prototypes using three visualization methods). *The aim of the first example is to demonstrate how the resulted plots of the projection should be interpreted and how the distance measure of the cluster prototype is "transformed" into Euclidian distance in the projected two-dimensional space.*

Table 2.1: Comparison of the performance of the mappings (Example 2.4).

Method	P	F	F^*	E
GK-PCA	0.1217	0.8544	0.7263	0.0000
GK-SAMMON	0.1217	0.8544	0.7263	0.0000
GK-FUZZSAM	0.0204	0.8495	0.8284	0.1177
FCM-FUZZSAM	0.0000	0.7468	0.7468	0.0000

This means that in the projected two-dimensional space, each cluster is represented by a single point, regardless of the form of the original cluster prototype, η_i*. In this example the result of Gustafson–Kessel algorithm is visualized, hence the distance norms defined by the inverse of the fuzzy covariance matrices are transferred to Euclidian distances with the presented FUZZSAM mapping. The application of the simple Euclidian distance measure increases the interpretability of the resulted plots. As Figure* 2.15 *shows in case of a properly selected cluster prototype the projected data will fall close to the projected cluster center represented by a point resulting in an approximately spherically distributed cluster (compare Figure* 2.15*c and Figure* 2.15*d). The numerical results summarized in Table* 2.1 *show that the presented FUZZSAM tool outperforms the linear method and the classical Sammon projection tools. The* P *errors of the membership values in between are much smaller, and the* F *and* F^* *cluster validity measures are similar when the projection is based on the presented FUZZSAM mapping.*

□

Benchmark Examples

The previous example showed that it is possible to obtain a good data structure by the presented mapping algorithm. However, the real advantage of the FUZZSAM algorithm, the visualization of higher-dimensional spaces was not shown. This will be done by the following real clustering problems. The first example is the visualization of the results of the clustering of the well-known Iris data, while the second

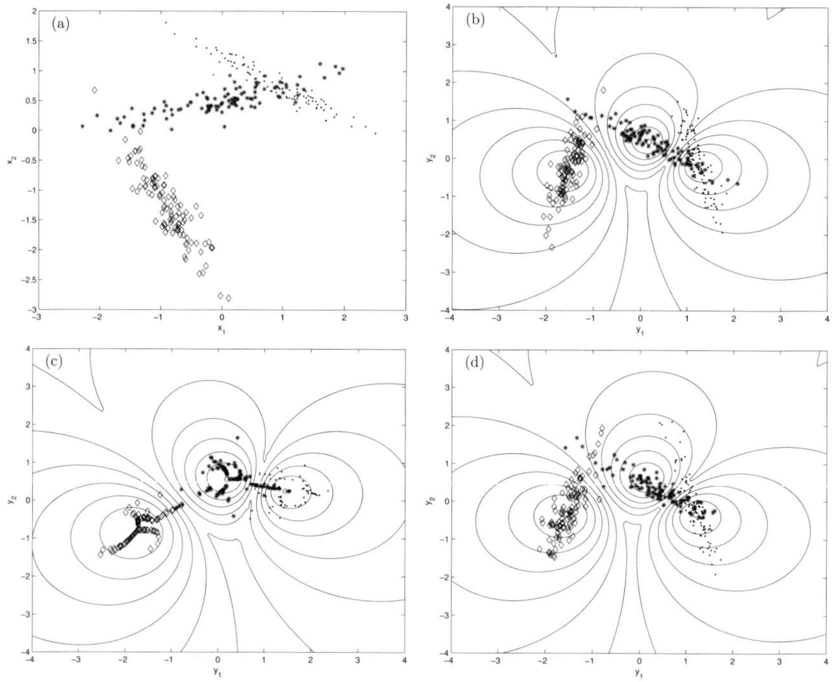

Figure 2.15: (a) Synthetic data in 2D, (b) PCA mapping of the data and the recalculated membership contours, (c) FUZZSAM projection of the results of GK clustering, (d) FUZZSAM projection of the results of FCM clustering.

one deals with the analysis of the Wisconsin Breast Cancer data, both coming from the UCI Repository of Machine Learning Databases (http://www.ics.uci.edu).

Example 2.5 (Iris data visualization). *The Iris data set contains measurements on three classes of Iris flower. The data set was made by measurements of sepal length and width and petal length and width for a collection of* 150 *irises. The problem is to distinguish the three different types (Iris setosa, Iris versicolor and Iris virginica). These data have been analyzed many times to illustrate various methods. To test the presented method the results of the clustering of the iris data were visualized by principal component analysis (PCA), the original Sammon mapping, and the modified method. The initial conditions in the applied Gustafson–Kessel fuzzy clustering algorithm were the following:* $c = 3$, $m = 2$, *and* $\alpha = 0.4$ *in the Sammon and FUZZSAM mapping algorithms. The results of the projections are shown on Figure* 2.16, *where the different markers correspond to different types of iris, and the level curves represent the recalculated membership degrees.*

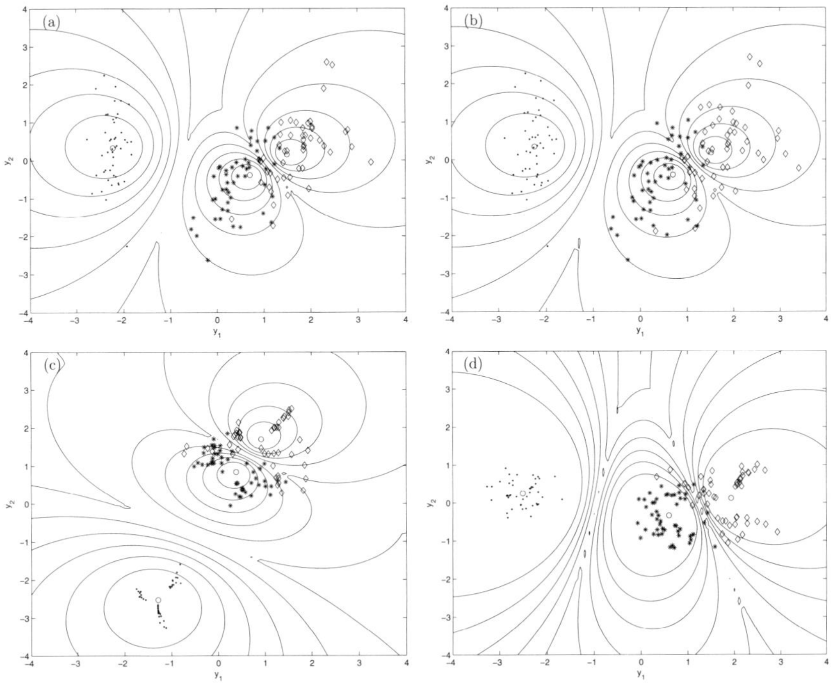

Figure 2.16: (a) PCA projection of the IRIS data and the recalculated membership contours, (b) SAMMON projection of the IRIS data and the recalculated membership contours., (c) FUZZSAM projection of the results of GK clustering of the IRIS data $m = 2$, (d) FUZZSAM projection of the results of FCM clustering of the IRIS data $m = 1.4$.

As Figure 2.16*c nicely illustrates, the data can be properly clustered by the GK algorithm. One of the clusters is well separated from the other two clusters. To illustrate how the fuzziness of the clustering can be evaluated from the resulted plot, Figure* 2.16*d shows the result of the clustering when* $m = 1.4$. *As can be seen in this plot the data points lie much closer to the center of the cluster and there are many more points in the first iso-membership curves. These observations are confirmed by the numerical data given in Table* 2.2.

□

Example 2.6 (Visualization of Wisconsin Breast Cancer data set). *The aim of the Wisconsin Breast Cancer classification problem is to distinguish between benign and malignant cancers based on the available nine features. This example is used to illustrate how the FUZZSAM algorithm can be used to visualize the clustering of nine-dimensional data, and how this visualization can be used to detect the number*

Table 2.2: Comparison of the performance of the mappings (Example 2.5).

Method	P	F	F^*	E
GK-PCA	0.1139	0.7262	0.6945	0.0100
GK-SAMMON	0.1153	0.7262	0.6825	0.0064
GK-FUZZSAM	0.0175	0.7262	0.7388	0.1481
GK-PCA-m=1.4	0.1057	0.9440	0.9044	0.0100
GK-SAMMON-m=1.4	0.1044	0.9440	0.8974	0.0064
GK-FUZZSAM-m=1.4	0.0011	0.9440	0.9425	0.0981

of clusters (Figure 2.17 *and Table* 2.3*). It can be seen from the values of partition coefficient F that two clusters fit much better to the data than four. It can also be observed in Figure* 2.17*a and d: three clusters out of four are overlapping and there are many points that belong to these three clusters with similar membership values (see the iso-membership curves). It can be seen in Figure* 2.17*d that more points lie closer to the cluster center in case of two than four clusters. In case of two clusters the performance of all methods is better than by four clusters (e.g., the values of P decrease), but the best is FUZZSAMM also in this case.*

Table 2.3: Comparison of the performance of the mappings (Example 2.6).

Method	P	F	F^*	E
FMC-PCA	0.0465	0.7696	0.8707	0.0678
FMC-SAMMON	0.0303	0.7696	0.8217	0.0262
FMC-FUZZSAM	0.0089	0.7696	0.7713	0.0826
FMC-PCA-c=2	0.0247	0.9191	0.9586	0.0678
FMC-SAMMON-c=2	0.0164	0.9191	0.9372	0.0262
FMC-FUZZSAM-c=2	0.0008	0.9191	0.9189	0.0517

□

2.2.3 Conclusions

This section has presented a tool that gives the user feedback about the result of fuzzy clustering. The FUZZSAM method generates two-dimensional informative plots about the quality of the cluster prototypes and the required number of clusters. This tool uses the basic properties of fuzzy clustering algorithms to map the cluster centers and the data such that the distances between the clusters and the data-points are preserved. The numerical results show superior performance over Principal Component Analysis and the classical Sammon projection tools.

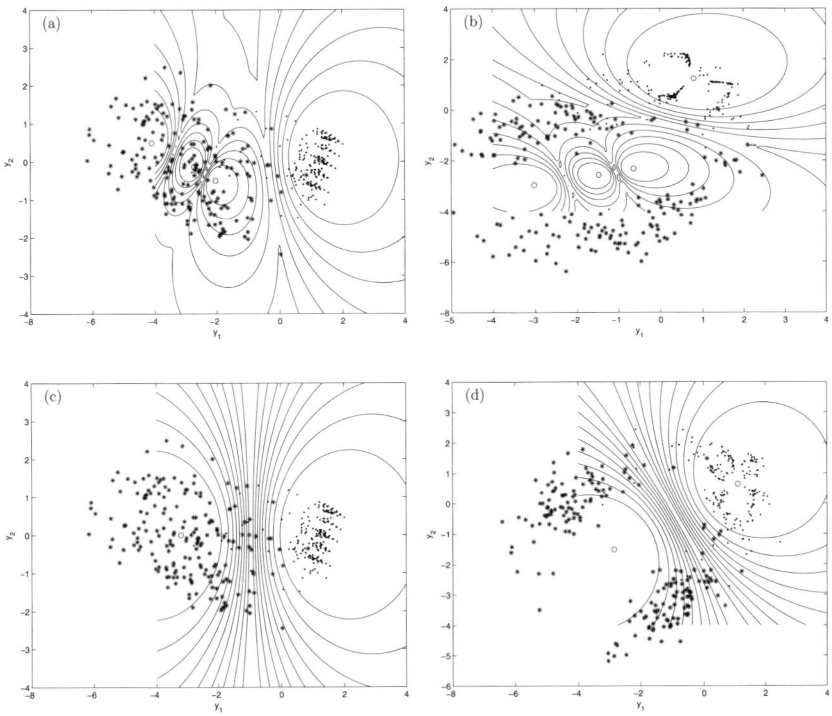

Figure 2.17: (a) PCA projection of the results of a FCM clustering of the WISCONSIN data $c = 4$, (b) FUZZSAM projection of the results of a FCM clustering of the WISCONSIN data $c = 4.$, (c) PCA projection of the results of a FCM clustering of the WISCONSIN data $c = 2$, (d) FUZZSAM projection of the results of a FCM clustering of the WISCONSIN data $c = 2$.

2.3 Fuzzy Self-Organizing Map based on Regularized Fuzzy c-Means Clustering

Since usually high-dimensional data are clustered, the resulted clusters are high-dimensional geometrical objects that are difficult to interpret. In this section a method is described whose aim is to increase the applicability of the classical fuzzy c-means algorithm by ordering the cluster centers (prototypes) in an easily visualizable low-dimensional space. In this approach, similarly to SOM, the cluster centers are distributed on a regular low-dimensional grid, and a regularization term is added to guarantee a smooth distribution for the values of the code vectors on the grid. The idea of the ordering of the clusters in a smaller dimensional space can be also found in [291], where the fuzzy c-means functional has been modified to detect smooth lines.

The key idea of this section is to give more insight to the result of clustering than what traditionally is given by the numerical values of the cluster centers and membership degrees, which are often difficult to analyze and interpret in the case of high-dimensional data and a large number of clusters. An extension of the fuzzy c-means algorithm is realized by adding a regularization term to the cost function. The algorithm can be used for simultaneously clustering and visualizing high-dimensional data. In addition, it is shown that special choices of the regularization term lead to clustering algorithms that can detect smooth lines or generate smooth and ordered feature maps which can easily be used in exploratory data analysis.

In the case of the generation of smooth feature maps, the cluster centers are arranged on a two-dimensional grid and the regularization is based on the measure of map smoothness expressed by the sum of the second-order partial derivatives of the cluster centers.

2.3.1 Regularized Fuzzy c-Means Clustering

Regularized Fuzzy c-Means Clustering Algorithm

The presented algorithm performs a topology preserving mapping from the high, n-dimensional space of the cluster centers onto a smaller, $q \ll n$-dimensional map [193]. The cluster centers of this q-dimensional map are connected to the adjacent cluster centers by a neighborhood relation, which dictates the topology of the map. For instance, Figure 2.18 shows a regular square grid, corresponding to $c = 9$

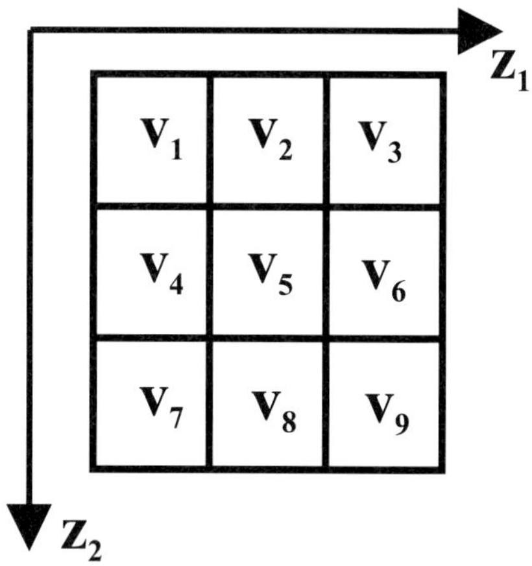

Figure 2.18: Example of cluster centers arranged in a two-dimensional space.

cluster centers. In this section, c is assumed to be known, based on prior knowledge, for instance. For methods to estimate or optimize c refer to [26]. This kind of arrangement of the clusters makes the result of the clustering interpretable only if the smoothness of the distribution of the cluster centers on this q-dimensional map is guaranteed (some illustrative examples will clarify this idea, see, e.g., Figure 2.19 and Figure 2.20). The neighborhood preserving arrangements of the cluster centers means the smoothness of the distribution of the cluster centers on this map. The aim of the visualization of the cluster centers is to provide easily interpretable maps of the variables, where every map is colored according to the numerical values of the cluster centers. In case of distance preserving arrangement it is expected that the resulted maps will be smooth, since clusters that are near to each other in the high-dimensional space of the original data have similar numerical values on each variable. Hence, the idea behind the presented algorithm is to cluster the data and simultaneously provide the smoothness of the resulted maps, of which the regularization indirectly ensures the neighborhood preserving properties of the arrangement.

The smoothness of the grid of the cluster centers can be measured by

$$S = \int \left\| \frac{\partial^2 \mathbf{v}}{\partial^2 \mathbf{z}} \right\| d\mathbf{z} \,. \tag{2.17}$$

This formula can be approximated by the summation of second-order differences, expressed in the following matrix multiplication form:

$$S \approx tr(\mathbf{V}\mathbf{G}^T\mathbf{G}\mathbf{V}^T) = tr(\mathbf{V}\mathbf{L}\mathbf{V}^T)\,, \tag{2.18}$$

where $\mathbf{G} = [\mathbf{G}_1^T, \cdots \mathbf{G}_q^T]^T$, where $\mathbf{G}_i$ denotes the second-order partial difference operator in the $i = 1, \ldots, q$th direction of the map. On the boarders, the operator is not defined. In case of the model depicted in Figure 2.18, these matrices are the following:

$$\mathbf{G}_1 = \begin{bmatrix} 0 & 0 & 0 & 0 & 0 & 0 & 0 & 0 & 0 \\ 1 & -2 & 1 & 0 & 0 & 0 & 0 & 0 & 0 \\ 0 & 0 & 0 & 0 & 0 & 0 & 0 & 0 & 0 \\ 0 & 0 & 0 & 0 & 0 & 0 & 0 & 0 & 0 \\ 0 & 0 & 0 & 1 & -2 & 1 & 0 & 0 & 0 \\ 0 & 0 & 0 & 0 & 0 & 0 & 0 & 0 & 0 \\ 0 & 0 & 0 & 0 & 0 & 0 & 0 & 0 & 0 \\ 0 & 0 & 0 & 0 & 0 & 0 & 1 & -2 & 1 \\ 0 & 0 & 0 & 0 & 0 & 0 & 0 & 0 & 0 \end{bmatrix}$$
$$\mathbf{G}_2 = \begin{bmatrix} 0 & 0 & 0 & 0 & 0 & 0 & 0 & 0 & 0 \\ 0 & 0 & 0 & 0 & 0 & 0 & 0 & 0 & 0 \\ 0 & 0 & 0 & 0 & 0 & 0 & 0 & 0 & 0 \\ 1 & 0 & 0 & -2 & 0 & 0 & 1 & 0 & 0 \\ 0 & 1 & 0 & 0 & -2 & 0 & 0 & 1 & 0 \\ 0 & 0 & 1 & 0 & 0 & -2 & 0 & 0 & 1 \\ 0 & 0 & 0 & 0 & 0 & 0 & 0 & 0 & 0 \\ 0 & 0 & 0 & 0 & 0 & 0 & 0 & 0 & 0 \\ 0 & 0 & 0 & 0 & 0 & 0 & 0 & 0 & 0 \end{bmatrix} \tag{2.19}$$

To obtain ordered cluster centers, this smoothness measure can be added to the original cost function of the clustering algorithm (1.16)

$$J(\mathbf{X}, \mathbf{U}, \mathbf{V}) = \sum_{i=1}^{c} \sum_{j=1}^{N} \mu_{i,k}^{m} d^2(\mathbf{x}_k, \mathbf{v}_i) + \varsigma tr(\mathbf{V}\mathbf{L}\mathbf{V}^T). \tag{2.20}$$

Taking into account constraints (1.12), (1.13) and (1.14), the minimum of (2.20) can be solved by a variety of methods. The most popular one is the alternating optimization algorithm [39] (see also Section 1.5.4). For fixed membership values, $\mathbf{U}$, the minimum of the cost function with respect to the cluster centers, $\mathbf{V}$, can be found by solving the following system of linear equations that can be followed from the Lagrange multiplier method:

$$\mathbf{v}_i \sum_{k=1}^{N} \mu_{i,k}^{m} + \varsigma \sum_{j=1}^{c} L_{i,j} \mathbf{v}_j = \sum_{k=1}^{N} \mu_{i,k} \mathbf{x}_k, \; i = 1, \ldots, c \tag{2.21}$$

where $L_{i,j}$ denotes the i, jth element of the $\mathbf{L} = \mathbf{G}^T\mathbf{G}$ matrix.

Based on the previous equation, the regularized FCM-AO algorithm is in Algorithm 2.3.1.

Relation to the Fuzzy Curve-Tracing Algorithm

The fuzzy curve-tracing algorithm [291] developed for the extraction of smooth curves from unordered noisy data can be considered as a special case of the previously presented algorithm. The neighboring cluster centers are linked to produce a graph according to the average membership values. After the loops in the graph have been removed, the data are then re-clustered using the fuzzy c-means algorithm, with the constraint that the curve must be smooth. In this approach, the measure of smoothness is also the sum of second-order partial derivatives for the cluster centers:

$$J(\mathbf{X}, \mathbf{U}, \mathbf{V}) = \sum_{i=1}^{c} \sum_{k=1}^{N} \mu_{i,k}^{m} d^2(\mathbf{x}_k, \mathbf{v}_i) + \varsigma \sum_{i=1}^{c} \left\| \frac{\partial^2 \mathbf{v}_i}{\partial^2 z} \right\|^2. \tag{2.25}$$

As this cost-function shows, the fuzzy curve-tracing applies a clustering algorithm that can be considered as a special one-dimensional case of the presented regularized clustering algorithm (2.20), where $\mathbf{G}_1$ is the second-order difference operator. If $c = 5$ and the clusters are arranged as can be seen, e.g., in Figure 2.18, then $\mathbf{G}_1$ and $\mathbf{L}$ are:

$$\mathbf{G}_1 = \begin{bmatrix} 0 & 0 & 0 & 0 & 0 \\ 1 & -2 & 1 & 0 & 0 \\ 0 & 1 & -2 & 1 & 0 \\ 0 & 0 & 1 & -2 & 1 \\ 0 & 0 & 0 & 0 & 0 \end{bmatrix} \quad hop\mathbf{L} = \begin{bmatrix} 1 & 2 & -1 & 0 & 0 \\ 2 & 5 & 4 & -1 & 0 \\ -1 & 4 & 6 & 4 & -1 \\ 0 & -1 & 4 & 5 & 2 \\ 0 & 0 & -1 & 2 & 1 \end{bmatrix} \tag{2.26}$$

Algorithm 2.3.1 (Regularized Fuzzy c-Means).

- **Initialization**

  Given a set of data $\mathbf{X}$ *specify* c, *choose a weighting exponent* $m > 1$ *and a termination tolerance* $\epsilon > 0$. *Initialize the partition matrix such that* (1.12), (1.13) *and* (1.14) *holds.*

- **Repeat for** $l = 1, 2, \ldots$

- **Step 1: Calculate the cluster centers.**

$$\mathbf{v}_i^{(l)} = \frac{\sum_{k=1}^{N} \left(\mu_{i,k}^{(l-1)}\right)^m \mathbf{x}_k - \varsigma \sum_{j=1, j\neq i}^{c} L_{i,j}\mathbf{v}_j}{\sum_{k=1}^{N} \left(\mu_{i,k}^{(l-1)}\right)^m + \varsigma L_{i,j}}, \; 1 \leq i \leq c. \quad (2.22)$$

- **Step 2 Compute the distances.**

$$d^2(\mathbf{x}_k, \mathbf{v}_i) = \left(\mathbf{x}_k - \mathbf{v}_i^{(l)}\right)^T \mathbf{A} \left(\mathbf{x}_k - \mathbf{v}_i^{(l)}\right) \; 1 \leq i \leq c, 1 \leq k \leq N. \quad (2.23)$$

- **Step 3: Update the partition matrix.**

$$\mu_{i,k}^{(l)} = \frac{1}{\sum_{j=1}^{c} \left(d(\mathbf{x}_k, \mathbf{v}_i)/d(\mathbf{x}_k, \mathbf{v}_j)\right)^{2/(m-1)}}, \quad (2.24)$$
$$1 \leq i \leq c, 1 \leq k \leq N$$

until $||\mathbf{U}^{(l)} - \mathbf{U}^{(l-1)}|| < \epsilon$.

Note that the elements of L are identical to the regularization coefficients derived "manually" in [291]. As the presented approach gives compact and generalized approach to the fuzzy curve-tracing algorithm, we can call our algorithm as Fuzzy Surface-Tracing (FST) method.

Relation to the Smoothly Distributed FCM Algorithm

The aim of this section is to illustrate that the smoothness of the grid can be defined on several ways. One possible implementation of smoothness (as used, for example, in spline theory) is achieved by demanding in an overall fashion that the value of a code vector must be close to the average value of its nearest neighbors on the grid. This interpretation has been used in [211]. Referring to Figure 2.18, this means that $\mathbf{G}_i$ becomes a first-order gradient operator [211]:

$$\mathbf{G}_1 = \begin{bmatrix} 0 & 0 & 0 & 0 & 0 & 0 & 0 & 0 & 0 \\ 1 & 1 & 0 & 0 & 0 & 0 & 0 & 0 & 0 \\ 0 & -1 & 1 & 0 & 0 & 0 & 0 & 0 & 0 \\ 0 & 0 & 0 & 0 & 0 & 0 & 0 & 0 & 0 \\ 0 & 0 & 0 & -1 & 1 & 0 & 0 & 0 & 0 \\ 0 & 0 & 0 & 0 & -1 & 1 & 0 & 0 & 0 \\ 0 & 0 & 0 & 0 & 0 & 0 & 0 & 0 & 0 \\ 0 & 0 & 0 & 0 & 0 & 0 & -1 & 1 & 0 \\ 0 & 0 & 0 & 0 & 0 & 0 & 0 & -1 & 1 \end{bmatrix}$$

$$\mathbf{G}_2 = \begin{bmatrix} 0 & 0 & 0 & 0 & 0 & 0 & 0 & 0 & 0 \\ 0 & 0 & 0 & 0 & 0 & 0 & 0 & 0 & 0 \\ 0 & 0 & 0 & 0 & 0 & 0 & 0 & 0 & 0 \\ -1 & 0 & 0 & 1 & 0 & 0 & 0 & 0 & 0 \\ 0 & -1 & 0 & 0 & 1 & 0 & 0 & 0 & 0 \\ 0 & 0 & -1 & 0 & 0 & 1 & 0 & 0 & 0 \\ 0 & 0 & 0 & -1 & 0 & 0 & 1 & 0 & 0 \\ 0 & 0 & 0 & 0 & -1 & 0 & 0 & 1 & 0 \\ 0 & 0 & 0 & 0 & 0 & -1 & 0 & 0 & 1 \end{bmatrix} \tag{2.27}$$

This type of penalizing is not always advantageous, since the cluster centers get attracted to each other and thereby to the center of the data, where the magnitude of this contraction is determined by the regulation parameter ς.

In the following the presented algorithm is used to trace a spiral in three-dimensional space and to show how the wine data can be visualized.

Example 2.7 (Tracing a Spiral in 3D based on regularized Fuzzy c-Means algorithm). *The aim of the first example is to illustrate how increasing the smoothness of the of the grid of the cluster centers can generate topology preserving arrangement of the clusters. The illustrative problem that the regularized fuzzy clustering algorithm has to handle is to trace a part of a spiral in* 3*D. For this purpose* 300 *datapoints around a* 3*D curvature have been clustered, where the datapoints were superposed by noise with* 0 *mean and variance* 0.2*. Both FCM and its regularized version are used to generate seven clusters that can be lined up to detect the curvature type structure of the data.*

As can be seen from the comparison of Figure 2.19 *and Figure* 2.20*, with the help of the presented regularized clustering algorithm the detected cluster centers are arranged in a proper order.*

In this example it was easy to qualify the result of the clustering, since the clusters and the data can be easily visualized in 3*D. This allows the easy tune of the only one parameter of the algorithm,* ς*, since with the increase of this parameter it is straightforward to validate the ordering of the resulted clusters (as it is*

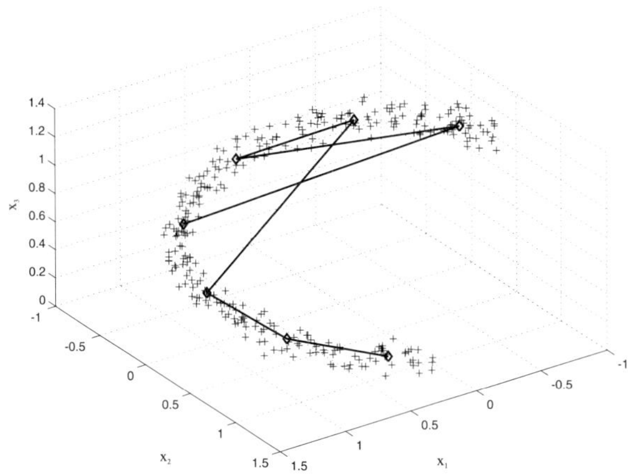

Figure 2.19: Detected clusters and the obtained ordering when the standard FCM algorithm is used.

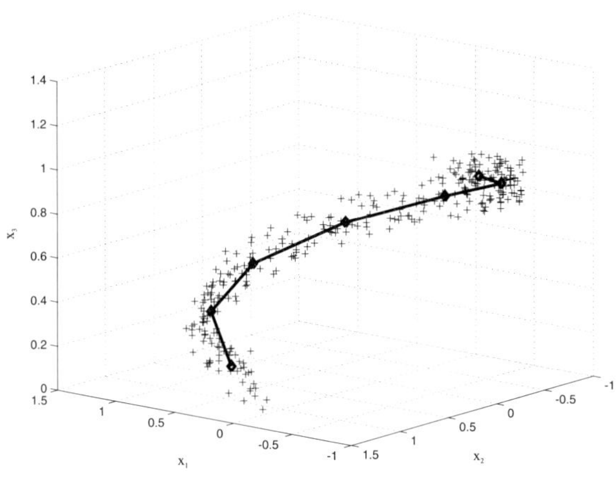

Figure 2.20: Detected clusters and the obtained ordering when the regularized FCM algorithm is used.

depicted in Figure 2.20). *This result shows that with the use of a properly weighted smoothing* $\varsigma = 2$, *it is possible to obtain proper arrangement of the cluster centers. The aim of the next example is to illustrate how this kind of arrangement of the cluster centers can be obtained and used for the analysis of high-dimensional data.*

□

Example 2.8 (Clustering and Visualization of the Wine data based on regularized Fuzzy c-Means clustering). *For the illustration of the results of the smoothed clustering, the MATLAB® implementation of the algorithm – which can be downloaded from the website:* www.fmt.vein.hu/softcomp *– generates independent plots for each variables. On these plots the numerical values of the cluster centers are color coded, see Figure* 2.21.

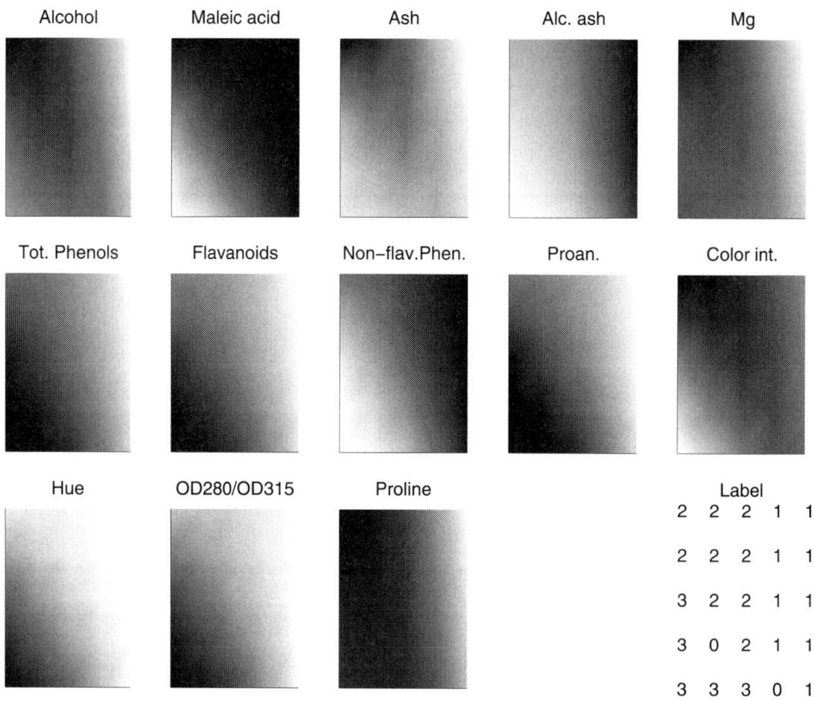

Figure 2.21: Results of regularized clustering.

For the illustration how these plots can be interpreted the clustering and the visualization of the well-known wine data is considered.

From Figure 2.4 *in Example* 2.1 *the different wines cannot be distinguished, and the relations between the variables and the origin of the wine are hard to get.*

With the use of regularized fuzzy clustering three different regions corresponding to the tree different types of wine can be obtained, see Figure 2.21. *The three regions can be seen on the maps of alcohol, color intensity, flavanoids, and ash. This means, these variables have distinguishable value, hence a classifier system should be based on these variables.*

For comparison, the maps obtained with the Kohonen's SOM can be seen in Figure 2.11 *in Example* 2.3. *The FCM based algorithms found the same clusters as Kohonen's SOM and the same relations reveal. The SOM and the presented clustering algorithm can be effectively used for "correlation hunting", that is, inspecting the possible correlations between vector components in the data. These procedures can also be useful for detecting the redundant features.*

□

2.3.2 Case Study

To illustrate the practical benefits of the presented approach, the monitoring of a medium and high-density polyethylene (MDPE, HDPE) plant is considered. HDPE is versatile plastic used for household goods, packaging, car parts and pipe. The plant is operated by TVK Ltd., which is the largest polymer production company (www.tvk.hu) in Hungary. A brief explanation of the Phillips license based low-pressure catalytic process is provided in the following section.

Process Description

Figure 2.22 represents the Phillips Petroleum Co. suspension ethylene polymerization process.

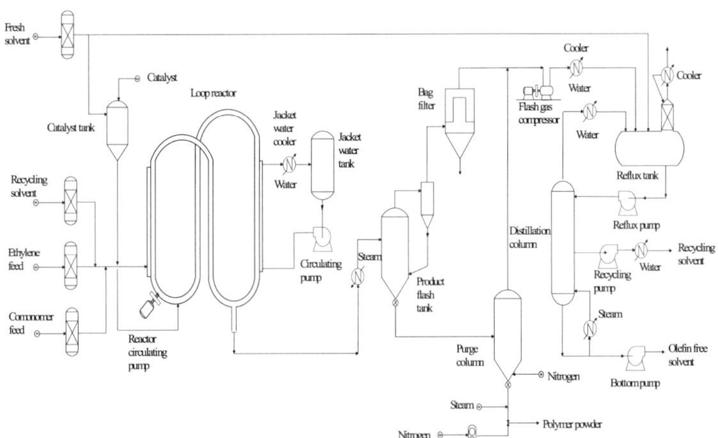

Figure 2.22: Scheme of the Phillips loop reactor process.

The polymer particles are suspended in an inert hydrocarbon. The melting point of high-density polyethylene is approximately 135° Celsius. Therefore, slurry polymerization takes place at a temperature below 135° Celsius; the polymer formed is in the solid state. The Phillips process takes place at a temperature between 85 110° Celsius. The catalyst and the inert solvent are introduced into the loop reactor where ethylene and an α-olefin (hexene) are circulating. The inert solvent (isobuthane) is used to dissipate heat as the reaction is highly exothermic. A cooling jacket is also used to dissipate heat. The reactor consists of a folded loop containing four long runs of pipe 1 m in diameter, connected by short horizontal lengths of 5 m. The slurry of HDPE and catalyst particles circulates through the loop at a velocity between 5–12 m/s. The reason for the high velocity is because at lower velocities the slurry will deposit on the walls of the reactor causing fouling. The concentration of polymer products in the slurry is 25–40% by weight. Ethylene, α-olefin comonomer (if used), an inert solvent, and catalyst components are continuously charged into the reactor at a total pressure of 450 psig. The polymer is concentrated in settling legs to about 60–70% by weight slurry and continuously removed. The solvent is recovered by hot flashing. The polymer is dried and pelletized. The conversion of ethylene to polyethylene is very high (95%–98%), eliminating ethylene recovery. The molecular weight of high-density polyethylene is controlled by the temperature of catalyst preparation. The main properties of polymer products (Melt Index (MI) and density) are controlled by the reactor temperature, monomer, comonomer and chain-transfer agent concentration.

Problem Description

The process has to produce about ten different product grades according to the market demand. Hence, there is a clear need to minimize the time of changeover because off-specification products may be produced during transition. The difficulty comes from the fact that there are more than ten process variables to consider. Measurements are available in every 15 seconds on process variables $\mathbf{x}_k$, which are the $x_{k,1}$ reactor temperature (T), $x_{k,2}$ ethylene concentration in the loop reactor ($C2$), $x_{k,3}$ hexene concentration ($C6$), $x_{k,4}$ the ratio of the hexene and ethylene inlet flowrate ($C6/C2in$), $x_{k,5}$ the flowrate of the isobutane solvent ($C4$), $x_{k,6}$ the hydrogen concentration (H2), $x_{k,7}$ the density of the slurry in the reactor (roz), $x_{k,8}$ polymer production intensity (PE), and $x_{k,9}$ the flowrate of the catalyzator (KAT). The product quality $\mathbf{y}_k$ is only determined later, in another process.

The interval between the product samples is between half an hour and five hours. The $y_{k,1}$ melt index (MI) and the $y_{k,2}$ density of the polymer (ro) are monitored by off-line laboratory analysis after drying the polymer that causes one hour time-delay. Since it would be useful to know if the product is good before testing it, the monitoring of the process would help in the early detection of poor-quality product. There are other reasons why monitoring the process is advantageous. Only a few properties of the product are measured and sometimes these are not sufficient to define entirely the product quality. For example, if only rheological properties of polymers are measured (melt index), any variation in end-use application that arises due to variation of the chemical structure (branching, composition, etc.) will not be captured by following only these product properties. In these cases the process data may contain more information about events with special causes that may effect the product quality.

Application to Process Monitoring

The SOM of the process has been applied to predict polymer properties from measured process variables and to interpret the behavior of the process. The process data analyzed in this book have been collected over three months of operation. The database contains the production of nine product grades with the distribution given in Table 2.4.

Table 2.4: Sample distribution of the products

Number of samples	Product grades
2	1
76	2
5	3
1	4
65	5
94	6
103	7
11	8
52	9

First we applied the classical FCM algorithm with $7 \times 7 = 49$ clusters. (The cause of this choice is to ease the visualization in a regular grid.) Then the cluster centers can be arranged in a grid along each dimension. These grids can also be called maps. The maps of the coordinates of the clusters are given in Figure 2.23.

From these maps the following relations can be seen:

- The hydrogen concentration and temperature are highly correlated with the melt index (MFI). Since the first two are state variables and they can be controlled, it can also be said that the hydrogen concentration and the reactor temperature determine the melt index.
- The hydrogen, C6-C2 concentrations, C6/C2 ratio, temperature have an effect on the density of polymer
- Production intensity is influenced by the flowrate of C4 and katalysator.

Unfortunately, it is hard to get more information from these images because there are many scattered regions with different colors so we cannot be sure that all available information has been detected.

The regularized FCM algorithm with random initialization shows better results (Figure 2.24). Here we can see the additional information that the C2, C6 concentration influence the melt index. The relations of the images are easier to interpret, because there are only a few regions with different colors.

Last we initialized the code vectors based on the first two principal component of the data and applied the regularized FCM algorithm. The resulting images

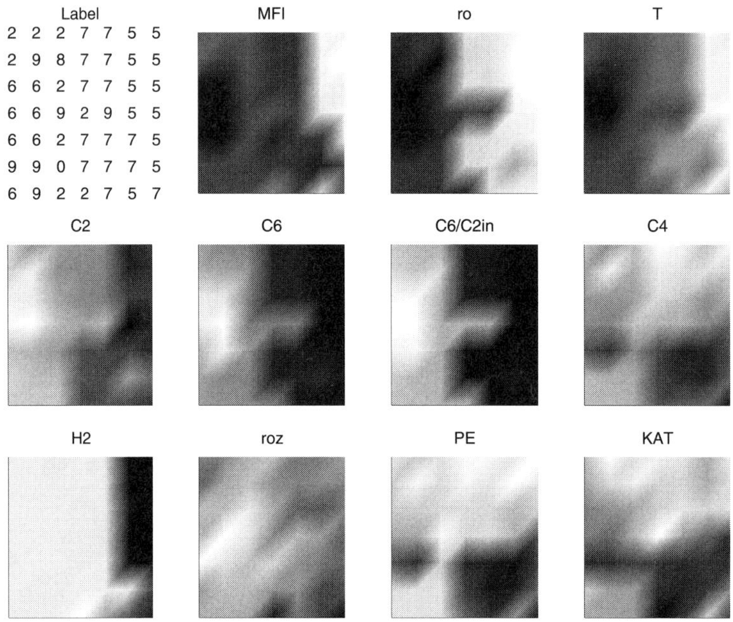

Figure 2.23: Map obtained by the FCM algorithm.

(Figure 2.25) are easier interpretable as before, but additional information was not revealed.

Based on this map the typical operating regions related to different product grades could be determined. With the regularized FCM algorithm it is possible to detect the typical operating regions related to the frequently produced products: 5, 7, 6, 9. Furthermore, the presented clustering algorithm is a good tool for hunting for correlation among the operating variables. For example, it can be easily seen that the melt index of the polymer (MFI) is highly correlated to the reactor temperature (T). Such extracted knowledge has been proven to be extremely useful for the support of the operators, as the extracted information can be used in the supervision of the process.

For comparison, the maps obtained by Kohonen's SOM algorithm are also presented. On the map (Figure 2.26) the color codes are the inverse of ours. The border of different regions are sharper on the SOM map because we used interpolated shading for the representation of the fuzziness of our map. The results show that both SOM and the presented regularized clustering algorithm results in similar component maps.

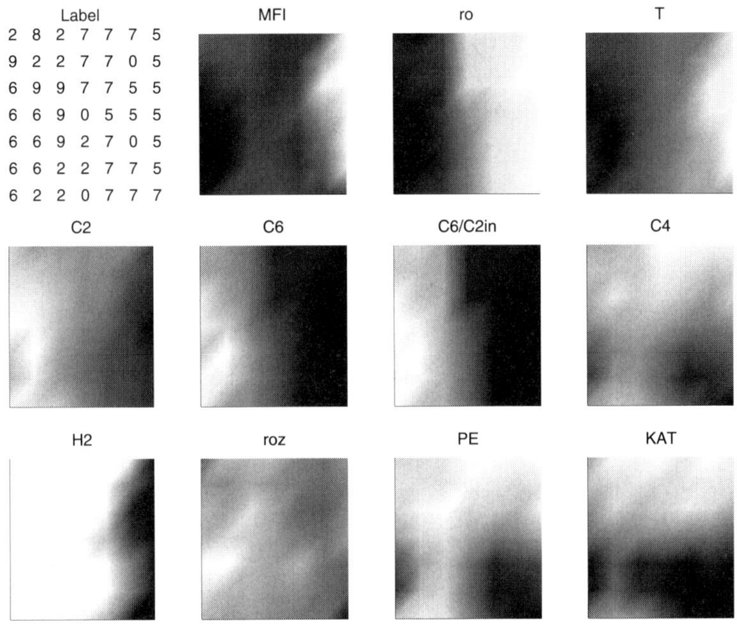

Figure 2.24: Map of code vectors obtained by randomly initialized regularized FCM.

2.3.3 Conclusions

Clustering is a useful tool to detect the structure of the data. If there are many clusters, it is hard to evaluate the results of the clustering. To avoid this problem, it is useful to arrange the clusters on a low-dimensional grid and visualize them. The aim was to develop an algorithm for this visualization task. With the introduction of a new regularization term into the standard fuzzy clustering algorithm it is possible to arrange the cluster prototypes on a grid. This type of arrangement of the code vectors helps greatly in the visualization of the results, as the regularization orders the similar cluster centers closer to each other. The color-coding of the numerical values of the clusters results in regions on the map that shows the relations among the variables. Hence, the presented algorithm can be used as an alternative of SOM.

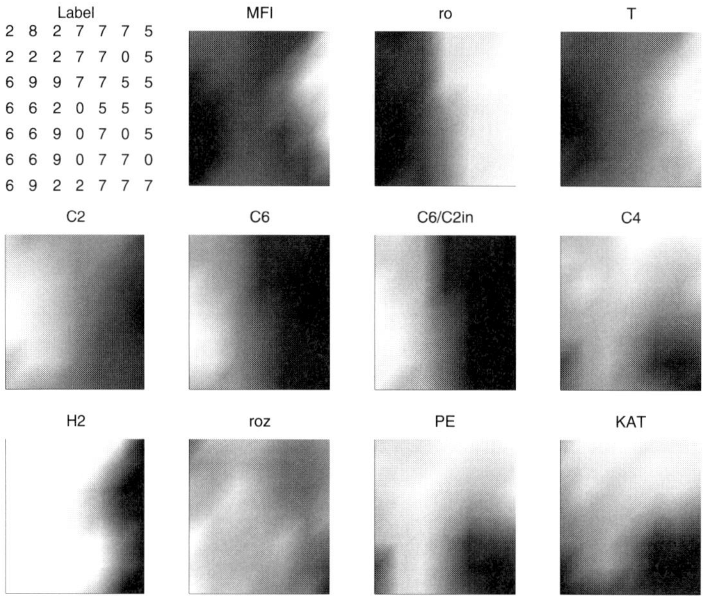

Figure 2.25: Map of cluster prototypes obtained by regularized FCM with initialization.

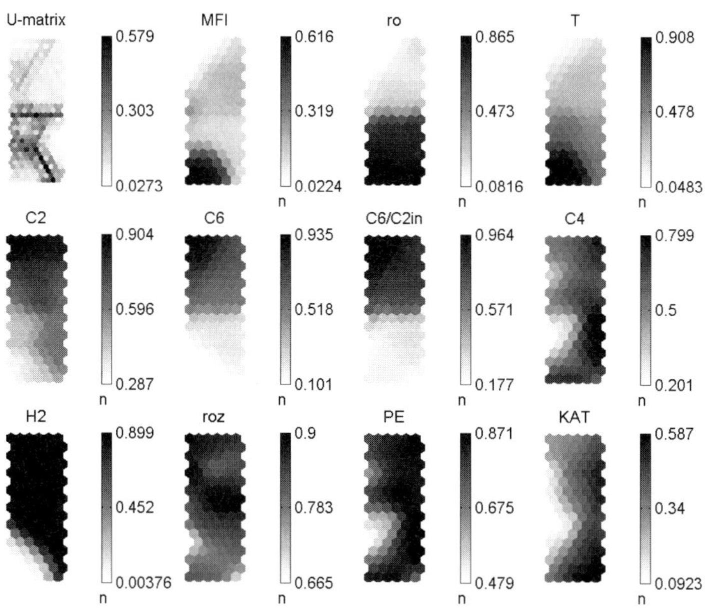

Figure 2.26: SOM map of the process data.

Chapter 3

Clustering for Fuzzy Model Identification – Regression

3.1 Introduction to Fuzzy Modelling

For many real world applications a great deal of information is provided by human experts, who do not reason in terms of mathematics but instead describe the system verbally through vague or imprecise statements like,

$$\textbf{If } \text{The } \textit{Temperature} \text{ is } \textit{Big} \textbf{ then } \text{The } \textit{Pressure} \text{ is } \textit{High}. \tag{3.1}$$

Because so much human knowledge and expertise is given in terms of verbal rules, one of the sound engineering approaches is to try to integrate such linguistic information into the modelling process. A convenient and common approach of doing this is to use fuzzy logic concepts to cast the verbal knowledge into a conventional mathematics representation (model structure), which subsequently can be fine-tuned using input-output data.

Fuzzy logic – first proposed by Lotfi Zadeh in 1965 [298] – is primarily concerned with the representation of the sort of imprecise knowledge which is common in natural systems. It facilitates the representation in digital computers of this kind of knowledge through the use of fuzzy sets. From this basis, fuzzy logic uses logical operators to collate and integrate this knowledge in order to approximate the kind of reasoning common in natural intelligence.

A fuzzy model is a computation framework based on the concepts of fuzzy sets, fuzzy if-then rules, and fuzzy reasoning. This section will present detailed information about particular fuzzy models which are used in this book. It will not attempt to provide a broad survey of the field. For such a survey the reader is referred to "An Introduction to Fuzzy Control" by Driankov, Hellendoorn, and Reinfrank [75] or "Fuzzy Control" by K.M. Passino and S. Yurkovic [212], or "A course in Fuzzy Systems and Control" by L.X. Wang [280].

Conventional set theory is based on the premise that an element either belongs to or does not belong to a given set. Fuzzy set theory takes a less rigid view and allows elements to have degrees of membership of a particular set such that elements are not restricted to either being in or out of a set but are allowed to be "somewhat" in. In many cases this is a more natural approach. For example, consider the case of a person describing the atmospheric temperature as being "hot". If one was to express this concept in conventional set theory one would be forced to designate a distinct range of temperatures, such as $25°C$ and over, as belonging to the set *hot*. That is:

$$hot = [25, \infty)°C.$$

This seems contrived because any temperature which falls just slightly outside this range would not be a member of the set, even though a human being may not be able to distinguish between it and one which is just inside the set.

In fuzzy set theory, a precise representation of imprecise knowledge is not enforced since strict limits of a set are not required to be defined, instead a *membership function* is defined. A membership function describes the relationship between a variable and the degree of membership of the fuzzy set that correspond to particular values of that variable. This degree of membership is usually defined in terms of a number between 0 and 1, inclusive, where 0 implies total absence of membership, 1 implies complete membership, and any value in between implies partial membership of the fuzzy set. This may be written as follows:

$$A(x) \in [0,1] \quad for \quad x \in U$$

where $A(\cdot)$ is the membership function and U is the *universe of discourse* which defines the total range of interest over which the variable x should be defined.

For example, to define membership of the fuzzy set, *hot*, a function which rises from 0 to 1 over the range $15°C$ to $25°C$ may be used, i.e.,

$$A(x) = \begin{cases} 0 & x < 15°C \\ \frac{x-15}{10} & 15 \geq x \geq 25°C \\ 1 & x > 25°C. \end{cases}$$

This implies that $15°C$ is not hot; $20°C$ is a bit hot; $23°C$ is quite hot; and $30°C$ is truly hot. Specific measurable values, such as 15, and 20 are often referred to as *crisp* values or *fuzzy singletons*, to distinguish them from *fuzzy* values, such as *hot*, which are defined by a fuzzy set. Fuzzy values are sometimes also called *linguistic* values.

As Figure 3.1 illustrates, this definition is more reflective of human or *linguistic* interpretations of temperatures and hence better approximates such concepts.

While seeming imprecise to a human being, fuzzy sets are mathematically precise in that they can be fully represented by exact numbers. They can therefore be seen as a method of tying together human and machine knowledge representations. Given that such a natural method of representing information in a computer

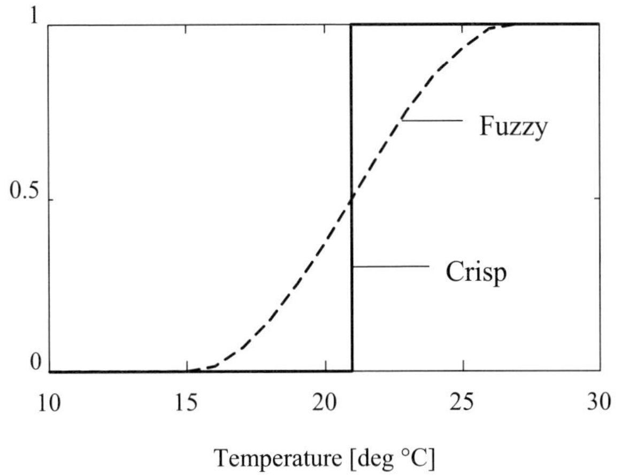

Figure 3.1: Representation of the high temperature.

exists, information processing methods can be applied to it by the use of fuzzy models.

The basic configuration of a fuzzy model is shown in Figure 3.2. As it is depicted

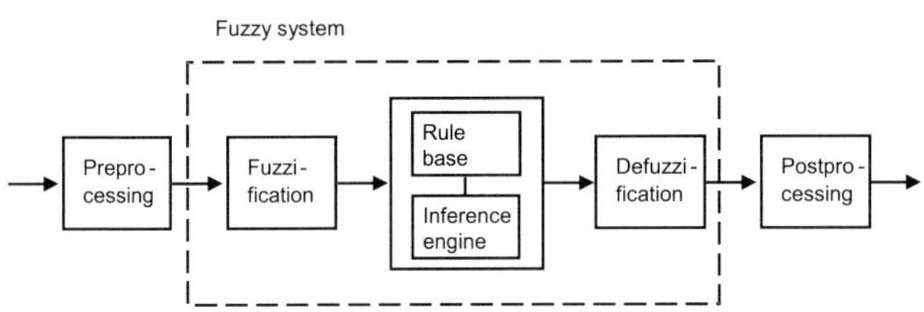

Figure 3.2: Structure of a fuzzy system.

in this figure, the fuzzy model involves the following components [288]:

- *Data preprocessing.* The physical values of the input of the fuzzy system may differ significantly in magnitude. By mapping these to proper normalized (but interpretable) domains via scaling, one can instead work with signals roughly of the same magnitude, which is desirable from an estimation point of view.

- *Fuzzification.* Fuzzification maps the crisp values of the preprocessed input of the model into suitable fuzzy sets represented by *membership functions* (MF). As the antecedent and consequent fuzzy sets take on linguistic meanings such as "high temperature" they are called linguistic labels of the sets of linguistic variables. For instance, if the linguistic variable is "temperature", several fuzzy sets can be defined for this variable, e.g., "low", "medium", "high", etc; see Figure 3.1. The degree of membership of a single crisp variable to a single fuzzy set could be evaluated using a membership function. A fuzzifier calculates the degree of membership of multiple crisp variables to multiple fuzzy sets in a one-to-many fashion. There are $n \geq 1$ crisp input variables and each crisp variables can belong to $M_i > 1 : \ i = 1 \dots n$ fuzzy sets.

For example, an air conditioning system might have two crisp input variables, temperature and humidity, i.e., $n = 2$. These might be transformed to two fuzzy variables consisting of the fuzzy sets $\{cold, cool, tepid, warm, hot\}$ and $\{dry, normal, hot\}$, respectively. This means that $M_1 = 5$ and $M_2 = 3$.

Systems of only one input variable are feasible, but it is quite apparent that if only one fuzzy set is defined for one particular input variable then no distinctions can be made in the rules on this variable and its inclusion in the fuzzy model is redundant. Therefore two or more fuzzy sets will usually be defined for each input variable. It has already been mentioned that the degree of membership of a crisp variable to a fuzzy set is defined by a membership function. In this work, triangular membership functions will be used exclusively. The triangular membership function and some other commonly used membership function shapes are shown in Figure 3.3.

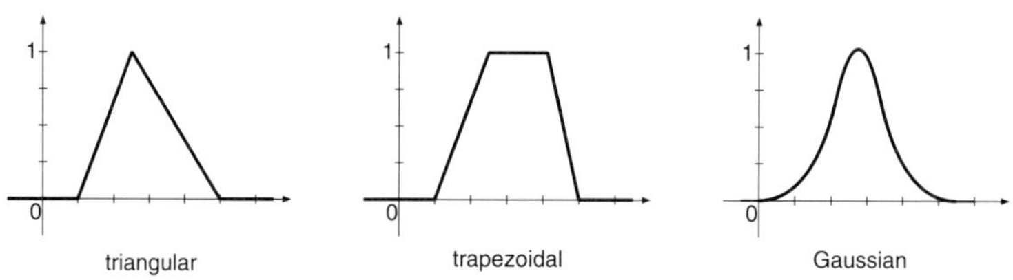

Figure 3.3: Three common membership function shapes.

- *Rule base.* The rule base is the cornerstone of the fuzzy model. The expert knowledge, which is assumed to be given as a number of if-then rules, is stored in a fuzzy rule base.

  In rule-based fuzzy systems, the relationship between variables are represented by means of If-Then rules of the following general form:

$$\textbf{If } \text{antecedent proposition } \textbf{then } \text{consequent proposition} \tag{3.2}$$

According to the consequent proposition and to the structure of the rule base, there are three distinct classes of fuzzy models:

  – *Fuzzy linguistic models* (*Mamdani models*) [184, 185] where both the antecedent and consequent are fuzzy propositions. Hence, a general rule of a linguistic or Mamdani fuzzy model is given by

$$R_j: \ \textbf{If } x_1 \text{ is } A_{1,j} \textbf{ and } \ldots \textbf{ and } x_n \text{ is } A_{n,j} \textbf{ then } y \text{ is } B_j \tag{3.3}$$

where R_j denotes the jth rule, $j = 1, \ldots, N_r$, and N_r is the number of the rules. The antecedent variables represent the input of the fuzzy system $\mathbf{x}$. $A_{i,j}$ and B_j are fuzzy sets described by membership functions $\mu_{A_{i,j}}(x_i) :\rightarrow [0, 1]$ and $\mu_{B_i}(y) :\rightarrow [0, 1]$.

- *Fuzzy relational models* are based on fuzzy relations and relational equations [297]. These models can be considered as a generalization of the linguistic model, allowing one particular antecedent proposition to be associated with several different consequent propositions via a fuzzy relation.
- *Takagi–Sugeno (TS) fuzzy models* where the consequent is a crisp function of the input variables, $f_j(\mathbf{x})$, rather than a fuzzy proposition [257].

$$R_j: \ \textbf{If } x_1 \text{ is } A_{1,j} \textbf{ and } \ldots \textbf{ and } x_n \text{ is } A_{n,j} \textbf{ then } y = f_j(\mathbf{x}). \tag{3.4}$$

This book deals with this type of fuzzy models.

- *Inference engine.* The inference mechanism or inference engine is the computational method which calculates the degree to which each rule *fires* for a given fuzzified input pattern by considering the rule and label sets. A rule is said to fire when the conditions upon which it depends occur. Since these conditions are defined by fuzzy sets which have degrees of membership, a rule will have a degree of firing or *firing strength*, β_j. The firing strength is determined by the mechanism which is used to implement the *and* in the expression (3.4); in this book the product of the degrees of membership will be used, that is:

$$\beta_j = \prod_{i=1}^{n} A_{i,j} \tag{3.5}$$

where $A_{i,j}$ defines the membership function on input i is used in rule j. Again, there are different methods for implementing each of the logical operators and the reader is referred to [75] for details on these.

- *Defuzzification.* A defuzzifier compiles the information provided by each of the rules and makes a decision from this basis. In linguistic fuzzy models the defuzzification converts the resulted fuzzy sets defined by the inference engine to the output of the model to a standard crisp signal. The method which is used in this book is the method commonly called the centre-of-gravity or *centroid* method. In case of TS fuzzy models it is described by the following equation:

$$y = \frac{\sum_{j=1}^{N_r} \beta_j f_j(\mathbf{x})}{\sum_{j=1}^{N_r} \beta_j}. \tag{3.6}$$

It can be seen that the centroid method of defuzzification takes a weighted sum of the designated consequences of the rules according to the firing strengths of the rules. There are numerous other types of defuzzifiers such as centre-of-sums, first-of-maxima, and middle-of-maxima [75].

- *Postprocessing.* The preprocessing step gives the output of the fuzzy system based on the crisp signal obtained after defuzzification. This often means the scaling of the output.

In their paper, "Fuzzy Basis Functions, Universal Approximation, and Orthogonal Least-Squares Learning" [279], Wang and Mendel define fuzzy basis functions as:

$$\beta_j(\mathbf{x}) = \frac{\prod_{i=1}^{n} A_{i,j}(x_i)}{\sum_{j=1}^{N_r}\prod_{i=1}^{n} A_{i,j}(x_i)}, \qquad y = \sum_{j=1}^{N_r} \beta_j(\mathbf{x})\theta_j$$

where $\beta_j(\mathbf{x})$ is the normalized firing strength of rule j, $A_{i,j}(x_j)$ represents a membership function, x_i is the ith input, and θ_j is the crisp rule consequent, $f_j(\mathbf{x}) = \theta_j$. Wang and Mendel prove that, given enough rules, this system can approximate any real continuous function to any given accuracy; this is stated as follows:

Theorem 3.1.1. *Given any continuous function $f(\cdot)$ on the compact set $U \subset R^n$ and an arbitrary constant $\epsilon > 0$, there exists a function $\hat{f}(\cdot)$, defined on the set of all fuzzy basis function expansions, such that:*

$$\min\,(x_1, \dots, x_n) \in U|\hat{f}(x_1, \dots, x_n) - f(x_1, \dots, x_n)| < \epsilon$$

where $\hat{f}(x_1, \dots, x_n)$ is the function implemented by the fuzzy basis function.

Although this is an interesting result, it should be noted that it is usually undesirable to have to define a separate set of membership functions for each rule. In addition, the theorem does not define the number of basis functions or rules required to achieve the desired accuracy (given by ϵ) – this number could be very large in some cases. Given that one of the most important features of fuzzy rules is that humans should be able to interpret them, a large number of rules could work against this purpose.

3.2 Takagi–Sugeno (TS) Fuzzy Models

This book mainly deals with a Takagi–Sugeno (TS) fuzzy model proposed by Takagi, Sugeno, and Kang [257, 254] to develop a systematic approach for generating fuzzy rules from a given input-output data set. In this section the structure of this model and the related modelling paradigms will be presented.

3.2.1 Structure of Zero- and First-order TS Fuzzy Models

The TS model is a combination of a logical and a mathematical model. This model is also formed by logical rules; it consists of a fuzzy antecedent and a mathematical function as consequent part. The antecedents of fuzzy rules partition the input space into a number of fuzzy regions, while the consequent functions describe the system behavior within a given region:

$$R_j : \textbf{If } z_1 \text{ is } A_{1,j} \textbf{ and } \ldots \textbf{ and } z_n \text{ is } A_{n,j} \textbf{ then} \\ y = f_j\left(q_1, \ldots, q_m\right) \tag{3.7}$$

where $\mathbf{z} = [z_1, \ldots, z_n]^T$ is the n-dimensional vector of the antecedent variables, $\mathbf{z} \in \mathbf{x}$, $\mathbf{q} = [q_1, \ldots, q_m]^T$ is the m-dimensional vector of the consequent variables $\mathbf{q} \in \mathbf{x}$, where $\mathbf{x}$ denotes the set of all inputs of the $y = f(\mathbf{x})$ model. $A_{i,j}(z_i)$ denotes the antecedent fuzzy set for the ith input. The antecedents of fuzzy rules partition the input space into a number of fuzzy regions, while the $f_j(\mathbf{q})$ consequent functions describe the system behavior within a given region.

The spirit of fuzzy inference systems resembles that of a 'divide and conquer' concept – the antecedent of fuzzy rules partition the input-space into a number of local fuzzy regions, while the consequents describe the behavior within a given region via various constituents [131].

With respect to the antecedent membership functions and the structure of the rules three typical ways of partitioning of the input space can be obtained. Figure 3.4 illustrates these partitions in a two-dimensional input space.

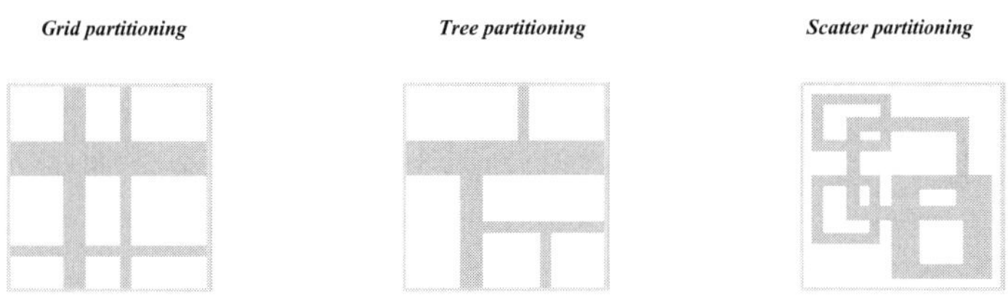

Figure 3.4: Various methods for partitioning the input space.

- **Grid partition** The conjunctive antecedent divides the antecedent space into a lattice of axis-orthogonal hyperboxes. In this case the number of rules needed to cover the entire domain is an exponential function of the input space dimension and of the fuzzy sets used in each variable. This partition method is often chosen in designing a fuzzy system which usually involves a few input variables. This partition strategy needs only a small number of membership functions for each input. However, it encounters problems when we have moderately large number of inputs. For instance, a fuzzy model with ten inputs and two membership functions on each input would result in $2^{10} = 1024$ fuzzy if-then rules. This problem, usually referred to as the *curse* *of dimensionality*, can be alleviated by other partition strategies introduced below.

- **Tree partition** This partition relieves the problem of the exponential increase of the number of rules. However, more membership functions for each input are needed to define these fuzzy regions and these membership functions are not interpretable as they do not bear clear linguistic meaning such as "small".
- **Scatter partition** By covering a subset of the whole input space that characterizes a region of possible occurrence of the input vectors, the scatter partition can be obtained that can also limit the number of rules to a reasonable amount. In this case the use of multivariate membership functions is the most general one, as there is no restriction on the shape of fuzzy sets. The boundaries between the fuzzy sets can be arbitrarily curved and opaque to the axes. This multidimensional membership function based fuzzy model is described in [1].

Usually, the f_j consequent function is a polynomial in the input variables, but it can be any arbitrary chosen function that can appropriately describe the output of the system within the region specified by the antecedent of the rule. When $f_j(\mathbf{q})$ is a first-order polynomial,

$$f_j(\mathbf{q}) = \theta_j^0 + \theta_j^1 q_1 + \cdots + \theta_j^m q_m = \sum_{l=0}^{m} \theta_j^l q_l \text{ , where } q_0 = 1 \tag{3.8}$$

the resulting fuzzy inference system is called *first-order Takagi–Sugeno* or simply *Takagi–Sugeno fuzzy model.* If $f_j(\mathbf{q})$ is a constant (fuzzy singleton) $f_j = \theta_j^0$ we have a zero-order Takagi–Sugeno or *Singleton fuzzy model*, which is a special case of the linguistic fuzzy inference systems [104] and the TS fuzzy model ($m = 0$).

Using fuzzy inference based upon *product-sum-gravity* at a given input [288], the final output of the fuzzy model, y, is inferred by taking the weighted average of the consequent functions as it is depicted in Figure 3.5:

$$y = \frac{\sum_{j=1}^{N_r} \beta_j(\mathbf{z}) f_j(\mathbf{q})}{\sum_{j=1}^{N_r} \beta_j(\mathbf{z})} \tag{3.9}$$

where the weight, $0 \leq \beta_j(\mathbf{z}) \leq 1$, represents the overall truth value (degree of fulfillment) of the ith rule calculated based on the degrees of membership

$$\beta_j(\mathbf{z}) = \prod_{i=1}^{n} A_{i,j}(z_i). \tag{3.10}$$

Figure 3.5 shows the fuzzy reasoning procedure for a TS fuzzy model.

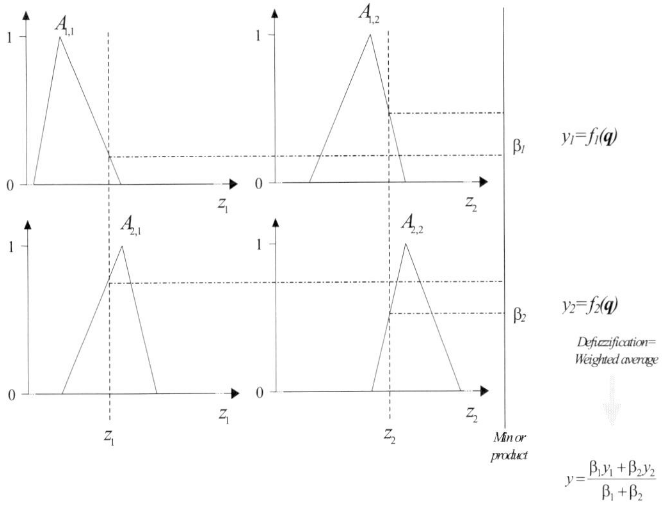

Figure 3.5: Inference method of the Takagi–Sugeno fuzzy model.

Example 3.1 (SISO TS fuzzy model). *This example presents a simple Takagi–Sugeno fuzzy model that has one input variable. The $y = f(x)$ model consists of three rules with local linear models on the consequent parts*

$$R_j : \textbf{ If } z_1 \textit{ is } A_{1,j} \textbf{ then } y = \theta_j^0 + \theta_j^1 q_1$$

where $x_1 = x$, $z_1 = x_1$, $q_1 = x_1$, $j = 1, \ldots, 3$.

As Figure 3.6 shows, when the operating regimes of these local models are defined by fuzzy sets instead of crisp ones, the resulted model behaves as a smooth nonlinear function.

□

As Figure 3.7 shows, the membership functions are arranged by Ruspini-type partition keeping the sum of the membership degrees equal to one

$$\sum_{i_l=1}^{i_l=M_l} A_{l,i_l}(z_l) = 1 \; l = 1, \ldots, n \tag{3.11}$$

where M_l represents the number of the fuzzy sets on the lth input domain. Hence, the triangular membership functions are defined by:

$$\begin{aligned} A_{l,i_l}(z_l) &= \frac{z_l - a_{l,i_j-1}}{a_{l,i_l} - a_{l,i_l-1}}, \; a_{l,i_l-1} \leq z_l < a_{l,i_l} \\ A_{l,i_l}(z_l) &= \frac{a_{l,i_l+1} - z_l}{a_{l,i_l+1} - a_{l,i_l}}, \; a_{l,i_l} \leq z_l < a_{l,i_l+1} \end{aligned} \tag{3.12}$$

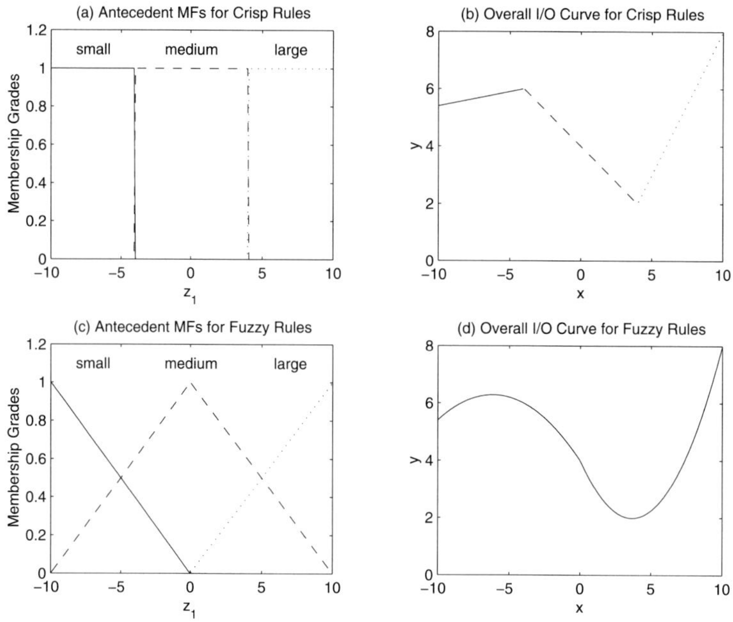

Figure 3.6: Example of a TS fuzzy model.

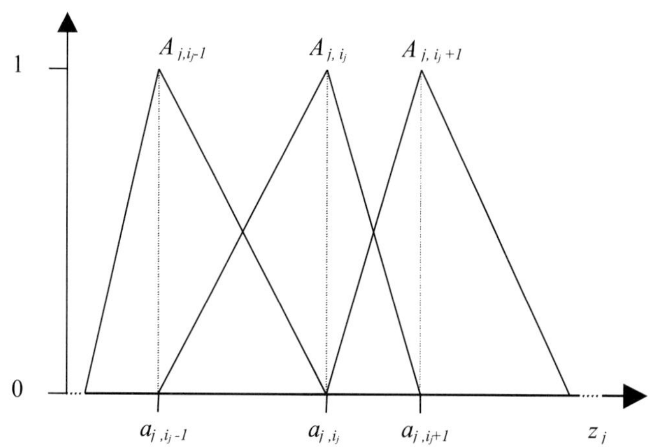

Figure 3.7: Ruspini parameterization of triangular membership functions.

where a_{l,i_l} cores of the adjacent fuzzy sets determine the support ($\text{sup}_{l,i_l} = a_{l,i_l+1} - a_{l,i_l-1}$) of a set.

$$a_{l,i_l} = \text{core}(A_{l,i_l}(z_l)) = \{z_l | A_{l,i_l}(z_l) = 1\} \,. \tag{3.13}$$

Example 3.2 (Ruspini partition of two-dimensional input space). *In multivariable case, the presented model obtains grid-type axes-parallel partition of the input space that helps to obtain an easily interpretable rule base. Figure* 3.8 *illustrates such partitioning of a two-dimensional input space when* $n = 2$, $M_1 = 5$, *and* $M_2 = 5$.

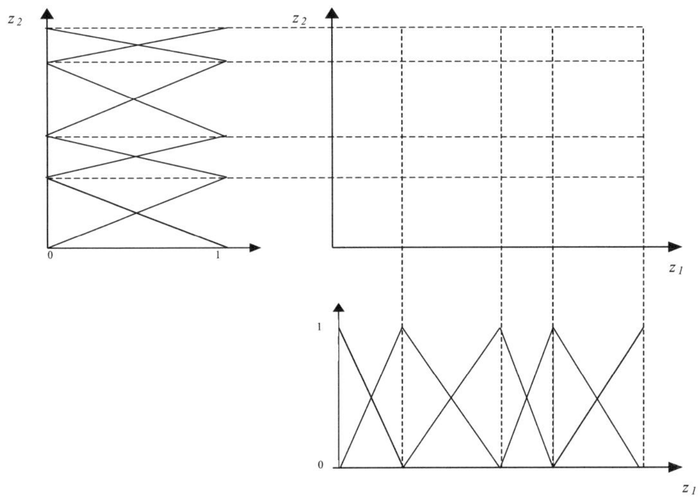

Figure 3.8: Partition of the input domain (when $n = 2$, $M_1 = 5$, $M_2 = 5$).

□

In the following, the mathematical formalism of TS fuzzy models for nonlinear regression is described. Consider the identification of an unknown nonlinear system

$$y = f(\mathbf{x}) \tag{3.14}$$

based on some available input-output data $\mathbf{x}_k = [x_{1,k}, \ldots, x_{n,k}]^T$ and y_k, respectively. The index $k = 1, \ldots, N$ denotes the individual data samples.

While it may be difficult to find a model to describe the unknown system globally, it is often possible to construct local linear models around selected operating points. The modelling framework that is based on combining local models valid in predefined operating regions is called *operating regime-based modelling* [197]. In this framework, the model is generally given by:

$$\hat{y} = \sum_{i=1}^{c} \phi_i(\mathbf{x}) \left(\mathbf{a}_i^T \mathbf{x} + b_i\right) \tag{3.15}$$

where $\phi_i(\mathbf{x})$ is the validity function for the ith operating regime and $\theta_i = [\mathbf{a}_i^T b_i]^T$ is the parameter vector of the corresponding local linear model. The operating regimes can also be represented by fuzzy sets in which case the Takagi–Sugeno fuzzy model is obtained [257]:

$$R_i: \ \textbf{If } \mathbf{x} \text{ is } \mathbf{A}_i(\mathbf{x}) \textbf{ then } \hat{y} = \mathbf{a}_i^T \mathbf{x} + b_i, \ [w_i] \qquad i = 1, \ldots, c\,. \tag{3.16}$$

Here, $\mathbf{A}_i(\mathbf{x})$ is a multivariable membership function, $\mathbf{a}_i$ and b_i are parameters of the local linear model, and $w_i \in [0, 1]$ is the weight of the rule. The value of w_i is usually chosen by the designer of the fuzzy system to represent the belief in the accuracy of the ith rule. When such knowledge is not available $w_i = 1, \forall\, i$ is used.

The antecedent proposition "$\mathbf{x}$ is $\mathbf{A}_i(\mathbf{x})$" can be expressed as a logical combination of propositions with univariate fuzzy sets defined for the individual components of $\mathbf{x}$, usually in the following conjunctive form:

$$R_i: \ \textbf{If } x_1 \text{ is } A_{i,1}(x_1) \textbf{ and } \ldots \textbf{ and } x_n \text{ is } A_{i,n}(x_n) \textbf{ then } \hat{y} = \mathbf{a}_i^T \mathbf{x} + b_i, \ [w_i]\,. \tag{3.17}$$

The *degree of fulfillment* of the rule is then calculated as the product of the individual membership degrees and the rule's weight:

$$\beta_i(\mathbf{x}) = w_i \mathbf{A}_i(\mathbf{x}) = w_i \prod_{j=1}^{n} A_{i,j}(x_j)\,. \tag{3.18}$$

The rules are aggregated by using the fuzzy-mean formula

$$\hat{y} = \frac{\sum\limits_{i=1}^{c} \beta_i(\mathbf{x}) \left(\mathbf{a}_i^T \mathbf{x} + b_i\right)}{\sum\limits_{i=1}^{c} \beta_i(\mathbf{x})}\,. \tag{3.19}$$

3.2.2 Related Modelling Paradigms

There are many well-known or developing modelling strategies that can be seen as special case of the previously presented fuzzy model. The remaining part of this section presents the connections with these methods to show the possible interpretations of the TS fuzzy models.

- Operating Regime Based Modelling

  As the combination of fuzzy sets partition the input space into a number of fuzzy regions and the consequent functions (local models) describe the system behavior within a given region, the TS fuzzy model can be seen as multiple model network [197]. The soft transition between the operating regimes is handled by the fuzzy system in elegant fashion [31]. This representation is appealing, since many systems change its behavior smoothly as a function of the operating point.

From (3.15) and (3.19) one can see that the TS fuzzy model is equivalent to the operating regime-based model when the validity function is chosen to be the normalized rule degree of fulfillment:

$$\phi_i(\mathbf{x}) = \frac{\beta_i(\mathbf{x})}{\sum\limits_{i=1}^{c} \beta_i(\mathbf{x})}. \tag{3.20}$$

In this chapter, Gaussian membership functions are used to represent the fuzzy sets $A_{i,j}(x_j)$:

$$A_{i,j}(x_j) = \exp\left(-\frac{1}{2}\frac{\left(x_j - v_{i,j}\right)^2}{\sigma_{i,j}^2}\right) \tag{3.21}$$

with $v_{i,j}$ being the center and $\sigma_{i,j}^2$ the variance of the Gaussian curve. This choice leads to the following compact formula for (3.18):

$$\beta_i(\mathbf{x}) = w_i \mathbf{A}_i(\mathbf{x}) = w_i \exp\left(-\frac{1}{2}\left(\mathbf{x} - \mathbf{v}_j^x\right)^T \left(\mathbf{F}_i^{xx}\right)^{-1} \left(\mathbf{x} - \mathbf{v}_j^x\right)\right). \tag{3.22}$$

The center vector is denoted by $\mathbf{v}_j^x = [v_{1,j}, \ldots, v_{n,j}]$ and $(\mathbf{F}_i^{xx})^{-1}$ is the inverse of the matrix containing the variances on its diagonal:

$$\mathbf{F}_i^{xx} = \begin{bmatrix} \sigma_{1,i}^2 & 0 & \cdots & 0 \\ 0 & \sigma_{2,i}^2 & \cdots & 0 \\ \vdots & \vdots & \ddots & \vdots \\ 0 & 0 & \cdots & \sigma_{n,i}^2 \end{bmatrix}. \tag{3.23}$$

- Piece-wise Models

  When the antecedent of a first-order TS fuzzy model consists of crisp sets or the fuzzy sets are defined by piece-wise linear membership functions, the resulted fuzzy model has piece-wise (linear or quadratic) behavior. Modelling techniques based on piece-wise linear models are widely applied for control relevant modelling [127]. Skeppstedt describes the use of local models for modelling and control purposes with hard transfer from one model to the next [248]. Pottman describes a multi model approach where the local models overlap [219]. These models can be effectively used in model based control [94].

When the rule consequent is a crisp number (Singleton) and the rule antecedent contains piece-wise linear membership functions, the resulted fuzzy model has piece-wise linear input-output behavior. Piece-wise linear multidimensional fuzzy sets can be obtained by Delaunay triangulation of characteristic points defined on the input space of the fuzzy model. This technique has already been suggested in the context of complexity reduction of fuzzy systems [242]. Moreover, Delaunay-based multivariable spline approximation from scattered samples of an unknown function has proven to be an effective tool for classification [67]. Recently, Delaunay Networks were introduced to represent interpolating models and controllers [270] and the integration of expert knowledge in these models has been also studied [269]. Fuzzy models based on this concept are pursued in [1].

- B-spline Networks

  The presented grid-type fuzzy model makes piece-wise polynomial approximation of the nonlinear system to be modeled. Piecewise polynomials are closely related to spline modelling and have been successfully applied to many real-life problems, e.g., in experimental pH control and for the prediction of the viscosity for an industrial polymerization reactor [175]. B-spline basis functions can be regarded as piece-wise polynomials, where B-spline basis functions are defined by a set of knots that represent piecewise polynomial intervals. The order of these local polynomials is defined by the order of the B-splines, denoted by k. A knot vector of a set of kth order basis functions is defined by

$$\mathbf{a}_i = \left[a_{i,1}, a_{i,2}, \ldots, a_{i,M_i+k-1}\right]^T \tag{3.24}$$

  where M_i is the number of the basis functions defined on the ith variable, and $a_{i,j}$ is the jth knot. The univariate basis functions are calculated using the following recurrence relationship [50]:

$$A_{i,j}^{k}\left(z_i\right) = \left(\frac{z_i - a_{i,j-k}}{a_{i,j-1} - a_{i,j-k}}\right) A_{i,j-1}^{k-1}\left(z_i\right) + \left(\frac{a_{i,j} - z_i}{a_{i,j} - a_{i,j-k+1}}\right) A_{i,j}^{k-1}\left(z_i\right) \tag{3.25}$$

$$A_{i,j}^{1}\left(z_i\right) = \begin{cases} 1, \text{ if } z_i \in \left[a_{i,j-1}, a_{i,j}\right] \\ 0, \text{ otherwise} \end{cases} \tag{3.26}$$

  where $A_{i,j}^{k}\left(z_i\right)$ is the jth univariate basis function of order k. Triangular membership functions are identical to second-order B-spline basis functions (3.12).

  Multi-dimensionality of the model is achieved by the means of the tensor product of the univariate B-splines. Given a set of B-splines defined over the input variables the multivariate B-spline can be defined as

$$\beta_j^k\left(\mathbf{z}\right) = \prod_{i=1}^{n} A_{i,j}^{k}\left(z_i\right). \tag{3.27}$$

  This multivariate B-spline is identical to the weight of the jth rule.

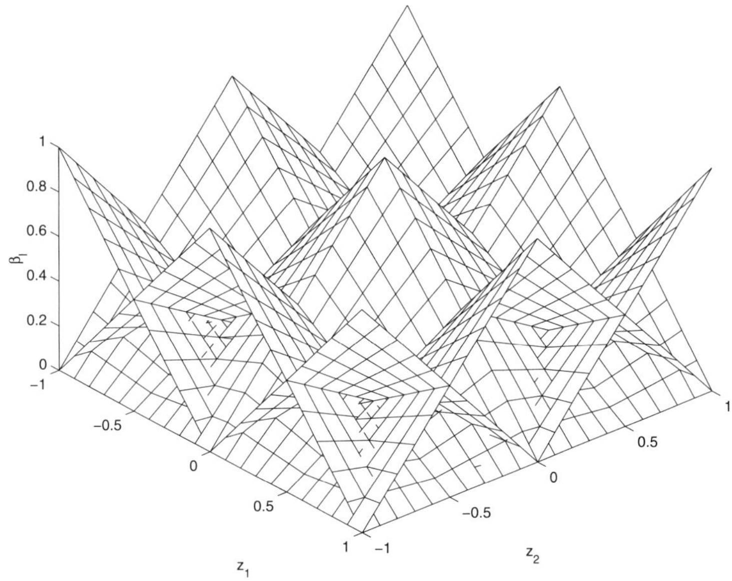

Figure 3.9: Example of multi(bi)variate B-spline (membership) functions.

Figure 3.9 shows an example of bivariate B-splines that are identical to the rule weights of nine rules in a fuzzy system with three membership functions defined on its two input domains.

- Radial Basis Function Networks

Basis function networks have been used for function approximation and modelling in various forms for many years. The original radial basis function methods come from interpolation theory [220], where a basis function is associated with each data point. Basis function nets have also received attention from neural network and the control community [182].

Radial Basis Functions Networks (RBFNs), as proposed in 1989 by Moody and Darken [195], are often considered to be a type of neural network in which each unit has only a local effect instead of a global effect as in multi-layer percetron based neural networks (MLPs). In a similar way to MLPs, RBFNs perform function approximation by superimposing a set of N_r Radial Basis Functions (RBFs) as follows:

$$\beta_j = \exp\left(-\sum_{i=1}^{n}\left(\frac{x_i - a_{i,j}}{\sigma_{i,j}}\right)^2\right) \tag{3.28}$$

$$y = \frac{\sum_{j=1}^{N_r}\beta_j\theta_j}{\sum_{j=1}^{N_r}\beta_j} \tag{3.29}$$

where $\beta_j : \ j = 1, \ldots, N_r$ is the firing strength of unit j, N_r is the number of RBFs, $x_i : \ i = 1, \ldots, n$ are the inputs, y is the output, and $a_{i,j} : \ i = 1, \ldots, n; j = 1, \ldots, N_r$, $\sigma_{i,j} : \ i = 1, \ldots, n; j = 1, \ldots, N_r$, and $\theta_j : \ j = 1, \ldots, N_r$ are free parameters which respectively determine the position, width, and height of the humps.

RBFNs can be trained in the same way as MLPs, i.e., they are initialized randomly and then minimized by gradient-descent. Alternatively, the position of the centres of the RBFs and their widths can be determined by a clustering algorithm and then the heights can be set by a least-squares type of algorithm [195]. Like MLPs, they have been proven to be universal approximators [112].

Jang has pointed out, under certain constraints, the radial basis function network (RBFN) is functionally equivalent to zero-order TS fuzzy model [22, 129] as

$$\begin{aligned} \beta_j &= \exp\left(-\sum_{i=1}^{n}\left(\frac{x_i - a_{i,j}}{\sigma_{i,j}}\right)^2\right) \\ &= \prod_{i=1}^{n} \underbrace{\exp\left(-\left(\frac{x_i - a_{i,j}}{\sigma_{i,j}}\right)^2\right)}_{A_{i,j}(x_i)}. \end{aligned} \tag{3.30}$$

Hunt has developed a generalized radial basis function network (GBFN) that is similar to the first-order TS fuzzy model [121]. These models are identical to TS fuzzy models under the following conditions.

- The number of the basis function units is equal to the number of the fuzzy if-then rules.
- Both the basis function network and the fuzzy inference system use the same method to calculate their overall outputs.

Thus, the model presented in Section 3.2.1 can be seen as RBFN if $q = \{0\}$ and GBFN if $q \neq \{0\}$, with piecewise linear basis functions.

The realization that fuzzy models are very similar to RBFN function approximators also means that methods which have been developed in fuzzy control, such as those analyzed in this work, can be applied to neural control.

3.3 TS Fuzzy Models for Nonlinear Regression

Fuzzy identification is an effective tool for the approximation of uncertain nonlinear systems on the basis of measured data [115]. Among the different fuzzy modelling techniques, the Takagi–Sugeno (TS) model [257] has attracted most attention. This model consists of if–then rules with fuzzy antecedents and mathematical functions in the consequent part. The antecedents fuzzy sets partition the input space into a number of fuzzy regions, while the consequent functions describe the system's behavior in these regions [254].

The construction of a TS model is usually done in two steps. In the first step, the fuzzy sets (membership functions) in the rule antecedents are determined. This can be done manually, using knowledge of the process, or by some data-driven techniques. In the second step, the parameters of the consequent functions are estimated. As these functions are usually chosen to be linear in their parameters, standard linear least-squares methods can be applied.

The bottleneck of the construction procedure is the identification of the antecedent membership functions, which is a nonlinear optimization problem. Typically, gradient-decent neuro-fuzzy optimization techniques are used [131], with all the inherent drawbacks of gradient-descent methods: (1) the optimization is sensitive to the choice of initial parameters and hence can easily get stuck in local minima; (2) the obtained model usually has poor generalization properties; (3) during the optimization process, fuzzy rules may loose their initial meaning (i.e., validity as local linear models of the system under study). This hampers the a posteriori interpretation of the optimized TS model. An alternative solution are gradient-free nonlinear optimization algorithms. Genetic algorithms proved to be useful for the construction of fuzzy systems [136, 230]. Unfortunately, the severe computational requirements limit their applicability as a rapid model-development tool.

Fuzzy clustering in the Cartesian product-space of the inputs and outputs is another tool that has been quite extensively used to obtain the antecedent membership functions [255, 29, 26]. Attractive features of this approach are the simultaneous identification of the antecedent membership functions along with the consequent local linear models and the implicit regularization [139].

By clustering in the product-space, multidimensional fuzzy sets are initially obtained, which are either used in the model directly or after projection onto the individual antecedent variables. As it is generally difficult to interpret multidimensional fuzzy sets, projected one-dimensional fuzzy sets are usually preferred. However, the projection and the approximation of the point-wise defined membership functions by parametric ones may deteriorate the performance of the model. This is due to two types of errors: the decomposition error and the approximation error. The decomposition error can be reduced by using eigenvector projection [26, 151] and/or by fine-tuning the parameterized membership functions. This fine-tuning, however, can result in overfitting and thus poor generalization of the identified model.

In this chapter, we propose to use the Gath–Geva (GG) clustering algorithm [93] instead of the widely used Gustafson–Kessel method [108], because with the GG method, the parameters of the univariate membership functions can directly be derived from the parameters of the clusters. Through a linear transformation of the input variables, the antecedent partition can be accurately captured and no decomposition error occurs. Unfortunately, the resulting model is not transparent as it is hard to interpret the linguistic terms defined on the linear combination of the input variables. To form an easily interpretable model that does not rely on transformed input variables, a new clustering algorithm is presented based on the Expectation Maximization (EM) identification of Gaussian mixture of models. Mixtures are used as models of data originating from several mixed populations. The EM algorithm has been widely used to estimate the parameters of the components in the mixture [44]. The clusters obtained by GG clustering are multivariate Gaussian functions. The alternating optimization of these clusters is identical to the EM identification of the mixture of these Gaussian models when the fuzzy weighting exponent $m = 2$ [40].

In this chapter, a new cluster prototype is introduced, that can easily be represented by an interpretable Takagi–Sugeno (TS) fuzzy model. Similarly to other fuzzy clustering algorithms, the alternating optimization method is employed in the search for the clusters. This new technique is demonstrated on the MPG (miles per gallon) prediction problem and another nonlinear benchmark process. The obtained results are compared with results from the literature. It is shown that with the presented modified Gath–Geva algorithm not only good prediction performance is obtained, but also the interpretability of the model improves.

3.3.1 Fuzzy Model Identification Based on Gath–Geva Clustering

The available data samples are collected in matrix $\mathbf{Z}$ formed by concatenating the regression data matrix $\mathbf{X}$ and the output vector $\mathbf{y}$:

$$\mathbf{X} = \begin{bmatrix} \mathbf{x}_1^T \\ \mathbf{x}_2^T \\ \vdots \\ \mathbf{x}_N^T \end{bmatrix}, \quad \mathbf{y} = \begin{bmatrix} y_1 \\ y_2 \\ \vdots \\ y_N \end{bmatrix}, \quad \mathbf{Z}^T = [\mathbf{X}\,\mathbf{y}]\,. \tag{3.31}$$

Each observation thus is an $n+1$-dimensional column vector

$$\mathbf{z}_k = [x_{1,k}, \ldots, x_{n,k}, y_k]^T = [\mathbf{x}_k^T \, y_k]^T.$$

Through clustering, the data set $\mathbf{Z}$ is partitioned into c clusters (see also Figure 1.17 in Section 1.6). The c is assumed to be known, based on prior knowledge, for instance (refer to [26] for methods to estimate or optimize c in the context of system identification). The result is a fuzzy partition matrix $\mathbf{U} = [\mu_{i,k}]_{c\times N}$, whose element $\mu_{i,k}$ represents the degree of membership of the observation $\mathbf{z}_k$ in cluster i.

Clusters of different shapes can be obtained by using an appropriate definition of cluster prototypes (e.g., linear varieties) or by using different distance measures. The Gustafson–Kessel (GK) clustering algorithm has often been applied to identify TS models. The main drawbacks of this algorithm are that only clusters with approximately equal volumes can be properly identified and that the resulted clusters cannot be directly described by univariate parametric membership functions.

To circumvent these problems, Gath–Geva algorithm [93] is applied. Since the cluster volumes are not restricted in this algorithm, lower approximation errors and more relevant consequent parameters can be obtained than with Gustafson–Kessel (GK) clustering. An example can be found in [26], p. 91. The clusters obtained by GG clustering can be transformed into exponential membership functions defined on the linearly transformed space of the input variables.

Probabilistic Interpretation of Gath–Geva Clustering

The Gath–Geva clustering algorithm can be interpreted in the probabilistic framework. Denote $p(\eta_i)$ the unconditional cluster probability (normalized such that $\sum_{i=1}^{c} p(\eta_i) = 1$), given by the fraction of the data that it explains; $p(\mathbf{z}|\eta_i)$ is the domain of influence of the cluster, and will be taken to be multivariate Gaussian $N(\mathbf{v}_i, \mathbf{F}_i)$ in terms of a mean $\mathbf{v}_i$ and covariance matrix $\mathbf{F}_i$. The Gath–Geva algorithm is equivalent to the identification of a mixture of Gaussians that model the $p(\mathbf{z}|\eta)$ probability density function expanded into a sum over the c clusters

$$p(\mathbf{z}|\eta) = \sum_{i=1}^{c} p(\mathbf{z}, \eta_i) = \sum_{i=1}^{c} p(\mathbf{z}|\eta_i) p(\eta_i) \tag{3.32}$$

where the $p(\mathbf{z}|\eta_i)$ distribution generated by the ith cluster is represented by the Gaussian function

$$p(\mathbf{z}|\eta_i) = \frac{1}{(2\pi)^{\frac{n+1}{2}} \sqrt{|\mathbf{F}_i|}} \exp\left(-\frac{1}{2} (\mathbf{z} - \mathbf{v}_i)^T (\mathbf{F}_i)^{-1} (\mathbf{z} - \mathbf{v}_i) \right) . \tag{3.33}$$

Through GG clustering, the $p(\mathbf{z}) = p(\mathbf{x}, y)$ joint density of the response variable y and the regressors $\mathbf{x}$ is modeled as a mixture of c multivariate $n+1$-dimensional Gaussian functions.

The conditional density $p(y|\mathbf{x})$ is also a mixture of Gaussian models. Therefore, the regression problem (3.14) can be formulated on the basis of this probability as

$$\begin{aligned} y = f(\mathbf{x}) &= E[y|\mathbf{x}] = \\ &= \int y p(y|\mathbf{x}) dy = \frac{\int y p(y, \mathbf{x}) dy}{p(\mathbf{x})} = \\ &= \sum_{i=1}^{c} \frac{\left[[\mathbf{x}^T\ 1] \theta_i \right] p(\mathbf{x}|\eta_i) p(\eta_i)}{p(\mathbf{x})} = \sum_{i=1}^{c} p(\eta_i|\mathbf{x}) \left[[\mathbf{x}^T\ 1] \theta_i \right] . \end{aligned} \tag{3.34}$$

Here, θ_i is the parameter vector of the local models to be obtained later on (Section 3.3.2) and $p(\eta_i|\mathbf{x})$ is the probability that the ith Gaussian component is generated by the regression vector $\mathbf{x}$:

$$p(\eta_i|\mathbf{x}) = \frac{\frac{p(\eta_i)}{(2\pi)^{n/2}\sqrt{|\mathbf{F}_i^{xx}|}} \exp\left(-\frac{1}{2} (\mathbf{x} - \mathbf{v}_i^x)^T (\mathbf{F}_i^{xx})^{-1} (\mathbf{x} - \mathbf{v}_i^x) \right)}{\sum_{i=1}^{c} \frac{p(\eta_i)}{(2\pi)^{n/2}\sqrt{|(\mathbf{F})_i^{xx}|}} \exp\left(-\frac{1}{2} (\mathbf{x} - \mathbf{v}_i^x)^T (\mathbf{F}_i^{xx})^{-1} (\mathbf{x} - \mathbf{v}_i^x) \right)} \tag{3.35}$$

where $\mathbf{F}^{xx}$ is obtained by partitioning the covariance matrix $\mathbf{F}$ as follows

$$\mathbf{F}_i = \begin{bmatrix} \mathbf{F}_i^{xx} & \mathbf{F}_i^{xy} \\ \mathbf{F}_i^{yx} & \mathbf{F}_i^{yy} \end{bmatrix} \tag{3.36}$$

where

- $\mathbf{F}_i^{xx}$ is the $n \times n$ submatrix containing the first n rows and columns of $\mathbf{F}_i$,
- $\mathbf{F}_i^{xy}$ is an $n \times 1$ column vector containing the first n elements of last column of $\mathbf{F}_i$,
- $\mathbf{F}_i^{yx}$ is an $1 \times n$ row vector containing the first n elements of the last row of $\mathbf{F}_i$, and
- $\mathbf{F}_i^{yy}$ is the last element in the last row of $\mathbf{F}_i$.

3.3.2 Construction of Antecedent Membership Functions

The 'Gaussian Mixture of Regressors' model [144] defined by (3.34) and (3.35) is in fact a kind of operating regime-based model (3.15) where the validity function is chosen as $\phi_i(\mathbf{x}) = p(\eta_i|\mathbf{x})$. Furthermore, this model is also equivalent to the TS fuzzy model where the rule weights in (3.16) are given by:

$$w_i = \frac{p(\eta_i)}{(2\pi)^{n/2}\sqrt{|\mathbf{F}_i^{xx}|}} \tag{3.37}$$

and the membership functions are the Gaussians defined by (3.22). However, in this case, $\mathbf{F}_i^{xx}$ is not necessarily in the diagonal form (3.23) and the decomposition of $\mathbf{A}_i(\mathbf{x})$ to the univariate fuzzy sets $A_{i,j}(x_j)$ given by (3.21) is not possible.

If univariate membership functions are required (for interpretation purposes), such a decomposition is necessary. Two different approaches can be followed.

The first one is an approximation, based on the axis-orthogonal projection of $\mathbf{A}_i(\mathbf{x})$. This approximation will typically introduce some decomposition error, which can, to a certain degree, be compensated by using global least-squares re-estimation of the consequent parameters. In this way, however, the interpretation of the local linear models may be lost, as the rule consequents are no longer local linearizations of the nonlinear system [6, 140].

The second approach is an exact one, based on eigenvector projection [26], also called the transformed input-domain approach [151]. Denote $\lambda_{i,j}$ and $\mathbf{t}_{i,j}$, $j = 1, \ldots, n$, the eigenvalues and the unitary eigenvectors of $\mathbf{F}_i^{xx}$, respectively. Through the eigenvector projection, the following fuzzy model is obtained in the transformed input domain:

$$R_i: \textbf{ If } \widetilde{x}_{i,1} \text{ is } A_{i,1}(\widetilde{x}_{i,1}) \textbf{ and } \ldots \textbf{ and } \widetilde{x}_{i,n} \text{ is } A_{i,n}(\widetilde{x}_{i,n}) \textbf{ then } \hat{y} = \mathbf{a}_i^T\mathbf{x} + b_i \tag{3.38}$$

where $\widetilde{x}_{i,j} = \mathbf{t}_{i,j}^T \mathbf{x}$ are the transformed input variables. The Gaussian membership functions are given by

$$A_{i,j}(\widetilde{x}_{i,j}) = \exp\left(-\frac{1}{2}\frac{\left(\widetilde{x}_{i,j} - \widetilde{v}_{i,j}\right)^2}{\widetilde{\sigma}_{i,j}^2}\right) \tag{3.39}$$

with the cluster centers $\widetilde{v}_{i,j} = \mathbf{t}_{i,j}^T \mathbf{v}_i^x$ and and variances $\widetilde{\sigma}_{i,j}^2 = \lambda_{i,j}^2$.

Estimation of Consequent Parameters

Two least-squares methods for the estimation of the parameters in the local linear consequent models are presented: weighted total least squares and weighted ordinary least squares (see also Section 1.6).

- **Ordinary Least-Squares Estimation**

  The ordinary weighted least-squares method can be applied to estimate the consequent parameters in each rule separately, by minimizing the following criterion:

$$\min_{\boldsymbol{\theta}_i} \frac{1}{N}\left(\mathbf{y} - \mathbf{X}_e\boldsymbol{\theta}_i\right)^T \boldsymbol{\Phi}_i \left(\mathbf{y} - \mathbf{X}_e\boldsymbol{\theta}_i\right) \tag{3.40}$$

  where $\mathbf{X}_e = [\mathbf{X}\ \mathbf{1}]$ is the regressor matrix extended by a unitary column and $\boldsymbol{\Phi}_i$ is a matrix having the membership degrees on its main diagonal:

$$\boldsymbol{\Phi}_i = \begin{bmatrix} \mu_{i,1} & 0 & \cdots & 0 \\ 0 & \mu_{i,2} & \cdots & 0 \\ \vdots & \vdots & \ddots & \vdots \\ 0 & 0 & \cdots & \mu_{i,N} \end{bmatrix}. \tag{3.41}$$

  The weighted least-squares estimate of the consequent parameters is given by

$$\boldsymbol{\theta}_i = \left(\mathbf{X}_e^T \boldsymbol{\Phi}_i \mathbf{X}_e\right)^{-1} \mathbf{X}_e^T \boldsymbol{\Phi}_i \mathbf{y}\,. \tag{3.42}$$

  When $\mu_{i,k}$ is obtained by the Gath–Geva clustering algorithm, the covariance matrix can directly be used to obtain the estimate instead of (3.42):

$$\begin{aligned} \mathbf{a}_i &= \left(\mathbf{F}^{xx}\right)^{-1} \mathbf{F}^{xy}\,, \\ b_i &= v_i^y - \mathbf{a}_i^T \mathbf{v}_i^x\,. \end{aligned} \tag{3.43}$$

  This follows directly from the properties of least-squares estimation [74].

- **Total Least-Squares Estimation**

  As the clusters locally approximate the regression surface, they are n-dimensional linear subspaces of the $(n+1)$-dimensional regression space. Consequently, the smallest eigenvalue of the ith cluster covariance matrix $\mathbf{F}_i$ is typically in orders of magnitude smaller than the remaining eigenvalues [26]. The corresponding eigenvector $\mathbf{u}_i$ is then the normal vector to the hyperplane spanned by the remaining eigenvectors of that cluster:

$$\mathbf{u}_i^T(\mathbf{z} - \mathbf{v}_i) = 0\,. \tag{3.44}$$

Similarly to the observation vector $\mathbf{z} = [\mathbf{x}^T\, y]^T$, the prototype vector and is partitioned as $\mathbf{v}_i = \left[(\mathbf{v}_i^x)^T\ v_i^y\right]$, i.e., into a vector $\mathbf{v}^x$ corresponding to the regressor $\mathbf{x}$, and a scalar v_i^y corresponding to the output y. The eigenvector is partitioned in the same way, $\mathbf{u}_i = \left[(\mathbf{u}_i^x)^T\ u_i^y\right]^T$. By using these partitioned vectors, (3.44) can be written as

$$\left[(\mathbf{u}_i^x)^T\ u_i^y\right]\left([\mathbf{x}^T\, y] - \left[(\mathbf{v}_i^x)^T\ v_i^y\right]\right)^T = 0 \tag{3.45}$$

from which the parameters of the hyperplane defined by the cluster can be obtained:

$$y = \underbrace{\frac{-1}{u_i^y}(\mathbf{u}_i^x)^T}_{\mathbf{a}_i^T}\mathbf{x} + \underbrace{\frac{1}{u_i^y}(\mathbf{u}_i)^T\mathbf{v}_i}_{b_i}\,. \tag{3.46}$$

Although the parameters have been derived from the geometrical interpretation of the clusters, it can be shown [26] that (3.46) is equivalent to the weighted total least-squares estimation of the consequent parameters, where each data point is weighted by the corresponding membership degree.

The TLS algorithm should be used when there are errors in the input variables. Note, however, that the TLS algorithm does not minimize the mean-square prediction error of the model, as opposed to the ordinary least-squares algorithm. Furthermore, if the input variables of the model locally are strongly correlated, the smallest eigenvector then does not define a hyperplane related to the regression problem; it may rather reflect the dependency of the input variables.

3.3.3 Modified Gath–Geva Clustering

As discussed in Section 3.3.2, the main drawback of the construction of interpretable Takagi–Sugeno fuzzy models via clustering is that clusters are generally axes-oblique rather than axes-parallel (the fuzzy covariance matrix $\mathbf{F}^{xx}$ has nonzero off-diagonal elements) and consequently a decomposition error is made in their projection. To circumvent this problem, we propose a new fuzzy clustering method in this section.

Expectation Maximization based Fuzzy Clustering for Regression

Each cluster is described by an input distribution, a local model and an output distribution:

$$\begin{aligned} p(\mathbf{x}, y) &= \sum_{i=1}^{c} p(\mathbf{x}, y, \eta_i) = \sum_{i=1}^{c} p(\mathbf{x}, y|\eta_i)p(\eta_i) \\ &= \sum_{i=1}^{c} p(y|\mathbf{x}, \eta_i)p(\mathbf{x}|\eta_i)p(\eta_i)\,. \end{aligned} \tag{3.47}$$

The input distribution, parameterized as an unconditional Gaussian [95], defines the domain of influence of the cluster similarly to the multivariate membership functions (3.22)

$$p(\mathbf{x}|\eta_i) = \frac{1}{(2\pi)^{\frac{n}{2}}\sqrt{|\mathbf{F}_i^{xx}|}} \exp\left(-\frac{1}{2}(\mathbf{x}-\mathbf{v}_i^x)^T(\mathbf{F}_i^{xx})^{-1}(\mathbf{x}-\mathbf{v}_i^x)\right). \tag{3.48}$$

The output distribution is

$$p(y|\mathbf{x}, \eta_i) = \frac{1}{\sqrt{2\pi\sigma_i^2}} \exp\left(-\frac{(y-\mathbf{x}^T\theta_i)^T(y-\mathbf{x}^T\theta_i)}{2\sigma_i^2}\right). \tag{3.49}$$

When the transparency and interpretability of the model is important, the cluster covariance matrix $\mathbf{F}^{xx}$ can be reduced to its diagonal elements similarly to the simplified axis-parallel version of the Gath–Geva clustering algorithm [119]:

$$p(\mathbf{x}_k|\eta_i) = \prod_{j=1}^{n} \frac{1}{\sqrt{2\pi\sigma_{i,j}^2}} \exp\left(-\frac{1}{2}\frac{(x_{j,k}-v_{i,j})^2}{\sigma_{i,j}^2}\right). \tag{3.50}$$

The identification of the model means the determination of the cluster parameters: $p(\eta_i), \mathbf{v}_i^x, \mathbf{F}_i^{xx}, \theta_i, \sigma_i$. Below, the expectation maximization (EM) identification of the model is presented, followed by a re-formulation of the algorithm in the form of fuzzy clustering.

The basics of EM are the following. Suppose we know some observed values of a random variable $\mathbf{z}$ and we wish to model the density of $\mathbf{z}$ by using a model parameterized by η. The EM algorithm obtains an estimate $\hat{\eta}$ that maximizes the likelihood $\mathcal{L}(\eta) = p(\mathbf{z}|\eta)$ by iterating over the following two steps:

- **E-step.** In this step, the current cluster parameters η_i are assumed to be correct, and based on them the posterior probabilities $p(\eta_i|\mathbf{x}, y)$ are computed. These posterior probabilities can be interpreted as the probability that a particular piece of data was generated by the particular cluster's distribution. By using the Bayes theorem, the conditional probabilities are:

$$p(\eta_i|\mathbf{x}, y) = \frac{p(\mathbf{x}, y|\eta_i)p(\eta_i)}{p(\mathbf{x}, y)} = \frac{p(\mathbf{x}, y|\eta_i)p(\eta_i)}{\sum_{i=1}^{c} p(\mathbf{x}, y|\eta_i)p(\eta_i)}\,. \tag{3.51}$$

- **M-step.** In this step, the current data distribution is assumed to be correct and the parameters of the clusters that maximize the likelihood of the data are sought. The new unconditional probabilities are:

$$p(\eta_i) = \frac{1}{N}\sum_{k=1}^{N} p(\eta_i|\mathbf{x}, y)\,. \tag{3.52}$$

The means and the weighted covariance matrices are computed by:

$$\mathbf{v}_i^x = \frac{\sum_{k=1}^{N} \mathbf{x}_k p(\eta_i|\mathbf{x}_k, y_k)}{\sum_{k=1}^{N} p(\eta_i|\mathbf{x}_k, y_k)}\,, \tag{3.53}$$

$$\mathbf{F}_i^{xx} = \frac{\sum_{k=1}^{N} (\mathbf{x}_k - \mathbf{v}_i^x)(\mathbf{x}_k - \mathbf{v}_i^x)^T p(\eta_i|\mathbf{x}_k, y_k)}{\sum_{k=1}^{N} p(\eta_i|\mathbf{x}_k, y_k)}\,. \tag{3.54}$$

In order to find the maximizing parameters of the local linear models, the derivative of the log-likelihood is set equal to zero:

$$\begin{aligned} 0 &= \frac{\partial}{\partial\theta_i}\ln\prod_{k=1}^{N} p(\mathbf{x}_k, y_k) = \sum_{k=1}^{N}\frac{\partial}{\partial\theta_i}\ln p(\mathbf{x}_k, y_k) \\ &= \frac{1}{Np(\eta_i)}\sum_{k=1}^{N} p(\eta_i|\mathbf{x}, y)\left(y_k - f_i(\mathbf{x}_k, \theta_i)\right)\frac{\partial f_i(\mathbf{x}_k, \theta_i)}{\partial\theta_i}\,. \end{aligned} \tag{3.55}$$

Here, $f_i(\mathbf{x}_k, \theta_i)$ represents the local consequent models, $f_i(\mathbf{x}_k, \theta_i) = \mathbf{a}_i^T\mathbf{x}_k + b_i$. The above equation results in weighted least-squares identification of the local linear models (3.42) with the weighting matrix

$$\mathbf{\Phi}_j = \begin{bmatrix} p(\eta_i|\mathbf{x}_1, y_1) & 0 & \cdots & 0 \\ 0 & p(\eta_i|\mathbf{x}_2, y_2) & \cdots & 0 \\ \vdots & \vdots & \ddots & \vdots \\ 0 & 0 & \cdots & p(\eta_i|\mathbf{x}_N, y_N) \end{bmatrix}. \tag{3.56}$$

Finally, the standard deviations σ_i are calculated. These standard deviations are parameters of the $p(y|\mathbf{x}, \eta_i)$ distribution functions defined by (3.49).

$$\sigma_i^2 = \frac{\sum_{k=1}^{N} (y_k - f_i(\mathbf{x}_k, \theta_i))^T (y_k - f_i(\mathbf{x}_k, \theta_i)) p(\eta_i|\mathbf{x}_k, y_k)}{Np(\eta_i)}\,. \tag{3.57}$$

Modified Gath–Geva Fuzzy Clustering for the Identification of TS Models

In this section, the EM algorithm is re-formulated to provide an easily implementable algorithm, similar to Gath–Geva clustering, for the identification of TS fuzzy models that do not use transformed input domains. See Algorithm 3.3.1.

Note that the distance measure (3.62) consists of two terms. The first one is the distance between the cluster centers and $\mathbf{x}$, while the second one quantifies the performance of the local linear models.

Example 3.3 (Automobile MPG prediction – a comparative study among clustering based techniques). *The example under consideration is the Automobile MPG (miles per gallon) prediction benchmark. The following methods were used and compared:*

1. *GG-TLS: Gath–Geva clustering with total least-squares estimation of the consequent parameters.*
2. *GG-LS: Gath–Geva clustering with weighted ordinary least-squares estimation of the consequent parameters.*
3. *EM-TI: The presented method with transformed input variables.*
4. *EM-NI: The presented method with the original input variables.*

As some clustering methods are sensitive to differences in the numerical ranges of the different features, the data can be normalized to zero mean and unit variance:

$$\tilde{z}_{j,k} = \frac{z_{j,k} - \bar{z}_j}{\sigma_j} \tag{3.64}$$

where $\bar{z}_j$ and σ_j are the mean and the variance of the given variable, respectively.

The goal is to predict the fuel consumption of an automobile on the basis of several given characteristics, such as the weight, model year, etc. The data set was obtained from the UCI Repository of Machine Learning Databases and Domain Theories (FTP address: ftp://ics.uci.edu/pub/machine-learning-databases/auto-mpg). After removing samples with missing values, the data set was reduced to 392 entries. This data set was divided into a training set and a test set, each containing 196 samples.

The performance of the models is measured by the root mean squared prediction error (RMSE):

$$RMSE = \sqrt{\frac{1}{N}\sum_{k=1}^{N}(y_k - \hat{y}_k)^2}\,.$$

The approximation power of the identified models is then compared with fuzzy models with the same number of rules obtained by the Fuzzy Toolbox for MATLAB®

Algorithm 3.3.1 (Gath–Geva Clustering for Takagi–Sugeno Models).

Initialization
Given the data set $\mathbf{Z}$, *specify* c, *choose the weighting exponent* $m = 2$ *and the termination tolerance* $\epsilon > 0$. *Initialize the partition matrix such that* (1.12), (1.13) *and* (1.14) *holds.*

Repeat *for* $l = 1, 2, \ldots$ (l *is the iteration counter*)

Step 1. Calculate the parameters of the clusters:

- *Centers of the membership functions:*

$$\mathbf{v}_i^{x\,(l)} = \sum_{k=1}^{N} \mu_{i,k}^{(l-1)} \mathbf{x}_k \Big/ \sum_{k=1}^{N} \mu_{i,k}^{(l-1)} . \tag{3.58}$$

- *Standard deviations of the Gaussian membership functions:*

$$\sigma_{i,j}^{2\,(l)} = \sum_{k=1}^{N} \mu_{i,k}^{(l-1)} (x_{j,k} - v_{j,k})^2 \Big/ \sum_{k=1}^{N} \mu_{i,k}^{(l-1)} . \tag{3.59}$$

- *Parameters of the local models:*

$$\theta_i = \left(\mathbf{X}_e^T \mathbf{\Phi}_i \mathbf{X}_e \right)^{-1} \mathbf{X}_e^T \mathbf{\Phi}_i \mathbf{y} , \tag{3.60}$$

where the weights are collected in the $\mathbf{\Phi}_i$ *matrix given by* (3.41).

- A priori *probabilities of the clusters:* $\alpha_i = \frac{1}{N} \sum_{k=1}^{N} \mu_{i,k}$.
- *Weights of the rules:*

$$w_i = \prod_{j=1}^{n} \frac{\alpha_i}{\sqrt{2\pi\sigma_{i,j}^2}} . \tag{3.61}$$

Step 2. Compute the distance measure $D_{i,k}^2$**:**

$$\frac{1}{D_{i,k}^2} = \prod_{j=1}^{n} \frac{\alpha_i}{\sqrt{2\pi\sigma_{i,j}^2}} \exp\left(-\frac{1}{2} \frac{(x_{j,k} - v_{i,j})^2}{\sigma_{i,j}^2} \right) \cdot \frac{1}{\sqrt{2\pi\sigma_i^2}} \exp\left(-\frac{(y_k - f_i(\mathbf{x}_k, \theta_i))^T (y_k - f_i(\mathbf{x}_k, \theta_i))}{2\sigma_i^2} \right) . \tag{3.62}$$

Step 3. *Update the partition matrix*

$$\mu_{i,k}^{(l)} = \frac{1}{\sum_{j=1}^{c} \left(D_{i,k}(\mathbf{z}_k, \eta_i) / D_{j,k}(\mathbf{z}_k, \eta_j) \right)^{2/(m-1)}} ,\ 1 \le i \le c,\ 1 \le k \le N . \tag{3.63}$$

until $||\mathbf{U}^{(l)} - \mathbf{U}^{(l-1)}|| < \epsilon$.

(the ANFIS model [130]*) and the Fuzzy Model Identification (FMID) Toolbox based on Gustafson–Kessel clustering* [26].

The inputs to the TS fuzzy model are: x_1*: Displacement,* x_2*: Horsepower,* x_3*: Weight,* x_4*: Acceleration and* x_5*: Model year. Originally, there were six features available. The first one, the number of cylinders, is neglected here because the clustering algorithms run into numerical problems on features with only a small number of discrete values.*

Fuzzy models with two, three and four rules were identified with the presented method. With the two rule-model, the presented clustering method achieved RMSE values of 2.72 *and* 2.85 *for the training and test data, respectively, which is nearly the same performance as with the three and four-rule models.*

The FMID Toolbox gives very similar results: RMSE values of 2.67 *and* 2.95 *for the training and test data. Considerably worse results where obtained with the ANFIS algorithm, which gave an overtrained model with the RMSE of* 1.97 *on the training data but* 91.35 *on the test data. These results indicate that the presented clustering method has very good generalization properties.*

For a further comparison we also give the results of a linear regression model given in [130]. *This linear model has seven parameters and six input variables (the previously given five variables and the number of cylinders). The training and test RMSE of this model are* 3.45 *and* 3.44*, respectively.*

Fuzzy models with only two input variables were also identified, where the selected features were taken from [130]*, where the following model structure was proposed:*

$$\text{MPG} = f\left(\text{Weight}, \text{Year}\right) . \tag{3.65}$$

As the Gath–Geva and EM-TI models capture correlations among the input variables, the TS fuzzy model extracted from the clusters should use multivariable antecedent membership functions:

$$R_i \quad : \quad \textbf{If } \mathbf{x} \textit{ is } \mathbf{A}_i \textbf{ then } \hat{y} = \mathbf{a}_i^T\mathbf{x} + b_i$$

or transformed input variables:

$$R_i \quad : \quad \textbf{If } \mathbf{t}_{i,1}^T\mathbf{x} \textit{ is } A_{i,1} \textbf{ and } \mathbf{t}_{i,2}^T\mathbf{x} \textit{ is } A_{i,2} \textbf{ then } \hat{y} = \mathbf{a}_i^T\mathbf{x} + b_i$$

where $i = 1, \ldots, c$*,* $\hat{y}$ *is the estimated MPG and* $\mathbf{x}^T = [\text{Weight}, \text{Year}]$*. These models cannot be easily analyzed, interpreted and validated by human experts, because the fuzzy sets (linguistic terms) are defined in a multidimensional or linearly transformed space. However, the presented EM-NI method (Modified Gath–Geva clustering) results in the standard rules with the original antecedent variables in the conjunctive form:*

$$R_i : \textbf{If } \text{Weight} \textit{ is } A_{i,1} \textbf{ and } \text{Year} \textit{ is } A_{i,2} \textbf{ then } \hat{y} = a_{i,1}\text{Weight} + a_{i,2}\text{Year} + b_i \quad [w_i] . \tag{3.66}$$

Table 3.1 *compares the prediction performance of the obtained models.*

Among the four presented approaches, only the total-least-squares identification is sensitive to the normalization of the data. Hence, in Table 3.1 *GG-TLS-N denotes the results obtained by making the identification with the use of normalized data.*

Table 3.1: Comparison of the performance of the identified TS models with two input variables.

Method	2 rules (train)	2 rules (test)	4 rules (train)	4 rules (test)
GG-TLS	3.34	3.43	5.58	5.71
GG-TLS-N	3.25	3.57	3.43	4.00
GG-LS	2.97	2.97	2.77	2.95
EM-TI	2.97	2.95	2.62	2.93
EM-NI	2.97	2.95	2.90	2.95
ANFIS	2.67	2.95	2.37	3.05
FMID	2.96	2.98	2.84	2.94

Normally, the model performance on the training data improves with the increasing number of clusters, while the performance on the evaluation data improves until the effect of over-fitting appears and then it starts degrading (bias-variance tradeoff). However, when the total-least squares (TLS) method is applied, the training error became larger with the increase of the model complexity. This is because the input variables of the model are strongly correlated and the smallest eigenvector does not define a hyperplane related to the regression problem, but it reflects the dependency of the input variables. Already for two clusters, the difference between the two small eigenvalues is very small (the eigenvalues are $[10.92, 2.08, 3.4 \cdot 10^5]$ *for the first cluster and* $[1.37 \cdot 10^5, 37.69, 4.93]$ *for the second one).*

The presented fuzzy clustering method showed a slightly better performance than the Gath–Geva algorithm. As these methods identify fuzzy models with transformed input variables, they have good performance because of the effective axes-oblique partition of the input domain, which can be seen in Figure 3.10.

The EM-NI algorithm presented in Section 3.3.3 *yields clusters that are not rotated in the input space (see Figure* 3.11*). These clusters can be projected and decomposed to easily interpretable membership functions defined on the individual features as shown in Figure* 3.12 *for the two-rule model and in Figure* 3.13 *for the four-rule model. This constraint, however, reduces the flexibility of the model, which can result in worse prediction performance. We use EMR-TI to demonstrate how much performance one has to sacrifice for the interpretability. For this example, the difference in performances turns out to be negligible (see Table* 3.1*).*

The Fuzzy Toolbox of MATLAB® (ANFIS, neuro-fuzzy model) [132] *and the Fuzzy Model Identification (FMID) Toolbox* [26] *were also used to identify fuzzy models for the MPG prediction problem. As can be seen from Table* 3.1*, the presented method obtains fuzzy models that have good performance compared to these alternative techniques.*

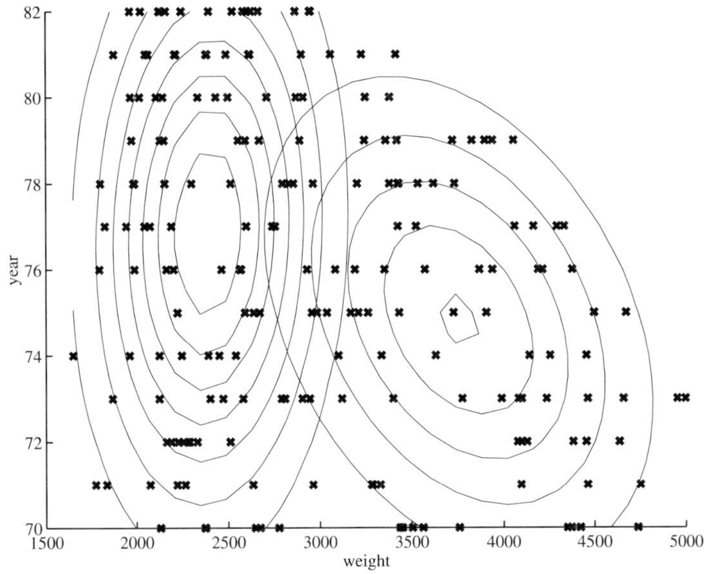

Figure 3.10: Clusters detected by GG clustering algorithm.

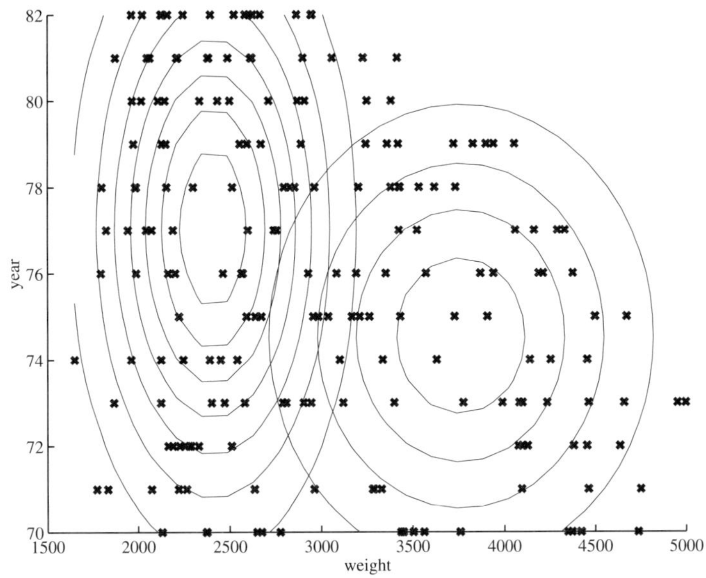

Figure 3.11: Clusters detected by the modified algorithm.

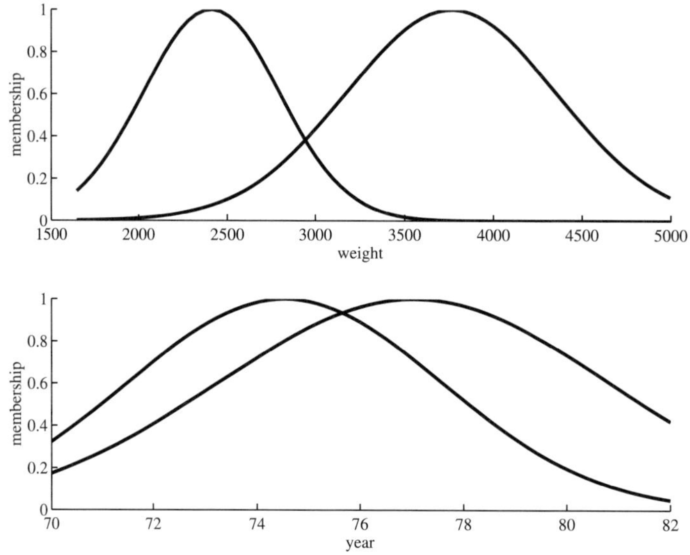

Figure 3.12: Membership functions obtained.

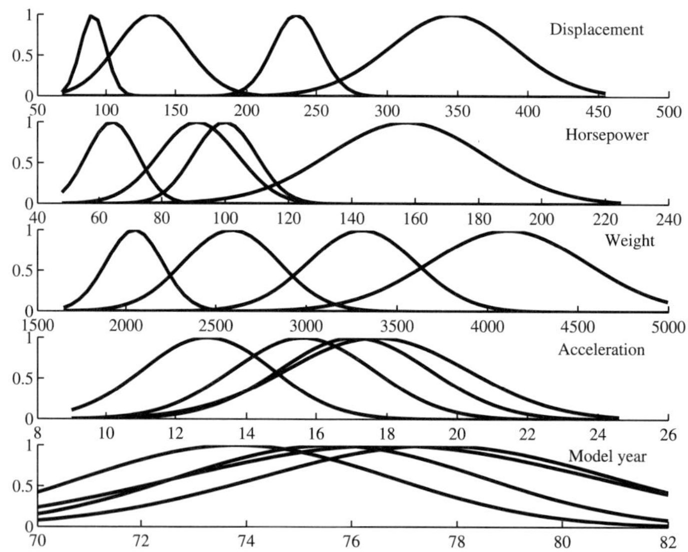

Figure 3.13: Membership functions of the TS model for MPG prediction based on five inputs.

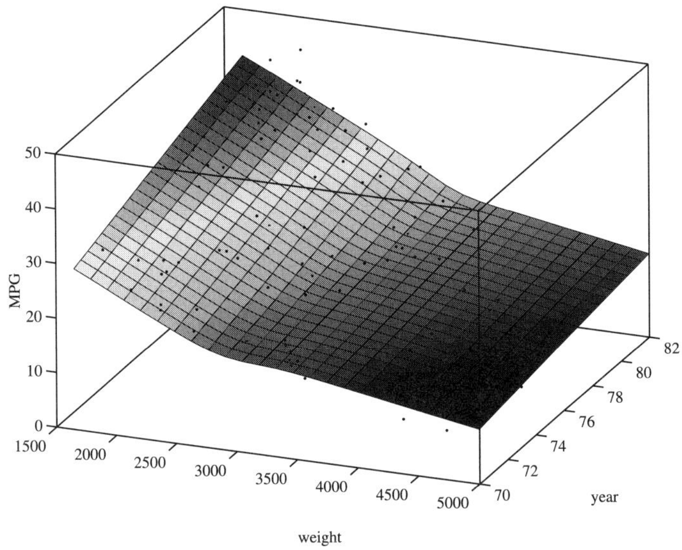

Figure 3.14: Prediction surface of the model.

The resulted model is also good at extrapolation. The prediction surface of the model with two inputs is shown in Figure 3.14. *If this surface is compared to the prediction surface of the ANFIS generated model (see* [130]*), one can see that the ANFIS model spuriously estimates higher MPG for heavy cars because of lack of data due to the tendency of manufacturers to begin building small compact cars during the mid 70s. As can be seen in Figure* 3.14*, the obtained EM-NI model does not suffer from this problem.*

□

3.3.4 Selection of the Antecedent and Consequent Variables

Using too many antecedent and consequent variables results in difficulties in the prediction and interpretability capabilities of the fuzzy model due to redundancy, non-informative features and noise. To avoid these problems in this section two methods are presented.

Selection of the Consequent Variables by Orthogonal Least Squares Method

As the fuzzy model is linear in the parameters θ_i, (3.14) is solved by least squares method (see (3.42)) that can be also formulated as:

$$\theta_i = \mathbf{B}^{+}\mathbf{y}\sqrt{\boldsymbol{\beta}_i} \tag{3.67}$$

where $\mathbf{B}^{+}$ denotes the Moore-Penrose pseudo inverse of $\mathbf{\Phi}_e\sqrt{\boldsymbol{\beta}_i}$.

The OLS method transforms the columns of $\mathbf{B}$ into a set of orthogonal basis vectors in order to inspect the individual contribution of each rule. To do this Gram-Schmidt orthogonalization of $\mathbf{B} = \mathbf{WA}$ is used, where $\mathbf{W}$ is an orthogonal matrix $\mathbf{W}^T\mathbf{W} = \mathbf{I}$ and $\mathbf{A}$ is an upper triangular matrix with unity diagonal elements. If $\mathbf{w}_i$ denotes the ith column of $\mathbf{W}$ and g_i is the corresponding element of the OLS solution vector $\mathbf{g} = \mathbf{A}\theta_i$, the output variance $(\mathbf{y}\sqrt{\boldsymbol{\beta}_i})^T(\mathbf{y}\sqrt{\boldsymbol{\beta}_i})/N$ can be explained by the regressors $\sum_{i=1}^{n_r} g_i \mathbf{w}_i^T \mathbf{w}_i/N$. Thus, the error reduction ratio, ϱ, due to an individual rule i can be expressed as

$$\varrho^i = \frac{g_i^2 \mathbf{w}_i^T \mathbf{w}_i}{(\mathbf{y}\sqrt{\boldsymbol{\beta}_i})^T(\mathbf{y}\sqrt{\boldsymbol{\beta}_i})}. \tag{3.68}$$

This ratio offers a simple mean for ordering the consequent variables, and can be easily used to select a subset of the inputs in a forward-regression manner.

Selection of the Scheduling Variables based on Interclass Separability

Feature selection is usually necessary. For this purpose, we modify the Fisher interclass separability method which is based on statistical properties of the data and has been applied for feature selection of labeled data [231]. The interclass separability criterion is based on the $\mathbf{F}_B$ between-class and the $\mathbf{F}_W$ within-class covariance matrices that sum up to the total covariance of the training data $\mathbf{F}_T$, where:

$$\mathbf{F}_W = \sum_{i=1}^{c} p(\eta_i)\mathbf{F}_i, \qquad \mathbf{F}_B = \sum_{i=1}^{c} p(\eta_i)\left(\mathbf{v}_i - \mathbf{v}_0\right)^T\left(\mathbf{v}_i - \mathbf{v}_0\right)$$
$$\mathbf{v}_0 = \sum_{i=1}^{c} p(\eta_i)\mathbf{v}_i. \tag{3.69}$$

The feature interclass separability selection criterion is a trade-off between $\mathbf{F}_W$ and $\mathbf{F}_B$:

$$J = \frac{\det \mathbf{F}_B}{\det \mathbf{F}_W}. \tag{3.70}$$

The importance of a feature is measured by leaving out the feature and calculating J for the reduced covariance matrices. The feature selection is made iteratively by leaving out the least needed feature.

Example 3.4. Automobile MPG prediction based on Orthogonal Least Squares and Interclass Separability methods

The presented fuzzy modelling approach is applied to a common benchmark problem, the Automobile MPG (miles per gallon) prediction case study [130] (*see Example* 3.3 *for more information*).

A TS-fuzzy model, that utilizes all the five information profile data about the automobiles, has been identified by the presented clustering algorithm. The inputs are: u_1*: Displacement,* u_2*: Horsepower,* u_3*: Weight,* u_4*: Acceleration and* u_5*: Model year. Originally, there are six features available. The first, the number of cylinders, was neglected because the clustering algorithm cannot handle discrete features with only four values* [1, 2, 3, 4].

Fuzzy models with two, three or four rules were identified with the presented method. The two rule-model gave almost the same performance on training and test data as a three and four rule-model and is therefore applied in the sequel. When $\mathbf{x}_k = \boldsymbol{\phi}_k = \boldsymbol{u}_k$*, the presented clustering method gave RMSE values of* 2.72 *and* 2.85 *training and test data.*

The FMID Toolbox gives very similar results; RMSE values of 2.67 *and* 2.95 *for train and test data. Bad results where obtained by the ANFIS algorithm when a neuro-fuzzy model with the same complexity was identified. The application of the ANFIS algorithm resulted in an overtrained model that had a RMSE performance of* 1.97 *on the training data but* 91.35 *on the test data. These results indicate that the presented clustering method has good generalization properties. For further comparison we also give the results of linear regression given by* [130]*. The linear model was based on seven parameters and six input variables (the previously presented five variables and the number of cylinders). The training and test RMS errors of this model were* 3.45 *and* 3.44*, respectively.*

The above presented model reduction techniques, OLS and FIS, are now applied to remove redundancy and simplify the model. First, based on the obtained fuzzy clusters, the OLS method is used for ordering of the input variables. The fuzzy model with two rules is applied. It turned out that for both clusters (local models) the "model year" and the "weight" become the most relevant input variables. Without re-estimation (re-clustering) of the model and additional removal of the other variables from the consequent part of the model, the modelling performance drops only to 3.75 *and* 3.55 *RMSE for the training and the test data, respectively. It is worth noting, that the same features turned out to be the most important in* [130]*, where a cross-validation based method has been used for input selection of neuro-fuzzy models.*

Based on the above considerations, the clustering has been applied again to identify a model based on the two selected attributes "weight" and "model year". This results in RMSE values of 2.97 *and* 2.95 *for training and test data, respectively. In addition, the Fisher Interclass separability method is applied and the second attribute "model year" could be removed from the antecedent of both rules. The final result is a model with RMSE values of* 2.97 *and* 2.98 *for training and test data, respectively. The other methods are now applied with the same attributes and their performances are summarized in Table* 3.2*.*

Because the obtained clusters are axis-parallel, they can be analytically projected and decomposed to easily interpretable membership functions defined on the individual features. The obtained fuzzy sets are shown in Figure 3.15*.*

Table 3.2: Performance of the identified TS models with two rules and only two input variables. HYBRID: Clustering + OLS + FIS, GG: Gath–Geva clustering, ANFIS: neuro-fuzzy model, FMID: Fuzzy Model Identification Toolbox

Method	train RMSE	test RMSE
HYBRID	2.97	2.95
GG	2.97	2.97
ANFIS	2.67	2.95
FMID	2.94	3.01

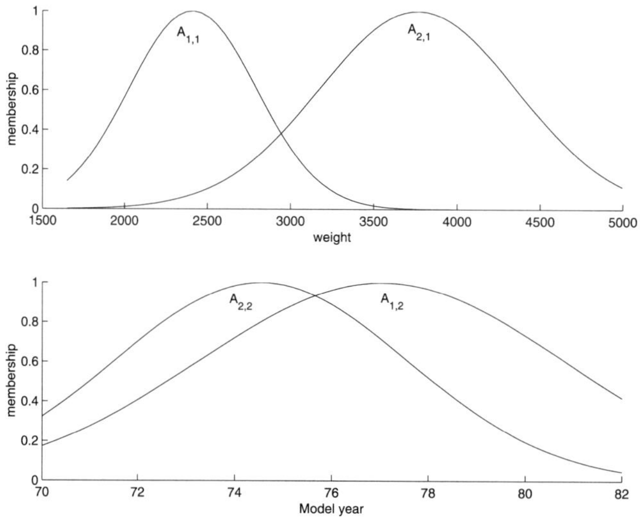

Figure 3.15: Membership functions obtained by the projection of the clusters.

The clusters are much more separated on the input "weight". Hence, the application of Fisher Interclass separability shows that it is enough to use only this input on the antecedent part of the model:

$$\begin{aligned} &R_1 : \textbf{ If } Weight \textit{ is } A_{i,1} \textbf{ then } MPG = \\ &[-0.0107 \;\; 1.0036]^T[Weight, Year]^T - 23.1652, \;[0.57] \\ &R_2 : \textbf{ If } Weight \textit{ is } A_{i,2} \textbf{ then } MPG = \\ &[-0.0038 \;\; 0.4361]^T[Weight, Year]^T - 1.4383, \;[0.43] \end{aligned} \tag{3.71}$$

Concluding, the obtained model has good approximation capabilities, similar to some other advanced methods, but is also very compact. Furthermore, the method seems to have good generalization properties because results on learning data are similar to those on training data. If the prediction surface of the obtained model is compared to the prediction surface of the ANFIS model (see [130]), one can see that the ANFIS approach spuriously estimates higher MPG for heavy cars because of lack of data due to the tendency of manufacturers to begin building small compact cars during the mid 70's, while the presented approach does not suffer from this problem.

□

3.3.5 Conclusions

We discussed the structure identification of Takagi–Sugeno fuzzy models and focused on methods to obtain compact but still accurate models. A new clustering method is presented to identify the local models and the antecedent part of the TS-fuzzy model. In addition two methods for the selection of the antecedent and consequent attributes have been presented. Those methods allow for the derivation of very compact models through subsequent ordering and removal of redundant and non-informative antecedent and consequent attributes. The approach is demonstrated by means of the identification of the MPG benchmark problem. The results show that the presented approach is able to identify compact and accurate models in a straightforward way. This method is attractive in comparison with other iterative schemes, like the one proposed in [230] that involves extensive intermediate optimization.

3.4 Fuzzy Regression Tree

The problems how to automatically partition the input space, how many fuzzy rules are really needed for properly approximating an unknown nonlinear system, and how to construct a fuzzy system from data examples automatically, mentioned above in this chapter, can be partially solved by the recent developments of hierarchical fuzzy systems. As a way to overcome the curse-of-dimensionality, it was suggested in [50] to arrange several low-dimensional rule base in a hierarchical structure, i.e., a tree, causing the number of possible rules to grow in a linear way with the number of inputs. But no method was given on how the rules could be determined automatically. In [228] the author described a new algorithm which derives the rules for hierarchical fuzzy associative memories that were structured as a binary tree. The basic idea is to use the benefits of the compact knowledge representation of decision trees to handle the structure selection problems. This proposal is motivated by the high performance and computational efficiency of the existing decision tree induction methods that are effective in the selection of the relevant features. The application of decision trees for the initialization of fuzzy and neural models has been already investigated by some researchers [16, 126, 167, 236, 238].

Decision trees are widely used in pattern recognition, machine learning and data mining applications thanks to the interpretable representation of the detected information [224]. Decision trees are hierarchical models consisting of nodes and branches. A decision tree is a directed acyclic graph in which the internal nodes represent tests on the attributes (input variables of the model), and the branches correspond to the outcomes of these tests. The leaf (terminal) nodes represent class labels or class distributions in the case of classification trees, or simple regression models (usually constants or linear models) in case of regression trees. The resulting knowledge, in the form of decision trees has been praised for comprehensibility. This appeals to a wide range of users who are interested in domain understanding, classification and/or regression capabilities, or the symbolic rules that may be extracted from the tree and subsequently used in a rule-based decision system.

Decision trees were popularized by Quinlan [224]. Before the last decade, only categorical information has been considered as input variables of decision trees. With the immanent need to handle numerical information too, various methods to create discrete partitions (and according predicates) for numerical attributes have been invented. The problem that arises when this discretization is done by means of partitions into crisp sets (intervals) is that small variations (e.g., noise) on the input side can cause large changes on the output. This is because in crisp decision trees, the test is based on Boolean logic and, as a result, only one branch can be followed after a test. In other words, only one branch will be valid according to the value of the attribute and the result of the test. Hence, one branch will be weighted by one, the other ones by zeros. An example can be seen in Figure 3.16. In the first node it has to be determined whether the value of the variable x_2 is

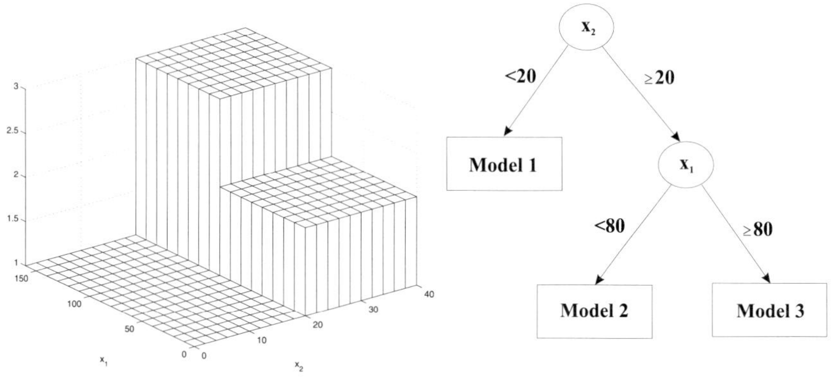

Figure 3.16: Crisp decision tree.

greater than 20 or not. If it is less than 20, Model 1 will be valid. If it is equal to or greater than 20, another test has to be executed, the other attribute, x_1 has to be analyzed. If it is less than 80, Model 2, in the other case Model 3 has to be used to estimate the output of the analyzed system. As this example illustrates, the disadvantage of classical crisp decision trees is their brittleness, i.e., a wrong path taken down the tree can lead to widely erroneous results. Furthermore, the crisp approach falls short also when the system under study does not exhibit crisp transitions. This entails the demand for admitting vagueness in the assignment of samples to predicates.

Using not only crisp but also fuzzy predicates, decision trees can be used to model vague decisions. The basic idea of fuzzy decision trees is to combine example based learning in decision trees with approximative reasoning of fuzzy logic [107, 133]. In fuzzy decision trees several branches originating from the same node can be simultaneously valid to a certain degree, according to the result of the fuzzy test (Figure 3.17). The path from the root node to a particular leaf model hence

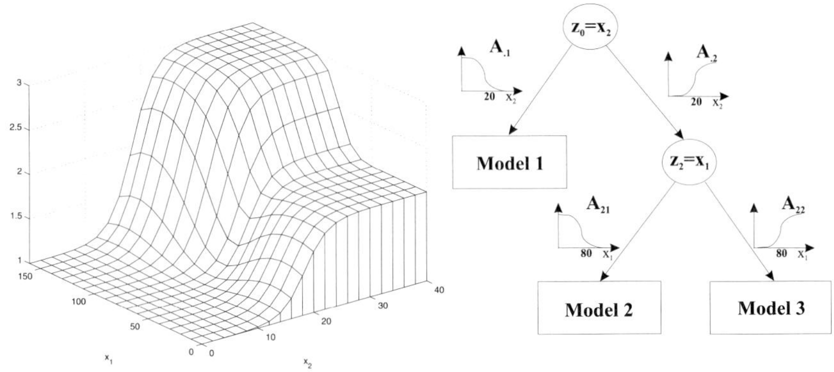

Figure 3.17: Fuzzy decision tree.

defines a fuzzy operating range of that particular model. The output of the tree is obtained by interpolating the outputs of the leaf models that are simultaneously active. This hybridization integrates the advantages of both methodologies compact knowledge representation of decision trees with the ability of fuzzy systems to process uncertain and imprecise information. Viewing fuzzy decision trees as a compressed representation of a (fuzzy) rule set, enables us to use decision trees not only for classification (classification tree), but also for approximation of continuous output functions (regression tree). Hence, classification and regression problems can be effectively solved with decision tree induction algorithms that are able to generate accurate and transparent models. Determination of the final structure and size of the tree is a challenging problem. A tree with many internal and terminal nodes causes difficulty in transparency and interpretability, demands more computing power and can also contain many branches without much benefit for the accuracy. The standard approach is the pruning of the tree. It can mean either an exhaustive search within the building of the tree to find the best variable and splitting value to minimize the classification/modelling error in the next step or it follows the building of the tree to find subtrees with relative small benefit. It is computationally demanding so it is worth eliminating it. This can be done by some heuristic.

In order to build a decision tree from data, one has to determine: **(i)** the structure of the tree and related to this, **(ii)** what tests should be evaluated in the nodes (e.g., which input variable have to be tested), and **(iii)** how the premise and **(iv)** consequent parameters should be identified. The data-driven determination of these structural parameters represents a complex mixed integer-continuous optimization problem.

Such problems can be solved by some computationally demanding stochastic optimization algorithms or by using some heuristic. The following strategies are commonly used for determination of these parameters:

- heuristic construction algorithms [204],
- nonlinear local optimization [53] and
- product space clustering [215].

Because the described type of models are linear in the parameter set of the local models they can be efficiently estimated from data. The three most common approaches are

- global least squares [26],
- local weighted least squares [26] and
- product space clustering [12].

There are good surveys in this topic, see, e.g., [118].

The locally linear model tree (LOLIMOT) algorithm combines a heuristic strategy for input space decomposition with a local linear least squares optimization [204]. In LOLIMOT radial basis functions are fitted to a rectangular partitioning of the input space performed by a decision tree with axis-orthogonal splits at the internal nodes. For each rectangle, the corresponding basis function has a standard deviation in each dimension that is equal to the width of the rectangle in that dimension multiplied by a constant (step **(iii)**). Locally weighted learning is used to train each local linear model (LLM) (step **(iv)**). The difficult part is deciding when to grow or prune the tree. In LOLIMOT, the greedy approach has been applied. It means that the worst performing model, hence with most modelling error has to be found and its operating regime has to be divided into parts with additional local models. The local loss function can be computed by weighting the squared model errors with the degree of validity of the corresponding local model. In the following steps all the identified local models have to be evaluated and compared. This approach is called greedy because it proceeds towards the worst performing model. These steps have to be carried out until reaching a predetermined modelling error or maximum size of the tree. The selection of which partition has to be splitted on which variable (step **(ii)**) is based on the local sum of squared errors loss function and not their mean is utilized for the comparison between the LLMs. Consequently, LOLIMOT preferably splits LLMs which contain more data samples. Thus, the local model quality depends on the training data distribution. This consequence is desired because more data allows to estimate more parameters.

The idea of LOLIMOT has the same background as the algorithm presented by Kubat [167] for the decision tree based identification of Radial Basis Function Networks since the corresponding basis function has a standard deviation in each dimension that is equal to the width of the rectangles of the partition. The incremental building of the tree makes splits, which always halve the rectangular partitioning and which are sensitive to the distribution of the data.

The aim is to present a slightly more sophisticated, but still heuristic algorithm that can make more effective partitioning than the recursive halving of the input variables. The main idea is to use fuzzy clustering for the effective partitioning of the input domains of decision trees. In [215] it has been shown that the results of the clustering coming in the form of a series of prototypes can be directly used to complete a quantization of the continuous attributes, and this quantization can be effectively used for the induction of a decision tree. In contrast with most discretization of continuous variables that deal with a single variable only, this approach concerns all the variables discussed at the same time. The discretization mechanism is straightforward: project the cluster prototypes on the respective variables (coordinate axes) and construct the discretization intervals. However, the projection of the clusters into the input variables and the approximation of the point-wise defined projected membership functions by parametric ones may significantly deteriorate the performance of the model.

Our approach differs from the previously presented methods in the following main issues:

Model-based clustering: In [12] we proposed a new supervised clustering algorithm that can be directly used for the identification of interpretable fuzzy models based on the Expectation Maximization (EM) identification of Gaussian mixture of models. The resulted clusters do not only represent the distribution of the data, but also the type of nonlinearity of the modeled system.

Incremental clustering: The clustering based quantization of the input variables of decision trees is independent from the decision tree induction algorithms, since the clustering is considered as a pre-processing step of the induction of the decision tree [215]. Contrary, in our approach the clustering is embedded into the incremental building of the tree according to the Algorithm 3.4.1.

In this section binary fuzzy regression trees are used. It means that only two branches belong to each internal node, and in every clustering step only two clusters were assumed. However, the presented algorithm is able to build trees with arbitrary number of branches per nodes. We used a binary tree since its compactness because the main goal of the research was to design an algorithm that is able to generate compact yet accurate models.

Algorithm 3.4.1 (Incremental Clustering for Fuzzy Regression Trees).

- **Step 1** *Cluster the data according to the scheduling variables into two clusters.*
- **Step 2** *Select the splitting variable based on the analysis of the separability of the clusters. This step enables to select a "scheduling" variable that is most representative to handle the nonlinearity of the system.*
- **Step 3** *Re-cluster the data according to the selected splitting variable exclusively. The result: two local linear models whose operating regime are represented by Gaussian membership functions. The parameters of these membership functions are given by the clustering algorithm.*
- **Step 4** *Greedy step Select the worst model based on the mean squared error of the local models. Remove this model, weight the data according to the weights (memberships) of this local model and jump to* **Step 1**.

3.4.1 Preliminaries

The aim of this section is to present the structure of the presented fuzzy regression tree and the notation followed in the section.

In rule-based fuzzy systems, the relationships between the input and output variables are represented by means of If-Then rules. In case of a fuzzy regression tree the rule antecedents are represented by the membership functions on the paths of the tree leading to the terminal nodes of the tree. These rule membership functions define the operating region of the rule in the n-dimensional feature space of the input variables. The consequences of the fuzzy rules are the terminal nodes of the tree. In case of regression trees these terminal nodes are local linear models describing the local linear behavior of the modeled nonlinear system in the operating region defined by the ascendents of the rules.

The terminal nodes (in Figure 3.18 denoted by rectangles) are numbered sequentially from left to right. Each terminal node corresponds to one fuzzy rule. The consequent of the ith rule is the function f_i, which may be either a constant (singleton or zero-order TS fuzzy model) or a (linear) function of the input variables (first- or higher-order TS fuzzy model). The antecedent of the rule is defined as the conjunction of the fuzzy assertions along the path leading from the root to that terminal node. The antecedent variables and fuzzy sets are labeled by index vectors p_j, denoting the path from the root to the particular node at level j.

The vectors are constructed recursively as follows:

$$p_0 = \emptyset \quad \text{empty vector, denotes the root node}$$
$$p_j = [p_{j-1}\ b_j]$$

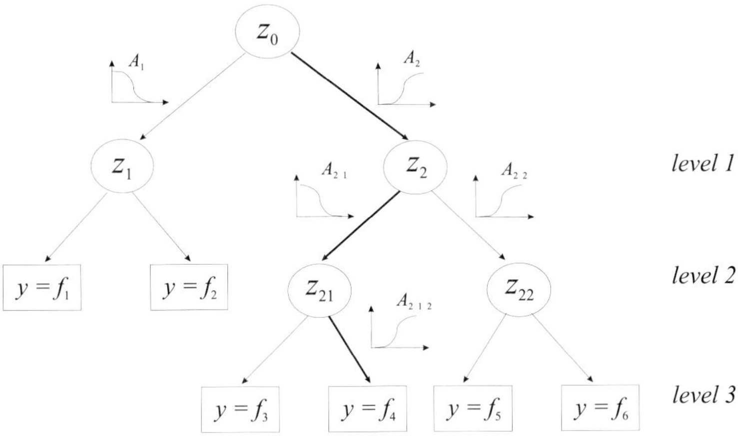

Figure 3.18: A fuzzy decision tree (to keep the figure simple, only some of the fuzzy sets are shown).

where b_j is the number of the respective branch from node p_{j-1} (numbered from left to right). Subscript p^i_j denotes the (sub)path from the root to a node at level j, which belongs to the rule (terminal node) i. The rules are then written as follows:

$$R_i : \textbf{ If } z_{p^i_0} \text{ is } \mathrm{A}_{p^i_1} \textbf{ and } z_{p^i_1} \text{ is } \mathrm{A}_{p^i_2} \textbf{ and } \dots \textbf{ and } z_{p^i_{L_i-1}} \text{ is } \mathrm{A}_{p^i_{L_i}} \textbf{ then } \mathbf{y} = \mathbf{f}_i$$

or in a more compact form:

$$R_i : \textbf{ If } \bigwedge_{j=1}^{L_i} \left(z_{p^i_{j-1}} \text{ is } \mathrm{A}_{p^i_j} \right) \textbf{ then } \mathbf{y} = \mathbf{f}_i$$

Here, L_i is the level of the ith terminal node. The antecedent variables z are instances of the model inputs x, selected by the tree construction algorithm.

The degree of fulfilment (weight of the rule) is computed by using the product conjunction operator:

$$w_i = \prod_{j=1}^{L_i} \mu_{\mathrm{A}_{p^i_j}} (z_{p^i_{j-1}})$$

and the interpolated output of the tree model is the weighted mean of the terminal node contributions:

$$\mathbf{y} = \sum_{i=1}^{K} w_i \mathbf{f}_i = \sum_{i=1}^{K} w_i \left[\mathbf{x}^T_k \; 1 \right] \boldsymbol{\theta}^T_i . \tag{3.72}$$

Example 3.5 (Fuzzy rule extraction from Fuzzy Decision Tree). *As an example, consider rule* R_4 *in Figure* 3.18 (*see the bold path*):

$$R_4 : \textbf{ If } z_0 \textit{ is } A_2 \textbf{ and } z_2 \textit{ is } A_{21} \textbf{ and } z_{21} \textit{ is } A_{212} \textbf{ then } \mathbf{y} = \mathbf{f}_4$$

whose degree of fulfillment is

$$w_4 = \mu_{A_2}(z_0) \cdot \mu_{A_{21}}(z_2) \cdot \mu_{A_{212}}(z_{21}).$$

□

The shape of the fuzzy set has to be decided before the building of the tree. In this section Gaussian fuzzy set has been applied. Consider the p_{j-1}^ith terminal node, its child's membership function that belongs to the ith (sub)path, $\mathrm{A}_{p_j^i}$ is

$$\mathrm{A}_{p_j^i} = \frac{\exp\left(-(z_{p_{j-1}^i} - v_{p_j^i})^2/2\sigma_{p_j^i}^2\right)}{\sum_i \exp\left(-(z_{p_{j-1}^i} - v_{p_j^i})^2/2\sigma_{p_j^i}^2\right)}. \tag{3.73}$$

Note that the model output is *linear* in the model parameters, $\boldsymbol{\theta}_i$, but is *non-linear* in the centers, $v_{p_{L_i}^i}$ and standard deviations, $\sigma_{p_{L_i}^i}$.

In the next section the building of such type of trees will be described in detail.

3.4.2 Identification of Fuzzy Regression Trees based on Clustering Algorithm

The aim of the identification of a fuzzy regression tree is the data-driven approximation of an unknown general nonlinear function $\mathbf{y}_k = f(\mathbf{x}_k)$ based on $k = 1, \ldots, N$ data sample, where $\mathbf{y}_k = [y_{1,k}, y_{2,k}, \ldots, y_{q,k}]^T$ is the output and $\mathbf{x}_k = [x_{1,k}, x_{2,k}, \ldots, x_{n,k}]^T$ is the input of the system.

Expectation-Maximization based Fuzzy Clustering for Regression Trees

The basic concept of this method is that the original data distribution, $p(\mathbf{x}_k, \mathbf{y}_k)$, $k = 1, \ldots, N$, should be described by hierarchical mixture of distributions according to the operating regimes of the local models. The operating regimes can be determined in several ways, this section proposes a method based on decision tree partitioning because of its advantages, e.g., compactness, interpretability and transparency.

The hierarchical structure of the model can be described in two steps. On the one hand, simple mixture of the presented Gaussian density models should be described (the first level of the regression tree) because its not evident how it can be done based on the whole dataset, on the other hand, it should be represented how the following levels can be formed known its parent's level parameters.

Simple Mixture of Gaussians

The corresponding model takes the form

$$p(\mathbf{x}_k, \mathbf{y}_k) = \sum_{i=1}^{K} p(\mathbf{x}_k, \mathbf{y}_k, \eta_{p_1^i}) = \sum_{i=1}^{K} p(\eta_{p_1^i}) p(\mathbf{x}_k, \mathbf{y}_k | \eta_{p_1^i}) \tag{3.74}$$

where the indexes of the terminal nodes, i, are related to the path leading to them, p_1^i means that the node belongs to the first level of the model, and K is the number of local models which is equal to the number of clusters, c in this level. The $p(\eta_{p_1^i})$ probabilities are the mixing coefficients, or prior probabilities, corresponding to the mixture components $p(\mathbf{x}_k, \mathbf{y}_k | \eta_{p_1^i})$ which can be divided into two parts: $p(\mathbf{x}_k, \mathbf{y}_k | \eta_{p_1^i}) = p(\mathbf{y}_k | \mathbf{x}_k, \eta_{p_1^i}) p(\mathbf{x}_k | \eta_{p_1^i})$, input and output distribution. Each component is an independent local linear model whose parameters should be determined. It can be done by the Expectation-Maximization algorithm similar to Hierarchical Latent Variable Models proposed by Tipping and Bishop in [47].

The prior expectations are given by the $p(\eta_{p_1^i})$ and the corresponding posterior probabilities, or responsibilities, are evaluated in the **E-step** using Bayes theorem in the form

$$p(\eta_{p_1^i} | \mathbf{x}_k, \mathbf{y}_k) = \frac{p(\eta_{p_1^i}) p(\mathbf{x}_k, \mathbf{y}_k | \eta_{p_1^i})}{\sum_{i=1}^{K} p(\eta_{p_1^i}) p(\mathbf{x}_k, \mathbf{y}_k | \eta_{p_1^i})}. \tag{3.75}$$

In the **M-step**, these posterior probabilities are used to obtain the 'new' values of the model parameters, using the following re-estimation formulas.

The probability of the p_1^ith mixture component or local model is

$$p(\eta_{p_1^i}) = \frac{1}{N} \sum_{k=1}^{N} p(\eta_{p_1^i} | \mathbf{x}_k, \mathbf{y}_k). \tag{3.76}$$

The input distribution, parameterized as an unconditional Gaussian, can be formed in a way similar to the multivariate membership functions

$$p(\mathbf{x}_k | \eta_{p_1^i}) = \frac{1}{(2\pi)^{\frac{n}{2}} \sqrt{\det(\mathbf{F}_{p_1^i})}} \exp\left(-\frac{1}{2} (\mathbf{x}_k - \mathbf{v}_{p_1^i})^T \left(\mathbf{F}_{p_1^i} \right)^{-1} (\mathbf{x}_k - \mathbf{v}_{p_1^i}) \right), \tag{3.77}$$

where the mean, $\mathbf{v}_{p_1^i}$, and the weighted covariance matrix, $\mathbf{F}_{p_1^i}$, are computed by

$$\mathbf{v}_{p_1^i} = \frac{\sum_{k=1}^{N} p(\eta_{p_1^i} | \mathbf{x}_k, \mathbf{y}_k) \mathbf{x}_k}{\sum_{k=1}^{N} p(\eta_{p_1^i} | \mathbf{x}_k, \mathbf{y}_k)}, \tag{3.78}$$

$$\mathbf{F}_{p_1^i} = \frac{\sum_{k=1}^{N} p(\eta_{p_1^i} | \mathbf{x}_k, \mathbf{y}_k)(\mathbf{x}_k - \mathbf{v}_{p_1^i})(\mathbf{x}_k - \mathbf{v}_{p_1^i})^T}{\sum_{k=1}^{N} p(\eta_{p_1^i} | \mathbf{x}_k, \mathbf{y}_k)}. \tag{3.79}$$

When the transparency and interpretability of the model is important, the cluster covariance matrix, $\mathbf{F}_{p_1^i}$ can be reduced to its diagonal elements, similar to the simplified axis-parallel version of the GG clustering algorithm

$$p(\mathbf{x}_k|\eta_{p_1^i}) = \sum_{l=1}^{n} \frac{1}{\sqrt{2\pi\sigma^2_{p_1^i,l}}} \exp\left(-\frac{1}{2}\frac{(x_{l,k} - v^2_{p_1^i,l})}{\sigma^2_{p_1^i,l}}\right). \tag{3.80}$$

The output distribution is

$$p(\mathbf{y}_k|\mathbf{x}_k, \eta_{p_1^i}) = \frac{1}{(2\pi)^{\frac{q}{2}}\sqrt{\det(\mathbf{P}_{p_1^i})}} \exp\left(-\frac{1}{2}(\mathbf{y}_k - \mathbf{f}_i)^T \left(\mathbf{P}_{p_1^i}\right)^{-1} (\mathbf{y}_k - \mathbf{f}_i)\right). \tag{3.81}$$

In order to obtain that, $\mathbf{f}_i$ has to be computed. It can be done by (3.72) if parameters of the local models are given. They can be computed by (3.85) where the $\mathcal{B}_{p_1^i}$ matrix contains the weighting values.

$$\mathbf{Y} = [\mathbf{y}_1, \mathbf{y}_2, \ldots, \mathbf{y}_N]^T \tag{3.82}$$

$$\mathbf{X} = [\mathbf{x}_1, \mathbf{x}_2, \ldots, \mathbf{x}_N]^T \tag{3.83}$$

$$\mathcal{B}_{p_1^i} = \begin{bmatrix} p(\eta_{p_1^i}|\mathbf{x}_1, \mathbf{y}_1) & 0 & \cdots & 0 \\ 0 & p(\eta_{p_1^i}|\mathbf{x}_2, \mathbf{y}_2) & \cdots & 0 \\ \vdots & \vdots & \ddots & \vdots \\ 0 & 0 & \cdots & p(\eta_{p_1^i}|\mathbf{x}_N, \mathbf{y}_N) \end{bmatrix} \tag{3.84}$$

$$\boldsymbol{\theta}_{p_1^i} = \left[\mathbf{X}^T \mathcal{B}_{p_1^i} \mathbf{X}\right]^{-1} \mathbf{X}^T \mathcal{B}_{p_1^i} \mathbf{Y}. \tag{3.85}$$

It corresponds to the local weighted least squares method (see also Section 1.6).

When $\boldsymbol{\theta}_{p_1^i}$ parameters are given, covariance of the modelling errors of the local models can be computed:

$$\mathbf{P}_{p_1^i} = \frac{\sum\limits_{k=1}^{N} (\mathbf{y}_k - \mathbf{f}_i)(\mathbf{y}_k - \mathbf{f}_i)^T p(\eta_{p_1^i}|\mathbf{x}_k, \mathbf{y}_k)}{\sum\limits_{k=1}^{N} p(\eta_{p_1^i}|\mathbf{x}_k, \mathbf{y}_k)}. \tag{3.86}$$

Hierarchical Mixture of Gaussians

According to the greedy approach, the worst performing model has to be found and its operating regime has to be divided further into c parts. The corresponding probability density can be written in the form

$$p(\mathbf{x}_k, \mathbf{y}_k, \eta_{p_{j-1}^i}) = p(\eta_{p_{j-1}^i}) p(\mathbf{x}_k, \mathbf{y}_k|\eta_{p_{j-1}^i}) = p(\eta_{p_{j-1}^i}) \sum_{i=1}^{c} p(\eta_{p_j^i|p_{j-1}^i}) p(\mathbf{x}_k, \mathbf{y}_k|\eta_{p_j^i}), \tag{3.87}$$

where $p(\eta_{p^i_j|p^i_{j-1}})$ denotes the conditional probability $p(\eta_{p^i_j}|\eta_{p^i_{j-1}})$. In this expression, $p(\mathbf{x}_k, \mathbf{y}_k|\eta_{p^i_j})$ again represents independent local linear models, and $p(\eta_{p^i_j|p^i_{j-1}})$ corresponds to sets of mixing coefficients, one for each p^i_{j-1}, which satisfy $\sum_{i'=i}^{i+c-1} p(\eta_{p^{i'}_j|p^i_{j-1}}) = 1$. Thus, each level of the hierarchy corresponds to a generative model, with lower levels giving more refined and detailed representations. According to it, (3.74) can be expressed in the form

$$p(\mathbf{x}_k, \mathbf{y}_k) = \sum_{i=1}^{K} p(\eta_{p^i_{j-1}}) \sum_{i'=i}^{i+c-1} p(\eta_{p^{i'}_j|p^i_{j-1}}) p(\mathbf{x}_k, \mathbf{y}_k|\eta_{p^i_j}). \tag{3.88}$$

The parameters of the local models can be determined by Expectation-Maximization method. This has the same form as the EM algorithm for a simple mixture (see above in this section), except that in the E-step, the posterior probability that model (p^i_j) generated data point $(\mathbf{x}_k, \mathbf{y}_k)$ is given by

$$p(\eta_{p^i_j}|\mathbf{x}_k, \mathbf{y}_k) = p(\eta_{p^i_{j-1}}|\mathbf{x}_k, \mathbf{y}_k) p(\eta_{p^i_j|p^i_{j-1}}, \mathbf{x}_k, \mathbf{y}_k) \tag{3.89}$$

in which

$$p(\eta_{p^i_j|p^i_{j-1}}|\mathbf{x}_k, \mathbf{y}_k) = \frac{p(\eta_{p^i_j|p^i_{j-1}}) p(\mathbf{x}_k, \mathbf{y}_k|\eta_{p^i_j})}{\sum_{i'=1}^{c} p(\eta_{p^{i'}_j|p^i_{j-1}}) p(\mathbf{x}_k, \mathbf{y}_k|\eta_{p^{i'}_j})}, \tag{3.90}$$

and

$$p(\eta_{p^i_j|p^i_{j-1}}) = \frac{\sum_{k=1}^{N} p(\eta_{p^i_j}|\mathbf{x}_k, \mathbf{y}_k)}{\sum_{k=1}^{N} p(\eta_{p^i_{j-1}}|\mathbf{x}_k, \mathbf{y}_k)}. \tag{3.91}$$

Every other equation from (3.77) to (3.86) is almost the same, just $\eta_{p^i_{j-1}}$ is replaced by $\eta_{p^i_j}$.

The applied greedy method means local identification of the model by least squares method. However, the above description of the model enables us to identify (or after the structure determination refine) the global model to obtain better results. Although it is computationally demanding because of nonlinear optimization, it can be worth executing it if the local models do not give acceptable results.

Note that the model has a hierarchical structure, therefore the probability of the (p^i_j)th local linear model, $p(\eta_{p^i_j})$ can be expressed in the following way:

$$p(\eta_{p^i_j}) = p(\eta_i \eta_{p^i_{j-1}}) = p(\eta_i|\eta_{p^i_{j-1}}) p(\eta_{p^i_{j-1}}). \tag{3.92}$$

Consider the p^i_{j-1}th node, its parameters are given, so is $p(\eta_{p^i_{j-1}})$, then its children's parameters, $p(\eta_i|\eta_{p^i_{j-1}})$ among them, can be computed by the presented algorithm. Hence, $p(\eta_{p^i_j})$ is given in a recursive form ($p(1)$ is equal to 1).

The probability above, $p(\mathbf{x}_k, \mathbf{y}_k, \eta_{p^i_j})$, means the probability that the kth data point, $(\mathbf{x}_k, \mathbf{y}_k)$, has been generated by the ith cluster in case of the (p^i_j)th node, and this probability and the distance between the point and the cluster are in inverse proportion

$$p(\mathbf{x}_k, \mathbf{y}_k, \eta_{p^i_j}) = \frac{1}{D^2_{p^i_j,k}}. \tag{3.93}$$

The identification of the model means the determination of the cluster parameters. It can be done an algorithm reformulated in the form of fuzzy clustering.

Modified Gath–Geva Clustering Algorithm for Identification of Fuzzy Regression Trees

Obtaining the parameters of the clusters: centers and standard deviations, and the parameters of the local models, the following cost function has to be minimized:

$$J(\mathbf{Y}, \mathbf{U}, \eta) = \sum_{i=1}^{K} \sum_{k=1}^{N} \left(\mu_{p^i_j,k}\right)^m D^2_{p^i_j,k}. \tag{3.94}$$

The used distance function enables us to identify the local linear models because it contains a term based on the model error:

$$\begin{aligned}\frac{1}{D^2_{p^i_j,k}} &= p(\eta_{p^i_j})p(\mathbf{x}_k|\eta_{p^i_j})p(\mathbf{y}_k|\mathbf{x}_k,\eta_{p^i_j})\\ &= p(\eta_{p^i_j}) \underbrace{\frac{1}{(2\pi)^{\frac{n}{2}}\sqrt{\det(\mathbf{F}_{p^i_j})}} \exp\left(-\frac{1}{2}(\mathbf{x}_k - \mathbf{v}_{p^i_j})^T \left(\mathbf{F}_{p^i_j}\right)^{-1} (\mathbf{x}_k - \mathbf{v}_{p^i_j})\right)}_{p(\mathbf{x}_k|\eta_{p^i_j})}\\ &\quad \underbrace{\frac{1}{(2\pi)^{\frac{q}{2}}\sqrt{\det(\mathbf{P}_{p^i_j})}} \exp\left(-\frac{1}{2}(\mathbf{y}_k - \mathbf{y}_k^{(p^i_j)})^T \left(\mathbf{P}_{p^i_j}\right)^{-1} (\mathbf{y}_k - \mathbf{y}_k^{(p^i_j)})\right)}_{p(\mathbf{y}_k|\mathbf{x}_k,\eta_{p^i_j})}. \end{aligned} \tag{3.95}$$

In case of this type of regression trees, the clusters are axis parallel:

$$p(\mathbf{x}_k|\eta_{p^i_j}) = \sum_{l=1}^{n} \frac{1}{\sqrt{2\pi\sigma^2_{p^i_j,l}}} \exp\left(-\frac{1}{2}\frac{(x_{l,k} - v^2_{p^i_j,l})}{\sigma^2_{p^i_j,l}}\right). \tag{3.96}$$

The clustering algorithm described in the following can be applied to minimize the cost function based on the Gath–Geva fuzzy clustering algorithm [93]. Let us assume that the p^i_{j-1}th node is the following.

Algorithm 3.4.2 (Gath–Geva Clustering for Fuzzy Regression Tree Induction).

Initialization

Given a set of data $\mathbf{Y} = [\mathbf{y}_k]_{q\times N}, \mathbf{X} = [\mathbf{x}_k]_{n\times N}$, *the number of clusters, c, choose a weighting exponent (usually* $m = 2$*) and a termination tolerance* $\epsilon > 0$. *Initialize the partition matrix (randomly),* $\mathbf{U} = [\mu_{p_j^{i'},k}]_{c\times N}$.

Repeat *for* $iter = 1, 2, \ldots$

Calculate the parameters of the clusters

- *Centers of the membership functions*

$$\mathbf{v}_{p_j^{i'}} = \sum_{k=1}^{N} \mu_{p_j^i,k}\, \mathbf{x}_k / \sum_{k=1}^{N} \mu_{p_j^i,k}, \quad i' = i, \ldots, i+c-1. \tag{3.97}$$

- *Standard deviations of the Gaussian membership functions:*

$$\sigma^2_{p_j^{i'},l} = \sum_{k=1}^{N} \mu_{p_j^{i'},k} \left(x_{l,k} - v_{p_j^{i'},l} \right)^2 / \sum_{k=1}^{N} \mu_{p_j^{i'},k}. \tag{3.98}$$

- *Parameters of the local models can be computed by* (3.85) *where the elements of the* $\mathcal{B}_{p_j^{i'}}$ *matrix are equal to the membership values:* $p(\eta_{p_j^{i'}}|\mathbf{x}_k, \mathbf{y}_k) = \mu_{p_j^{i'},k}$.
- *Covariance of the modelling errors of the local models:*

$$\mathbf{P}_{p_j^{i'}} = \frac{\sum_{k=1}^{N} (\mathbf{y}_k - \mathbf{f}_{i'})(\mathbf{y}_k - \mathbf{f}_{i'})^T \mu_{p_j^{i'},k}}{\sum_{k=1}^{N} \mu_{p_j^{i'},k}}. \tag{3.99}$$

- A priori *probability of the cluster can be computed in the following way:* $p(\eta_{p_j^{i'}}) \equiv p(\eta_{p_j^{i'}}, \eta_{p_{j-1}^i}) = p(\eta_{p_j^{i'}}|\eta_{p_{j-1}^i})p(\eta_{p_{j-1}^i})$, *and it can be expressed by* (3.76) *and* (3.91)*:* $p(\eta_{p_j^{i'}}) = \frac{1}{N}\sum_{k=1}^{N} p(\eta_{p_j^{i'}}|\mathbf{x}_k, \mathbf{y}_k)$, *so*

$$p(\eta_{p_j^{i'}}) = \frac{1}{N} \sum_{k=1}^{N} \mu_{p_j^{i'},k}. \tag{3.100}$$

Compute the distance $D^2_{p_j^{i'},k}$ **by** (3.95)

Update the partition matrix by (3.89) **and** (3.90)

$$\mu_{p_j^{i'},k} = \frac{\prod_{j'=1}^{j-1} \mu_{A_{p_{j'}^i}}(z_{p_{j'-1}^i})}{\sum_{d=i}^{i+c-1} \left(D_{p_j^{i'},k} / D_{p_j^d,k} \right)^{2/(m-1)}}, \tag{3.101}$$
$$i' = i, \ldots, i+c-1; 1 \le k \le N.$$

until $||\mathbf{U}^{(iter)} - \mathbf{U}^{(iter-1)}|| < \epsilon$.

Selection of the Splitting Variable

There are input variables, $\mathbf{x}_k, k = 1, \ldots, N$. According to the applied modelling technique, the regression tree, in case of the p^i_{j-1}th node one input variable has to be selected based on which it can be detected which operating regime the input data point is located in and which local model has to be used. We cluster the data based on all the input variables and we select the splitting variable from the input ones according to which the two clusters are the most separate. This variable is denoted by z^i_{j-1}. The splitting variable can be selected based on the cluster centers and standard deviations.

The result of the clustering algorithm in case of the p^i_{j-1}th node is the set $\mathbf{v}_{p^{i'}_j}, \sigma^2_{p^{i'}_j, l}$. These parameters are in connection with the modelling error because one term of the cost function is based on the modelling error (see (3.95)).

Choose the splitting variable based on the following heuristic criterium:

$$S(p^i_{j-1}, l) = \frac{1}{1 + (v_{p^i_j, l} - v_{p^{i+1}_j, l})^2 + (\sigma_{p^i_j, l} - \sigma_{p^{i+1}_j, l})^2}. \tag{3.102}$$

The value of $S(p^i_{j-1}, l)$ is in the range of 0 and 1. The more separate the clusters are according to the lth variable the smaller the value of $S(p^i_{j-1}, l)$ is. The p^i_{j-1}th splitting variable is chosen to be the variable that possesses the smallest $S(p^i_{j-1}, \cdot)$ value.

Fine Tuning

In the previous step the p^i_{j-1}th splitting variable, $z_{p^i_{j-1}}$ can be chosen. In this step the parameters of the $(p^{i'}_j)$th local linear models and their operating regime, $A_{p^{i'}_j}, i' = i, \ldots, i + c - 1$, have to be identified. (The parameters identified in the previous step are good choices for initial values in this step. This step can be regarded as fine tuning of the parameters.) It can be done by using the modified Gath–Geva algorithm presented in the previous section. The only difference is that the clustering procedure is applied only to the splitting variable and not all the input variables. The results of this step are $\boldsymbol{\theta}_{p^{i'}_j}$ parameter sets, $v_{p^{i'}_j}$ and $\sigma^2_{p^{i'}_j}$ for the operating regimes (see (3.73)).

Example 3.6 (Comparative study to demonstrate the effectiveness of GG clustering based Fuzzy Regression Tree induction). *The accuracy and transparency of the presented algorithm are shown based on five datasets, three real life and two synthetic ones. All datasets have been used before, most of them are originated from well-known data repositories. The result of the presented Fuzzy Regression Tree (FRT) algorithm has been compared with other decision tree or fuzzy model identification methods, like CART algorithm* [53]*; Fuzzy Model Identification (FMID)*

Algorithm 3.4.3 (The Whole Greedy Algorithm for FRT Induction).

1. *Modelling steps*

   Step 1 *Cluster the data according to the scheduling variables into two clusters.*

   Step 2 *Choose the splitting variable.*

   Step 3 *Re-cluster the data according to the splitting variable exclusively. The result: two (local linear) models and weights for the operating regime.*

2. *Greedy steps Choose the worst model based on the following criterium: the mean squared error of the* (i, p^i_{j-1})*th local model is*

$$mse(i, p^i_{j-1}) = \underbrace{\frac{\sum_{k=1}^{N} \beta_{i,p^i_{j-1}} \left(\mathbf{y}_k - [\mathbf{x}_k^T \; 1] \boldsymbol{\theta}^T_{i,p^i_{j-1}} \right)^2}{N}}_{\text{accuracy}} \underbrace{\frac{\sum_{k=1}^{N} \beta_{i,p^i_{j-1}}}{N}}_{\text{ratio in the dataset}} . \tag{3.103}$$

   Find the maximum of the mse values, remove this model, and jump to Step 1.

Toolbox based on Gustafson–Kessel clustering [26]; *GUIDE (linear regression tree)* [180]; *and Scalable Linear Regression Tree Algorithm (SECRET) and its modified version with oblique splits (SECRET(O))* [71]. *The last three methods can use constant regressors (special, zero-order linear regressor) or linear ones. This section compares the result of the presented method, which is based on linear regressors, with the results of the last three methods listed above based on linear regressors as well (the methods based on constant regressors give worse results). It is important also in comparability point of view.* 10 *fold cross validation method was used to avoid uncertain error based on sampling. The presented technique does not include pruning method but the other ones need pruning to avoid overfitting. In these cases the training set was used to fit and to prune the tree.*

Real life datasets:

- Abalone *Dataset from UCI machine learning repository used to predict the age of abalone from physical measurements. Contains* 4177 *cases with* 8 *attributes (*1 *nominal and* 7 *continuous).*
- MPG *See Example* 3.3.
- Kin8nm *Data containing information on the forward kinematics of an* 8 *link robot arm from the DVELVE repository. Contains* 8192 *cases with* 8 *continuous attributes.*

Synthetic datasets:

- Fried *Artificial dataset used by Friedman* [88] *containing* 10 *continuous attributes with independent values uniformly distributed in the interval* [0, 1]. *The value of the output variable is obtained with the equation:*

$$y = 10\sin(\pi x_1 x_2) + 20(x_3 - 0.5)^2 + 10x_4 + 5x_5 + \sigma(0,1). \tag{3.104}$$

- 3DSin *Artificial dataset containing* 2 *continuous predictor attributes uniformly distributed in interval* [−3, 3], *with the output defined as*

$$y = 3\sin(x_1)\sin(x_2). \tag{3.105}$$

3000 *data points were generated using these equations.*

As some clustering methods are sensitive to differences in the numerical ranges of the different features, the data can be normalized to zero mean and unit variance

$$\tilde{z}_{j,k} = \frac{z_{j,k} - \bar{z}_j}{\sigma_j} \tag{3.106}$$

where $\bar{z}_j$ *and* σ_j *are the mean and the variance of the given variable, respectively. The performance of the models is measured by the root mean squared prediction error* (*RMSE*)

$$RMSE = \sqrt{\frac{1}{N}\sum_{k=1}^{N}(y_k - \hat{y}_k)^2}. \tag{3.107}$$

Note that the algorithm deals with normalized data but the RMSE values have been computed by using denormalized data. Note that two initialization techniques were tried by the presented FRT method: random and using of the result of the more robust Gustafson–Kessel (*GK*) *fuzzy clustering method* [108]. *The experience is that the RMSE values by GK initialization were always worse than by random initialization. Its cause is that the GK algorithm takes account only of the distribution of the data, not the modelling error, and it forces the FRT algorithm to the wrong way. Another important remark is that the minimum number of datapoints in a node is to be considered for splitting to* 5% *of the size of the dataset. It can be important for the clustering because there should be enough data to compute the covariance matrices accurately.*

The results are shown in Table 3.3. *It can be seen that FRT gives always better results than CART or FMID methods. It gives by far the best results by MPG and Fried datasets. In case of the other datasets the FRT method gives slightly worse results than the current best algorithm. The aim of the presented method is not only to give accurate but also interpretable models. According to that, the number*

Table 3.3: Comparison of the prediction performance of different algorithms. (Where two values are given, the first is the training, the second is the test error. Numbers in brackets are the number of models.)

Dataset	FRT	CART	FMID		GUIDE	SECRET	SECRET(O)
Abalone	2.18 / 2.19 (4)	1.19 / 2.87 (664.8)	2.20 (4)	2.19 (12)	2.15	2.16	2.18
MPG	2.85 / 2.91 (2)	1.69 / 3.04 (51.7)	3.07 (4)	3.03 (12)	5.91	3.99	4.09
Kin8nm	0.15 / 0.15 (20)	0.09 / 0.23 (453.9)	0.20 (4)	0.20 (12)	0.15	0.15	0.13
3DSin	0.18 / 0.18 (12)	0.09 / 0.17 (323.1)	0.50 (4)	0.31 (12)	0.21	0.20	0.15
Fried	0.69 / 0.70 (15)	0.84 / 2.12 (495.6)	2.41 (4)	2.41 (12)	1.10	1.12	1.23

of leaves was forced into the range 2 and 22. The proper number of the leaves is chosen from the value by which the test error begins to increase or, if it decreases monotone in the analyzed range, reaches an acceptable small value.

GUIDE, SECRET and SECRET(O) generated trees with around 75 *nodes* [71], *trees generated by CART algorithm can be even much larger (see the numbers in brackets in Table* 3.3*), but the presented FRT algorithm has good generalization capability and can give accurate results within the limits of perspicuity. E.g., in case of the MPG problem a lot less leaves* (2) *are enough to give excellent result. It is also a considerable result compared with methods proposed in* [12], *EM-TI, EM-NI, ANFIS* [130] *or FMID* [26], *and contains less membership functions because the antecedent part of the fuzzy rules does not test all of the attributes, only those that can be found in the nodes.*

One generated FRT tree can be seen in Figure 3.19 *with* 12 *leaves (not depicted in the figure, only nodes that test an attribute). The label of axis x is the number of the tested attribute, the label of axis y means the order of generation of the nodes. In this problem, there are two predictor variables (see* (3.105)*). It can be seen how the algorithm divides the input space, e.g., the node* 4, 10 *and* 11 *test the same attribute, x_2. Node* 10 *divides the 'right side', node* 11 *the 'left side' of the input space. It is possible to develop the algorithm in this point of view: if it is the case, the algorithm will get back to the parent node (here node* 4*) and will increase the number of clusters (in this case up to* 4*). In this way the depth of the tree can be reduced, and more compact trees can be generated.*

Another generated FRT tree can be seen in Figure 3.20 *with* 15 *leaves. The presented method generated a well-arranged tree, despite, e.g., the CART algorithm (approximately* 495 *leaves after pruning). The presented method is able to clearly determine the attributes on which the output variable depends* (3.104)*, the tree does not contain nodes that tests any of the $x_6, \ldots, x_{10}$ attributes (they do not have an effect on the dependent variable). Their parameters in the linear models are two-three orders of magnitude smaller than the ones of $x_1, \ldots, x_5$.*

□

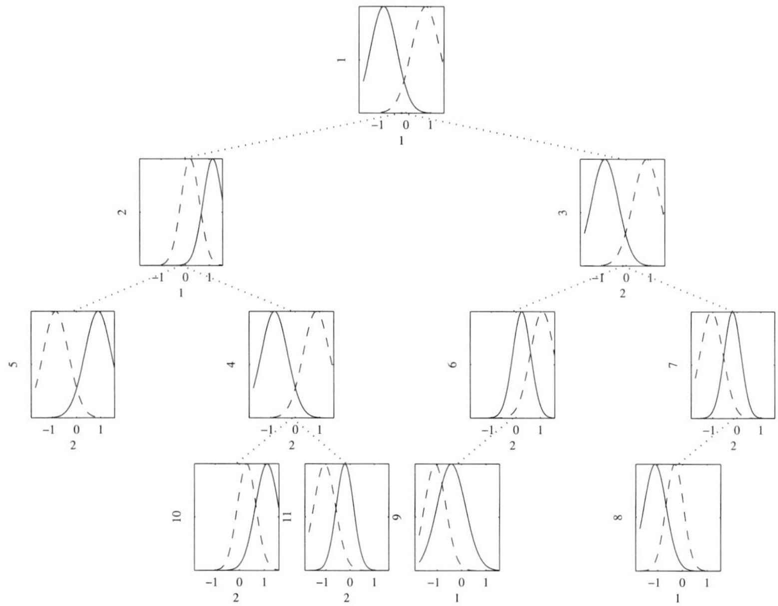

Figure 3.19: Result of FRT algorithm in 3DSin prediction.

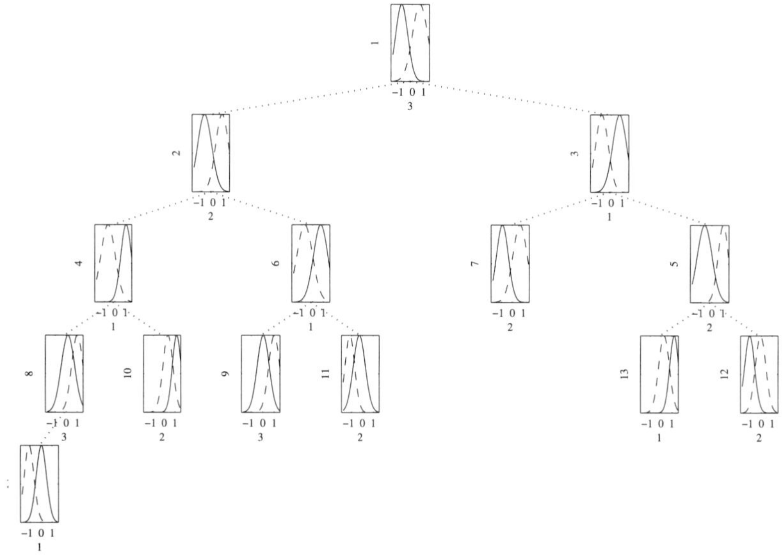

Figure 3.20: Result of FRT algorithm with 15 leaves in case of dataset by Friedman.

3.4.3 Conclusions

A novel representation of hierarchical Takagi–Sugeno fuzzy models has been presented in this section. The hierarchical structure is represented by a binary fuzzy tree. The model tree is identified by a supervised clustering algorithm embedded to a greedy incremental model building algorithm. The analysis of the clusters can be used for the selection of the relevant scheduling variables that are needed to handle the nonlinearity of the system. The presented fuzzy regression tree induction method is evaluated in some well-known synthetic (Fried, 3DSin) and real life (Abalone, MPG, Kin8nm) benchmark applications. Simulation results show that the identified hierarchical TS-FS models are effective for the identification of linear/nonlinear systems. When compared to other fuzzy model identification (FMID, ANFIS, EM) and regression tree building (CART, GUIDE, SECRET, SECRET(0)) algorithms, the hierarchical TS fuzzy model exhibits competing results with a high accuracy and smaller size of architecture. The preliminary results are encouraging.

3.5 Clustering for Structure Selection

3.5.1 Introduction

Real-world data analysis, data mining and modelling problems typically involve a large number of potential inputs. The set of inputs actually used by a model should be reduced to the necessary minimum, especially when the model is nonlinear and contains many parameters. Therefore, effective methods for input selection (also called structure selection) are very important for any modelling exercise [109]. For dynamic systems, the input-selection problem also includes the choice of the model's order (number of lagged inputs and outputs used as regressors) and the number of pure time delays (see Section 4.3 for more details).

A large array of structure-selection methods, like the analysis of correlations or principal component analysis, have been introduced for linear models. However, they usually fail to discover the significant inputs in real-world data, which are almost always characterized by nonlinear dependencies. Relatively few methods are available for structure selection in nonlinear models. These methods can generally be divided into two main groups:

- Model-free methods, which do not need to develop models in order to find significant inputs. They use the available data only and are based on statistical tests or properties of functions. An example is the method of Lipschitz coefficients [114] which exploits the continuity property of nonlinear functions.
- Model-based methods, using a particular model structure in order to find the significant inputs. The Akaike's information criterion (AIC, [179]) is often used for linear models. Models with different sets of input variables are

compared and the model that minimizes the AIC is selected. Input selection for nonlinear fuzzy systems is often based on various heuristic criteria [64, 130, 178, 232]. A common drawback of these methods is the need to develop fuzzy models in order to compare them and select the relevant inputs. Such an approach is very time consuming and the result may be biased by the choice of the fuzzy model type and structure.

The only model-based methods that provide the (theoretic) possibility to handle mixed continuous and discrete (categorical) inputs in an effective way are classification and regression trees [53]. However, due to their greedy nature (variables are selected and split sequentially, based on a local measure of performance or information gain) they do not always perform as expected. (Examples provided below in this section illustrate this point.)

A model-free method is presented here that should overcome some of the deficiencies of regression trees [97]. No specific model structure is assumed and the relevant inputs are found directly from the input-output data by using exhaustive search in the space of all possible model input combinations. For each input combination, the total variance of the output variable is computed over all subsets of discrete value combinations. For continuous variables, the data samples are first partitioned into fuzzy subsets by means of clustering and the variance is replaced by 'fuzzy variance'. Finally, models with the lowest variance are selected as candidates for further (model-based) analysis. In this way, the potential sets of inputs are ordered according to their relative power in explaining the output. The final interpretation of the result and the definitive selection of the model structure can be done by the user.

3.5.2 Input Selection for Discrete Data

Let us start with introducing the approach for discrete-valued data. Given is a set of multivariate input data stored in an $N \times n$ matrix $\mathbf{X}$ where N is the number of data samples and n the number of variables. The corresponding output data are contained in an $N \times 1$ vector $\mathbf{y}$.

We decompose this data set into disjoint subsets, such that the variance of the predicted variable within the individual subsets is minimal. The underlying assumption is that such a structure will allow us to construct a prediction model that minimizes the prediction error.

For the purpose of this decomposition, suppose that each independent variable $x_i, i = 1, \ldots, n$ takes on values (attributes) from a finite set $A_i = \{a_i^1, \ldots, a_i^{c_i}\}$. For the complete set of variables, define the index set $W = \{1, 2, \ldots, n\}$. Consider a partitioning W into two disjoint index subsets U and V, such that $U \bigcup V = W$ and $U \bigcap V = \emptyset$. The complete data set (X, y) can now be decomposed into subsets, such that in each of these subsets, the variables x_i are constant: $x_i = a_i^j, i \in U, j \in \{1, 2, \ldots, c_i\}$. We call such a subset a *data segment* and denote it by X_{a_U} (a_U is the set of constant values taken on by x_i, for all i):

$$X_{a_U} = \{(x_{kV}, y_k)|a_U, k \in \{1, 2, \ldots, N\}\}. \tag{3.108}$$

We can now search for such a partitioning of W into U and V, such that the variance of y_k within the individual segments is minimal. Each segment is then represented by the mean value of the corresponding y_k values. Note that a part of the variance in y will typically be caused by the dependence of y on x_V, which are not taken into account in the decomposition. Therefore, it is meaningful to look for decompositions with minimal variance, in which no important inputs are missing.

The complete set of possible values a_U is the Cartesian product $A = \times_{i \in U} A_i$. The number of segments, n_s, in the given partition

$$n_s = |A| = \prod_{j \in U} c_j \tag{3.109}$$

is thus growing exponentially with the cardinality of U. To keep the computational load limited, only partitions up to a certain pre-defined cardinality of U can be considered. As $|U| + |V| = |W| = const$, more variables in U will generally result in smaller data segments. Note that some of the segments X_{a_U} may be empty, simply because the corresponding values of x_U are not present in the data set.

Illustrative example: Consider a data set with three independent variables: f, g and h. The complete set of variables, $x_W = \{f, g, h\}$, can thus be partitioned in the following six pairs of subsets:

$$\begin{array}{lll lll}
x_{U_1} & = & \{\text{ f }\} & x_{V_1} & = & \{\text{ g, h }\} \\
x_{U_3} & = & \{\text{ h }\} & x_{V_3} & = & \{\text{ f, g }\} \\
x_{U_5} & = & \{\text{ g, h }\} & x_{V_5} & = & \{\text{ f }\} \\
x_{U_2} & = & \{\text{ g }\} & x_{V_2} & = & \{\text{ f, h }\} \\
x_{U_4} & = & \{\text{ f, g }\} & x_{V_4} & = & \{\text{ h }\} \\
x_{U_6} & = & \{\text{ f, h }\} & x_{V_6} & = & \{\text{ g }\}
\end{array}$$

Suppose that variables f, g and h can take on values from the following sets:

$$\begin{array}{lll}
A_f & = & \{0.5, 0.8, 1.1\}, \\
A_g & = & \{0, 1, 2, 3, 4\}, \\
A_h & = & \{1.5, 2.0, 2.5\}.
\end{array}$$

We can now segment the data for each of the above partitions. Partition (U_2, V_2), for instance, will result into five data segments corresponding to $g = 0, 1, 2, 3, 4$. Each of these segments will contain some independent data x_{kV_2} from the set $A_f \times A_h$ and the corresponding outputs y_k. Partition (U_4, V_4), will result into $15 (= 3 \times 5)$ data segments, each containing some independent data x_{kV_4} from the set A_h and the corresponding outputs y_k. Note that these segments will be smaller than the segments obtained with (U_2, V_2).

The obvious drawback of this approach is that it cannot be directly applied to continuous-valued data. Instead, the data first have to be quantized in some way. To avoid this step, which may introduce unwanted bias by an unsuitable choice of the quantization levels and boundaries, we propose to fuzzify the decomposition process and use fuzzy clustering to partition the continuous data into subsets.

3.5.3 Fuzzy Clustering Approach to Input Selection

Instead of quantizing the input data into hard subsets, fuzzy c-means (FCM) clustering [42] is used to partition all the candidate regressors into fuzzy subsets. To determine the proper number of clusters c_j for variable j, the Xie-Beni index [285] is employed. The cluster centers are randomly initialized. As a result, for each input x_j, the cluster centers $\mathbf{v}_j \in \mathbb{R}^{1\times c_j}$ and the partition matrix $\mathbf{U}^j = [0,1]^{c_j \times N}$ are obtained.

The fuzzy counterparts of the discrete segments (1) are fuzzy subsets obtained by the Cartesian product-space conjunction of the cluster membership degrees:

$$\beta_{i,k} = \prod_{j=1}^{n_i} \mu^j_{I_{ji}k},\ i = 1,2,\ldots,n_s,\ k = 1,2,\ldots,N, \tag{3.110}$$

where n_i is the number of inputs in the considered structure, $I = \times_{j\in U}(1,\ldots,c_j)$ is the set of all possible ordered n-tuples of cluster indices, whose total number is:

$$n_s = |I| = \prod_{j=1}^{n_i} c_j. \tag{3.111}$$

We normalize $\beta_{i,k}$ such that the sum over all the segments equals to one:

$$\gamma_{i,k} = \frac{\beta_{i,k}}{\sum_{i'=1}^{n_s} \beta_{i,k}}. \tag{3.112}$$

To assess the relative power of the inputs to explain the output, the weighted mean of each segment is computed

$$b_i = \sum_{k=1}^{n_s} \gamma_{i,k} y_k \tag{3.113}$$

and the total fuzzy variance σ_t is obtained by

$$\sigma_t = \sum_{k=1}^{N} \left(u_k - \sum_{i=1}^{n_s} \gamma_{i,k} b_i \right). \tag{3.114}$$

Often, σ_t will have comparable values for several model structures, some of them containing more inputs than others. In order to distinguish among such structures, the variance can be modified to account for the number of inputs. There are many ways to realize this modification. In our experiments, the following *variance-complexity trade-off* (VCT) measure proved useful and it is therefore used as a ranking criterion:

$$VCT = \sigma_t n_i. \tag{3.115}$$

The interpretation of this measure is discussed in the next section by the examples.

Remarks:

1. Each data segment can be interpreted as a fuzzy if-then rule and the set of all segments as a fuzzy rule base of the same form as the singleton fuzzy model [26]. The total fuzzy variance is equivalent to the sum of squared prediction errors over the data. Searching for the optimal partition corresponds to selecting the optimal rule base among all possible ones.
2. To reduce the computational complexity, the successive computation of the product in (3.110) can be truncated whenever the current value of $\beta_{i,k}$ becomes zero or lower than a certain threshold.
3. The final selection of the model structure should be done by the user, who will typically try to find a tradeoff between the performance and complexity of the model. Usually, it is feasible to select a small set of 'good' model structures and compare their performance in modelling. We illustrate this point in the last example of the next section.

3.5.4 Examples

In this section we present two examples to demonstrate the effectiveness of the presented approach and we compare it with the regression-tree method [53] implemented in the Statistics Toolbox of MATLAB®.

Example 3.7 (Input selection by a nonlinear regression problem based on Fuzzy c-Means clustering). *We first consider the well-known example taken from* [88] *where it was included to illustrate the performance of the multivariate adaptive regression splines technique. Later, it was used in other publications as well* [145]. *Consider the following nonlinear function of five variables* x_1 *through* x_5*:*

$$y = 10\sin(\pi x_1 x_2) + 20(x_3 - 0.5)^2 + 10x_4 + 5x_5. \tag{3.116}$$

A set of 1000 *samples are randomly generated for* x_1 *to* x_5 *according to the uniform distribution* $U[0,1]$ *and the corresponding output* y *is computed. In addition, five dummy inputs* x_6 *to* x_{10} *are added to the set of candidate inputs (these data are also generated randomly according to* $U[0,1]$*).*

Applying the regression-tree method to this data typically results in the correct variables x_1 *through* x_5 *being selected as model inputs, but in addition one or two dummy variables are selected as well. This is clearly undesirable, as it can lead to over-fitting and it can also have a negative influence on the decision-making or control based on such a model.*

With the presented method, the 10 *candidate inputs are first clustered one by one. The number of clusters can be determined by using the Xie-Beni index or set to a predefined value. The later option is preferred in this particular case, as it is easy to see that the data are uniformly distributed (check a histogram) and hence no natural cluster structure is present.*

Based on the obtained partition matrices, the total fuzzy variance is computed for each possible subset of inputs. The numbers obtained are sorted in ascending order and the corresponding model structures are sorted accordingly. Figure 3.21 *shows the variance-complexity tradeoff criterion for the first* 70 *structures, numbered sequentially in the order of increasing* σ_t. *The total number of structures that were compared is* 847.

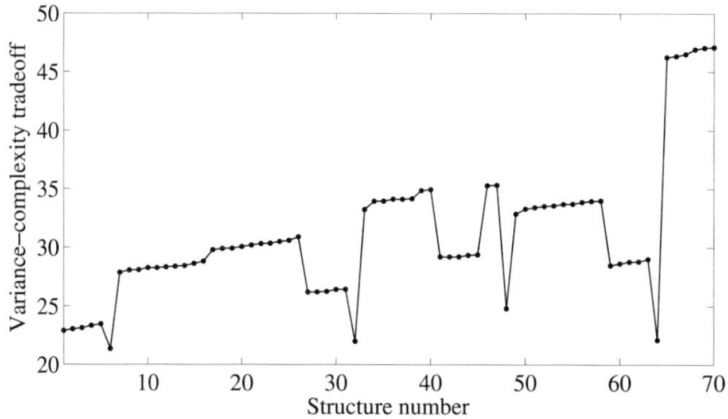

Figure 3.21: The variance-complexity tradeoff criterion for function (3.116)

Table 3.4 *shows the numerical results for several selected structures, including also the total variance* σ_t. *The first local (and global) minimum of* VCT *(structure number* 6*) refers to the model with the correct inputs* x_1 *through* x_5. *Another local minimum is found for structure number 32 in which variable* x_3 *is dropped, which can be expected since* x_3 *has the least relative contribution in* (3.116). *A similar observation can be made for structure* 64.

It is important to realize that the absolute values of the local minima on the VCT *curve (Figure* 3.21*) should not be compared one to another. For small values of* σ_t, *the number of inputs* n_i *will dominate the formula* (3.115), *which to a certain degree also happens in this example. Note, that the* VCT *index for structure* 64 *is only marginally larger than for structure* 6, *while the corresponding* σ_t *values differ considerably. Therefore, we use* VCT *only to compare structures with similar* σ_t *values (such as structures* 1 *through* 6 *in Table* 3.4*).*

□

structure	selected inputs	σ_t	VCT
1	x_1, x_2, x_3, x_4, x_5, x_8	3.82	22.90
2	x_1, x_2, x_3, x_4, x_5, x_{10}	3.84	23.05
5	x_1, x_2, x_3, x_4, x_5, x_9	3.91	23.46
6	x_1, x_2, x_3, x_4, x_5	4.27	21.34
7	x_1, x_2, x_4, x_5, x_6, x_8	4.64	27.86
32	x_1, x_2, x_4, x_5	5.50	21.98
64	x_1, x_2, x_4	7.36	22.09

Table 3.4: Summary of the results for function (3.116)

Example 3.8 (Input selection by continuous and discrete variables based on FCM). *This example demonstrates how the presented algorithm can handle data sets containing both continuous and discrete inputs. Consider the following function*

$$y = \begin{cases} x_1^2 + \epsilon & \text{for } x_2 = 1, \\ x_4^2 + x_3 + \epsilon & \text{for } x_2 = 0, \end{cases} \tag{3.117}$$

in which the output switches between two nonlinear functions depending on the value of the random discrete regressor $x_2 \in \{0, 1\}$. The remaining variables are randomly generated according to the uniform distribution $U[0, 1]$. Two additional random dummy inputs x_5 and x_6 are included in the set of candidate inputs. The noise term ϵ is a normally distributed random variable: $\epsilon \sim N(0, 0.1)$.

Similarly to the previous example, the regression-tree method selects the correct variables x_1 through x_4, but in addition one of the dummy variables is selected in about 20% of runs (recall that the input data are random).

With the presented algorithm, the number of clusters in the continuous variables is set to 3, while the discrete variable clearly has two distinct values. Figure 3.22 shows the variance-complexity tradeoff criterion for the first 30 structures, numbered sequentially in the order of increasing σ_t. The total number of structures compared was 62.

Table 3.5 shows the numerical results for several selected structures. The first local (and global) minimum of VCT (structure number 3) refers to the model with the correct inputs x_1 through x_4.

□

3.5.5 Conclusions

A model-free method for structure selection has been presented in data containing both continuous and categorical variables. The relevant inputs are found by an exhaustive search in the space of all possible input combination after partitioning the continuous variables into fuzzy subsets. Models are ranked according to the

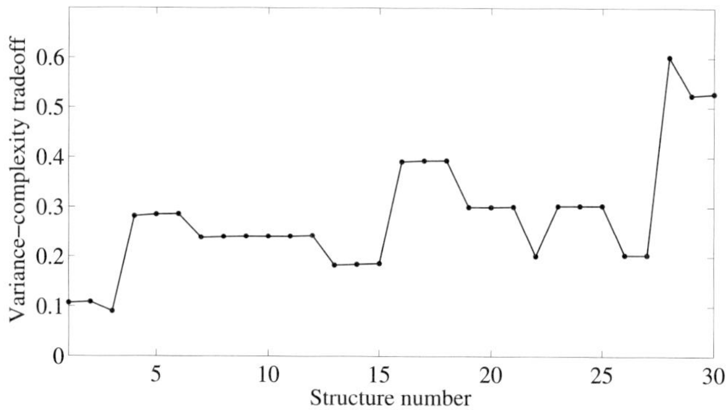

Figure 3.22: The variance-complexity tradeoff criterion for function (3.117)

structure #	selected inputs	σ_t	VCT
1	x_1, x_2, x_3, x_4, x_6	0.021	0.107
2	x_1, x_2, x_3, x_4, x_5	0.022	0.109
3	x_1, x_2, x_3, x_4	0.023	0.090
4	x_2, x_3, x_4, x_5, x_6	0.056	0.282
5	x_1, x_2, x_4, x_5, x_6	0.057	0.285

Table 3.5: Summary of the results for function (3.117)

resulting (fuzzy) variance of the output variable. A ranked list of candidate model structures is obtained and the final selection of the structure is left to the user. Two examples were presented to show that the method is able to find the correct model structures for simulated benchmarks. The presented method performs better than regression trees which in many cases overfit the data by selecting a larger set of inputs (despite pruning with the help of cross-validation). A clear limitation of our method is that it can only handle models with a limited number of potential inputs. Due to the exhaustive search, it will take too long to evaluate model structures with more than 15 candidate inputs. The method is fully automatic (no user interaction is needed) if the Xie-Beni index is used to determine the number of clusters for each variable. In some cases it might be preferable to set the number of clusters manually; this can be done after analyzing the data by means of histograms, for instance. The algorithm can be modified in several ways, for instance by using product-space clustering rather than clustering the individual input variables. While clusters in the product space of the inputs can better account for dependencies among the variables, computationally, this approach would be much more involved, as for each candidate structure, the clustering would have to be run again.

Chapter 4

Fuzzy Clustering for System Identification

In this chapter we deal with fuzzy model identification, especially by dynamical systems. In practice, there is a need for model-based engineering tools, and they require the availability of suitable dynamical models. Consequently, the development of a suitable nonlinear model is of paramount importance. Fuzzy systems have been effectively used to identify complex nonlinear dynamical systems. In this chapter we would like to show how effectively clustering algorithms can be used to identify a compact Takagi–Sugeno fuzzy model to represent single-input single-output *and also* multiple-input multiple-output dynamical systems.

The algorithm mentioned above uses only data from the modeled system, but if there is a priori information about the analyzed system and there are less defined parts as well, it is worth using both information sources, i.e., a white box structure based on mechanistic relationships and black-box substructures to model less defined parts based on measured data. These hybrid model structures are called semi-mechanistic models. It is shown that certain types of white-box models can be efficiently incorporated into a Takagi–Sugeno fuzzy rule structure. Next, the presented models are identified from learning data and special attention is paid to transparency and accuracy aspects. For the identification of the semi-mechanistic fuzzy model, a new fuzzy clustering method is presented. Subsequently, model reduction is applied to make the TS models as compact as possible.

Until this point, the order of the input-output model is assumed to be known. However, it is often not known beforehand and has to be determined somehow. Selecting the order of an input-output model of a dynamical system can be the key step toward the goal of system identification. The false nearest neighbors algorithm (FNN) is a useful tool for the estimation of the order of linear and nonlinear systems. Advanced FNN uses nonlinear input-output data based models. To increase the efficiency of this method, we present a clustering-based algorithm.

Clustering is applied to the product space of the input and output variables. The model structure is then estimated on the basis of the cluster covariance matrix eigenvalues. The main advantage of the proposed solution is that it is model-free. This means that no particular model needs to be constructed in order to select the order of the model, while most other techniques are 'wrapped' around a particular model construction method. This saves the computational effort and avoids a possible bias due to the particular construction method used.

The approach mentioned above can be used not only for input-output systems but autonomous systems as well. It is a challenging problem by chaotic systems and it can be the key step toward the analysis and prediction of nonlinear and chaotic time-series generated by these systems. The embedding dimension can be selected by the presented clustering based algorithm. The intrinsic dimension of the reconstructed space can then be estimated based on the analysis of the eigenvalues of the fuzzy cluster covariance matrices (the 'flatness' of the clusters), while the correct embedding dimension is inferred from the prediction performance of the local models of the clusters. The main advantage of the proposed solution is that three tasks are simultaneously solved during clustering: selection of the embedding dimension, estimation of the intrinsic dimension, and identification of a model that can be used for prediction.

4.1 Data-Driven Modelling of Dynamical Systems

In this section, before the introduction of fuzzy models of dynamical systems, the widely used dynamical model structures will be reviewed.

When the information necessary to build a fundamental model of dynamical processes is lacking or renders a model that is too complex for an on-line use, empirical modelling is a viable alternative. Empirical modelling and identification is a process of transforming available input-output data into a functional relation that can be used to predict future trends. While fundamental physical process models are mostly developed in continuous time, computer-based process control systems operate in discrete time: measurements are made and control actions are taken at discrete time instants. In addition, the available input-output data used for model identification are only available in discrete time instants. Hence, the identification of a discrete input-output dynamical model is based on the observed inputs $\{u(k)\}_k$ and outputs $\{y(k)\}_k$ [247],

$$\{u(k)\}_k = [u(1), u(2), \ldots, u(k)] \;, \tag{4.1}$$

$$\{y(k)\}_k = [y(1), y(2), \ldots, y(k)] \;. \tag{4.2}$$

Our aim is to find a relationship between past observations and future output

$$\hat{y}(k) = g(\{y(k-1)\}_k, \{u(k-1)\}_k, \theta) + e(k) \tag{4.3}$$

where θ is a finite-dimensional parameter vector and the additive term $e(k)$ accounts for the fact that the next output $y(k)$ will not be an exact function of the past data. However, the goal is that $e(k)$ remains small. Instead of using the whole previous input-output sequence, $\{u(k-1)\}_k$ and $\{y(k-1)\}_k$, a finite-dimensional regression vector has to be used. This can be done if (4.3) is written as a concentration of two mappings: one that takes the increasing number of past inputs and outputs and maps them into a finite-dimensional vector $\boldsymbol{\phi}(k)$ of a fixed dimension,

$$\boldsymbol{\phi}(k) = \varphi(\{y(k-1)\}_k, \{u(k-1)\}_k, \eta) \tag{4.4}$$

where $\boldsymbol{\phi}(k)$ input vector of the model is a subset of the previous input and output variables, generated by the $\varphi(.)$ model: where η denotes the parameter vector of the φ function that generates the regressors of the $f(.)$ model

$$\hat{y}(k) = f(\boldsymbol{\phi}(k), \theta)\,. \tag{4.5}$$

Following this scheme, the nonlinear system identification of dynamical systems involves the following tasks [179]:

- **Structure selection** How to choose the nonlinear mapping $f(\cdot)$ and the regression vector $\boldsymbol{\phi}(k)$?
- **Input sequence design** Determination of the input sequence $u(k)$ which is injected into the plant to generate the output sequence $y(k)$ that can be used for identification (see [172] for more details).
- **Noise modelling** Determination of the dynamic model which generates the noise $e(k)$.
- **Parameter estimation** Estimation of the model parameters from the dynamic plant data $u(k)$ and $y(k)$ and the noise $e(k)$.
- **Model validation** Comparison of the output of the plant and the model based on data not used in model development.

As the structure selection task has the largest effect on the performance of the model, a small review about regressor selection for dynamic systems is provided in the following subsection.

As the nomenclature of nonlinear dynamical models is based on the terminology used to categorize linear input-output models, in the following the linear empirical model structures are discussed that can be summarized by the general family [179, 247]

$$A(q)\,y(k) = \frac{B(q)}{F(q)}u(k) + \frac{C(q)}{D(q)}e(k) \tag{4.6}$$

where q denotes the shift operator. For instance, $A(q)$ is a polynomial in q^{-1}. This model can be given in a "pseudo-linear" regression form

$$\hat{y}(k) = \theta^T \boldsymbol{\phi}(k) \tag{4.7}$$

where the regressors, i.e., the components of $\boldsymbol{\phi}(k)$, can be given by

- $u(k-i)$, $i=1,\dots,n_b$, control signals (associated with the B polynomial)
- $y(k-i)$, $i=1,\dots,n_a$, measured process outputs (associated with the A polynomial),
- $\hat{y}_u(k-i)$ simulated outputs from past $u(k)$ (associated with the F polynomial),
- $e(k-i)=y(k-i)-\hat{y}(k-i)$ prediction errors (associated with the C polynomial),
- $e_u(k-i)=y(k-i)-\hat{y}_u(k-i)$ simulation errors (associated with the D polynomial).

Based on these regressors, different types of model structures can be constructed. For instance, the simplest linear dynamical model is the finite impulse response (FIR) model

$$\hat{y}(k)=B(q)u(k)+e(k)=b_1u(k-1)+\cdots+b_nu(k-n_b)+e(k)\,. \tag{4.8}$$

In this equation, the corresponding predictor $\hat{y}(k)=B(q)u(k)$ is thus based on the $\phi(k)=[u(k-1),\dots,u(k-n_b)]$ regression vector.

Other special cases of (4.6) are known as the *Box–Jenkins (BJ)* model ($A=1$), the *ARMAX*model ($F=D=1$), the *output-error (OE)* model ($A=C=D=1$) and the *ARX* model ($F=C=D=1$).

Following this nomenclature of linear models, it is natural to construct similar nonlinear model structures:

- *NFIR, Nonlinear Finite Impulse Response models*, in this case the regressor vector is composed as

$$\phi(k)=[u\,(k-1)\,,\dots,u\,(k-n_b)]\,.$$

- *NARX, Nonlinear AutoRegressive with eXogenous input models*, which use regressors

$$\phi(k)=[y(k-1),\dots,y(k-n_a),u\,(k-1)\,,\dots,u\,(k-n_b)].$$

- *NOE, Nonlinear Output Error Models*, which use

$$\phi(k)=[\hat{y}(k-1),\dots,\hat{y}(k-n_b),u\,(k-1)\,,\dots,u\,(k-n_b)].$$

- *NBJ, Nonlinear Box–Jenkins models*, where the regressors are past inputs, past estimated outputs, estimation errors using past outputs, and the estimation errors using past estimated outputs

$$\begin{aligned}\phi(k)=[&\hat{y}(k-1),\dots,\hat{y}(k-n_a),u(k-1),\dots,u(k-n_b),\\ &\varepsilon_u(k-1),\dots,\varepsilon_u(k-n_u),\varepsilon(k-1),\dots,\varepsilon(k-n_e)].\end{aligned}$$

In the soft-computing community, the NARX model is called a series-parallel model, while the NOE is referred to as a parallel model [201]. The NOE, and NBJ models are recurrent models, because they use the estimated output that constitutes a feedback. This makes the identification of these models difficult. Because the NARX model structure is non-recursive, its parameters are easy to estimate. Therefore, the NARX model is frequently used for identification. The identified NARX or series-parallel model is often used and tested as an NOE or parallel model, when the past outputs of the nonlinear model are used instead of the real plant outputs. By using this approach a multi-step-ahead prediction can be made, using the former predictions of the system. This procedure is often called 'free run' simulation. This 'free run' is a very rigorous test of the predictive power of the model, because in this way small errors can accumulate to major ones.

Many general empirical model structures, especially those employing a linear parameterization, lead to a large number of parameters to be estimated. An excessive number of unknown coefficients leads to an ill-conditioned estimation problem causing numerical difficulties and high sensitivity to noise and other identification errors. In the following some special model structure will be presented to avoid this dimensionality problem:

Lowering of the regressor dimension through input projection: The $\varphi(.)$ model (4.4) is used to transform the original input-space of the dynamical model into a lower-dimensional space. This transformation is parameterized by an η parameter vector and can be designed based on a priori knowledge about the dynamics of the system. The projection can also be determined on the basis of certain statistics computed using the data. Proposed methods are mostly extensions of the principal component analysis [72] and partial least squares. For instance, in the PLS neural network proposed by Qin and McAvoy, one-dimensional linear projections are decided one at a time based on the residuals resulting after fitting a neural network to the projected data [223]. Methods to obtain nonlinear projections have also been published, but the computational complexities associated with these extensions have made it very difficult to apply them to realistic problems [72]. However, based on prior knowledge, semi-physical regressors can also be defined [276, 277].

ANOVA decomposition: The other option to reduce the search-space of the identification problem is to decompose the general structure into a collection of simpler sub-structures. For instance, consider the Gabor–Kolmogorov or ANalysis Of VAriance (ANOVA) decomposition of a general nonlinear function

$$f(\boldsymbol{\phi}) = f_0 + \sum_{i=1}^{n} f_i(\phi_i) + \sum_{i=1}^{n} \sum_{j=i+1}^{n} f_{i,j}(\phi_i, \phi_j) + \cdots + f_{1,2,\ldots,n}(\phi_1, \ldots, \phi_n) \quad (4.9)$$

where the $f(\boldsymbol{\phi})$ function is simply an additive decomposition of simpler subfunctions; in which f_0 is a bias term and $f_i(\phi_i), f_{i,j}(\phi_i, \phi_j), \ldots$ represent univariate, bivariate, ... components. Any function and hence any reasonable dynamical system can be represented by this decomposition. In most of the systems certain input interactions will be redundant and hence components in the ANOVA decomposition can be ignored resulting in a more parsimonious representation.

By using this approach, the following well-known model classes can be constructed.

- *NAARX Nonlinear Additive AutoRegressive models with eXogenous input models* are defined as [214]

$$\hat{y}(k) = \sum_{i=1}^{n_a} f_i(y(k-i)) + \sum_{j=1}^{n_b} g_j(u(k-j)) + e(k) \tag{4.10}$$

  where the functions f_i and g_i are scalar nonlinearities. As can be seen, this model does not permit "cross terms" involving products of input and output values at different times.

- *Volterra models* are defined as multiple convolution sums

$$\begin{aligned}\hat{y}(k) = y_0 &+ \sum_{i=1}^{n_b} b_i u(k-i) \\ &+ \sum_{i=1}^{n_b}\sum_{j=1}^{n_b} b_{i,j} u(k-i)u(k-j) + \cdots + e(k)\,.\end{aligned} \tag{4.11}$$

- *Polynomial ARMA models* are superior to Volterra series models in the sense that the number of parameters needed to approximate a system is generally much less with polynomial models [116] because of the use of previous output values.

$$\begin{aligned}\hat{y}(k) = y_0 &+ \sum_{i=1}^{n_a} a_{1,i} y(k-i) + \sum_{i=1}^{n_b} b_{1,i} u(k-i) \\ &+ \sum_{i=1}^{n_a}\sum_{j=1}^{i} a_{1,ij} y(k-i)y(k-j) \\ &+ \sum_{i=1}^{n_b}\sum_{j=1}^{i} b_{2,ij} u(k-i)u(k-j) + \ldots + e(k)\,.\end{aligned} \tag{4.12}$$

Block-oriented modelling: Nonlinear effects encountered in some industrial processes, such as distillation columns, pH-neutralization processes, heat-exchangers, or electro-mechanical systems, can be effectively modeled as a combination of a nonlinear static element and a linear dynamic part [83, 218]. Because of the static nature of the nonlinearities, the problem of the nonlinear behavior of the system can be effectively removed from the control problem, allowing the use of simple linear algorithms instead of computationally intensive nonlinear programming ones [89, 205, 206].

According to the order of the static and dynamic blocks, three main block-oriented model structures can be defined.

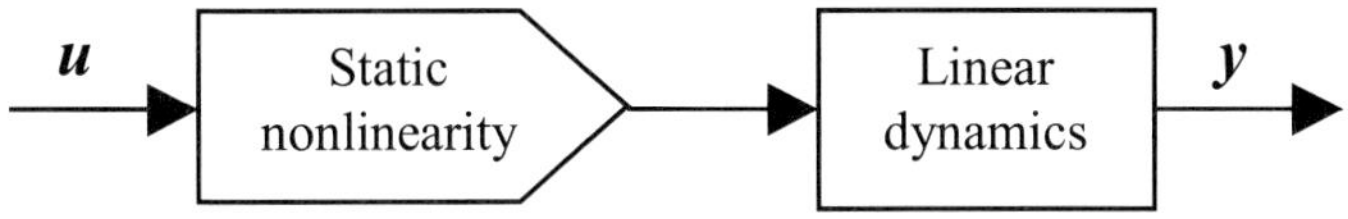

Figure 4.1: Hammerstein model structure.

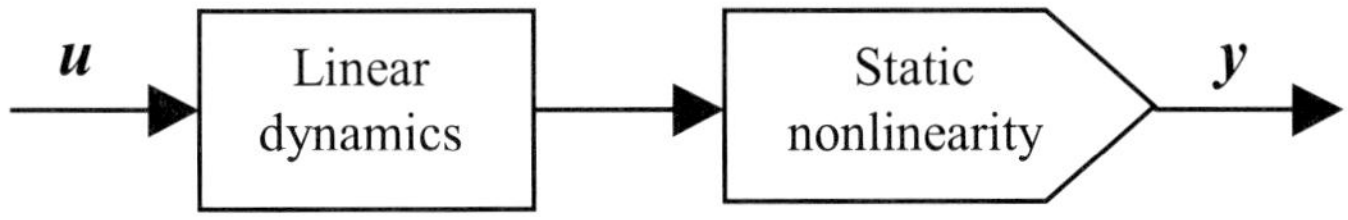

Figure 4.2: Wiener model structure.

- *Hammerstein models*

  A special case of the NAARX model is the *Hammerstein* model, where the same static nonlinearity f is defined for all of the delayed control inputs [80, 83]:
$$\hat{y}(k) = \sum_{i=1}^{n_a} a_i y(k-i) + \sum_{j=1}^{n_b} b_j f(u(k-j)) + e(k)\,. \tag{4.13}$$

  As it is depicted in Figure 4.1 this model can be obtained as a series combination of a memoryless nonlinearity and linear dynamics.

- *Wiener models*

  When the nonlinear element follows the linear block (see Figure 4.2), the resulting model is called the Wiener model [81, 302] that can be formulated as
$$\hat{y}(k) = f\left(\sum_{i=1}^{n_a} a_i y(k-i) + \sum_{j=1}^{n_b} b_j u(k-j)\right) + e(k)\,. \tag{4.14}$$

- *Feedback block-oriented models*

  The basic feedback block-oriented model structure as defined by the diagram shown in Figure 4.3 consists of a linear dynamic model in the forward path with a static nonlinearity in the feedback path [218].

  This model is also a member of the NAARX model family:
$$\hat{y}(k) = \sum_{i=1}^{n_a} a_i (y(k-i)) + \sum_{j=1}^{n_b} b_j u(k-j) + \sum_{j=1}^{n_b} f(y(k-j)) + e(k)\,. \tag{4.15}$$

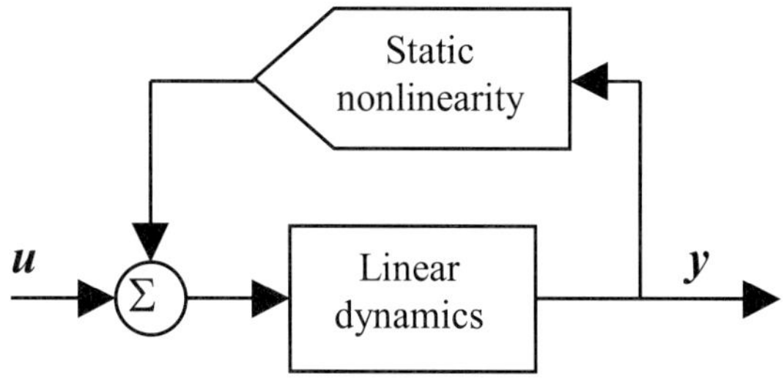

Figure 4.3: Feedback-block-oriented model structure.

The previously presented empirical model structures are frequently applied in the soft-computing community. Artificial neural networks are the most popular framework for nonlinear empirical model development [122, 201]. Usually, these models are applied in the NARX structure, although techniques based on Hammerstein [81, 89, 253] and Wiener models [81] and polynomial ARMAX models also have been presented [37, 116].

4.1.1 TS Fuzzy Models of SISO and MIMO Systems

Fuzzy models of dynamical single input-single output (SISO) and multiple input-multiple output (MIMO) systems are usually based on Nonlinear AutoRegressive with eXogenous input (NARX) model structure . These models can be represented by the following nonlinear vector function:

$$\mathbf{y}(k+1) = f\left(\mathbf{y}(k), \ldots, \mathbf{y}(k-n_a+1), \mathbf{u}(k-n_d), \ldots, \mathbf{u}(k-n_b-n_d+1)\right) \quad (4.16)$$

where, f represents the nonlinear model, $\mathbf{y} = \left[y_1, \ldots, y_{n_y}\right]^T$ is an n_y-dimensional output vector, $\mathbf{u} = [u_1, \ldots, u_{n_u}]^T$ is an n_u-dimensional input vector, n_a and n_b are maximum lags considered for the outputs and inputs, respectively, and n_d is the minimum discrete dead time.

While it may not be possible to find a model that is universally applicable to describe the unknown $f(.)$ system, it would certainly be worthwhile to build local linear models for specific operating points of the process. The modelling framework that is based on combining a number of local models, where each local model has a predefined operating region in which the local model is valid is called *operating regime based model* [197]. In case of MIMO NARX models, the operating regime is formulated as:

$$\mathbf{y}(k+1) = \sum_{i=1}^{c} \beta_i(\mathbf{z}(k)) \left(\sum_{j=1}^{n_a} \mathbf{A}_j^i \mathbf{y}(k-j+1) + \sum_{j=1}^{n_b} \mathbf{B}_j^i \mathbf{u}(k-j-n_d+1) + \mathbf{c}^i \right) \quad (4.17)$$

where the $\beta_i(\mathbf{z}(k))$ function describes the operating regime of the $i = 1, \ldots, c$th local linear ARX model, where $\mathbf{z} = [z_1, \ldots, z_{n_z}]^T$ is a "scheduling" vector, which is usually a subset of the previous process inputs and outputs,

$$\mathbf{z}(k) = \left\{y_1(k), \ldots, y_{n_y}(k - n_a + 1), u_1(k - n_d), \ldots, u_{n_u}(k - n_b - n_d + 1)\right\}.$$

The local models are defined by the $\boldsymbol{\theta}_i = \{\mathbf{A}_j^i, \mathbf{B}_j, \mathbf{c}^i\}$ parameter set. As n_a and n_b denote the maximum lags considered for the previous outputs and inputs, and n_d is the minimum discrete dead time, the lags considered for the separate input-output channels can be handled by zeroing the appropriate elements of the $\mathbf{A}_j^i$ and $\mathbf{B}_j^i$ matrices. If there is no *a priori* knowledge about the order of the nonlinear system, the model orders and the time delays can be directly estimated from input-output data. The next chapter of this book will deal with this problem and give a detailed overview.

The main advantage of this framework is its transparency, because the $\beta_i(\mathbf{z}(k))$ operating regimes of the local models can be represented by fuzzy sets [31]. This representation is appealing, since many systems change behaviors smoothly as a function of the operating point, and the soft transition between the regimes introduced by the fuzzy set representation captures this feature in an elegant fashion. Hence, the entire global model can be conveniently represented by Takagi–Sugeno fuzzy rules [257]:

$$\begin{aligned} R_i \quad : \quad & \textbf{If } z_1 \text{ is } A_{i,1} \textbf{ and } \ldots \textbf{ and } z_{n_z} \text{ is } A_{i,n} \textbf{ then} \\ & \mathbf{y}^i(k+1) = \sum_{j=1}^{n_a} \mathbf{A}_j^i \mathbf{y}(k-j+1) + \sum_{j=1}^{n_b} \mathbf{B}_j^i \mathbf{u}(k-j-n_d+1) + \mathbf{c}^i, \ [w_i] \end{aligned} \tag{4.18}$$

where $A_{i,j}(z_j)$ is the ith antecedent fuzzy set for the jth input and $w_i = [0,1]$ is the weight of the rule that represents the desired impact of the rule. The value of w_i is often chosen by the designer of the fuzzy system based on his or her belief in the goodness and accuracy of the ith rule. When such knowledge is not available w_i is set as $w_i = 1, \forall\, i$.

The one-step-ahead prediction of the MIMO fuzzy model, $\mathbf{y}(k+1)$, is inferred by computing the weighted average of the output of the consequent multivariable models,

$$\mathbf{y}(k+1) = \sum_{i=1}^{c} \beta_i(\mathbf{z}(k))\mathbf{y}^i(k+1) \tag{4.19}$$

where c is the number of the rules, and β_i is the weight of the ith rule,

$$\beta_i(\mathbf{z}(k)) = \frac{w_i \prod_{j=1}^{n} A_{i,j}(z_j)}{\sum_i^c w_i \prod_{j=1}^{n} A_{i,j}(z_j)}. \tag{4.20}$$

To represent the $A_{i,j}(z_j)$ fuzzy set, in this chapter Gaussian membership function is used as in the previous chapters

$$A_{i,j}(z_j) = \exp\left(-\frac{1}{2}\frac{(z_j - v_{i,j})^2}{\sigma_{i,j}^2}\right) \tag{4.21}$$

where $v_{i,j}$ represents the center and $\sigma_{i,j}^2$ the variance of the Gaussian function.

The presented fuzzy model can be seen as a multivariable linear parameter varying system model (LPV), where at the $\mathbf{z}$ operating point, the fuzzy model represents the following LTI model

$$\mathbf{y}(k+1) = \sum_{j=1}^{n_a} \mathbf{A}_j(\mathbf{z}(k))\mathbf{y}(k-j+1) + \sum_{j=1}^{n_b} \mathbf{B}_j(\mathbf{z}(k))\mathbf{u}(k-j-n_d+1) + \mathbf{c}(\mathbf{z}(k)) \tag{4.22}$$

with

$$\begin{aligned} \mathbf{A}_j(\mathbf{z}(k)) &= \sum_{i=1}^{c} \beta_i(\mathbf{z}(k))\mathbf{A}_j^i, \quad \mathbf{B}_j(\mathbf{z}(k)) = \sum_{i=1}^{c} \beta_i(\mathbf{z}(k))\mathbf{B}_j^i, \\ \mathbf{c}(\mathbf{z}(k)) &= \sum_{i=1}^{c} \beta_i(\mathbf{z}(k))\mathbf{c}^i. \end{aligned} \tag{4.23}$$

Example 4.1 (Identification of a TS fuzzy model for a SISO Nonlinear System). *The system under study is a second-order nonlinear system*

$$y(k) = f(y(k-1), y(k-2)) + u(k) \tag{4.24}$$

where

$$f(y(k-1), y(k-2)) = \frac{y(k-1)y(k-2)\,[y(k-1) - 0.5]}{1 + y^2(k-1) + y^2(k-2)}. \tag{4.25}$$

We approximate the nonlinear component f of the plant with a fuzzy model. Following the approach in [293], 400 *simulated data points were generated from the plant model:* 200 *samples of the identification data were obtained with a random input signal $u(k)$ uniformly distributed in* $[-1.5, 1.5]$, *followed by* 200 *samples of evaluation data obtained using a sinusoidal input signal* $u(k) = \sin(2\pi k/25)$, $k = 1001, \ldots, 1200$. *The simulated data are shown in Figure* 4.4. *The input of the model is* $\mathbf{z}_k = [y(k-1), y(k-2)]$.

Table 4.1 *compares the performance (mean squared error) of the models identified with these techniques. The nomenclature can be found in Example* 3.3 *(Section* 3.3.3*).*

From the prediction surface and the operating regimes of the local linear models of the fuzzy model (Figure 4.5 *and Figure* 4.6*), one can see that the presented EM-NI method results in almost optimal antecedent and consequent parameters.*

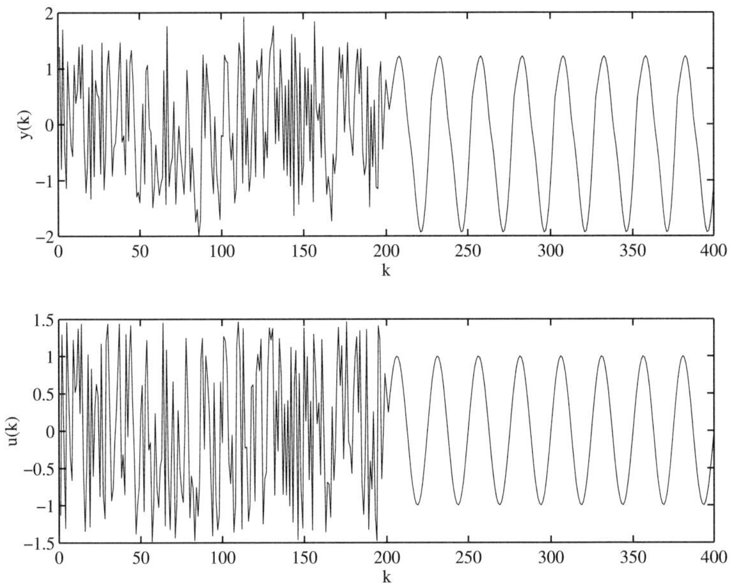

Figure 4.4: Simulated output of the plant and the corresponding input signal.

Table 4.1: Fuzzy models of the nonlinear dynamic plant.

	4 rules		12 rules	
Method	Training set	Test set	Training set	Test set
GG-TLS	$4.6\ 10^{-3}$	$2.1\ 10^{-3}$	$3.7\ 10^{-4}$	$2.9\ 10^{-4}$
GG-LS	$4.6\ 10^{-3}$	$2.0\ 10^{-3}$	$3.7\ 10^{-4}$	$2.9\ 10^{-4}$
EM-TI	$4.6\ 10^{-3}$	$2.0\ 10^{-3}$	$2.4\ 10^{-4}$	$4.1\ 10^{-4}$
EM-NI	$4.4\ 10^{-3}$	$3.5\ 10^{-3}$	$3.4\ 10^{-4}$	$2.3\ 10^{-4}$

We compare our results with those obtained by the optimal rule selection approach proposed by Yen and Wang [293]. Their method uses various information criteria to successively select rules from an initial set of 36 *rules in order to obtain a compact and accurate model. The initial rule base was obtained by partitioning each of the two inputs into six equally distributed fuzzy sets. The rules were selected in an order determined by an orthogonal transform. When linear rule consequents were used, the optimal fuzzy model with* 24 *rules achieved the mean squared error of* $2.0\ 10^{-6}$ *on the training data and* $6.4\ 10^{-4}$ *on the evaluation data.*

Based on this comparison, we can conclude that the presented modelling approach is capable of obtaining good accuracy, while using fewer rules than other approaches presented in the literature.

□

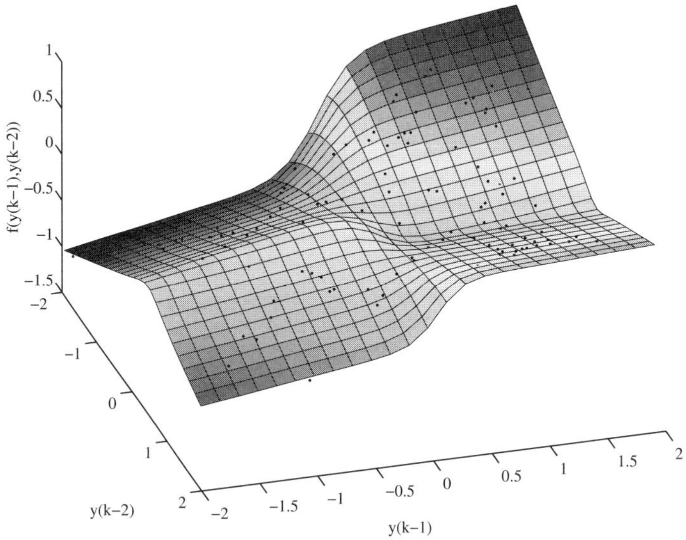

Figure 4.5: Surface plot of the TS model. Available data samples are shown as black dots.

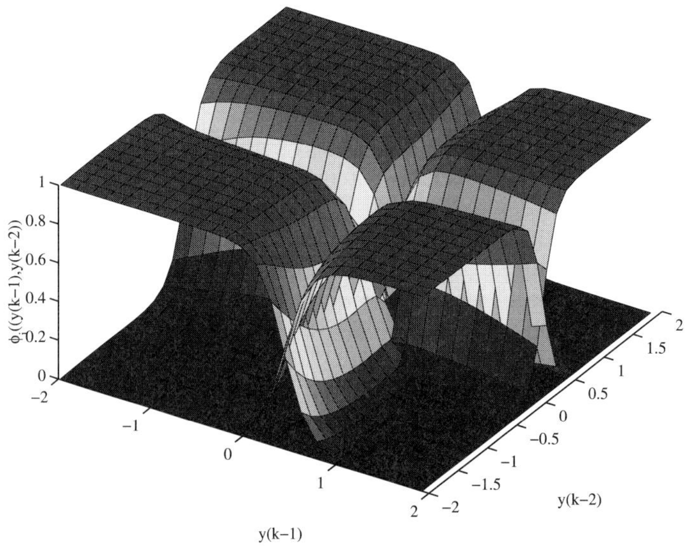

Figure 4.6: Operating regimes obtained.

4.1.2 Clustering for the Identification of MIMO Processes

Identification of the Local MIMO ARX Models

The fuzzy model can be formulated in the following compact form:

$$\mathbf{y}(k+1)^T = \sum_{i=1}^{c} \beta_i(\mathbf{z}(k)) \left[\boldsymbol{\phi}(k)\, \mathbf{I}_{1\times n_y}\right] \boldsymbol{\theta}_i^T + \mathbf{e}(k), \tag{4.26}$$

where $\boldsymbol{\phi}(k)$ is the regressor vector,

$$\boldsymbol{\phi}(k) = [\mathbf{y}(k)^T, \dots, \mathbf{y}(k-n_y+1)^T, \mathbf{u}(k-n_d)^T, \dots, \mathbf{u}(k-n_u-n_d+1)^T], \tag{4.27}$$

$\boldsymbol{\theta}_i$ is the parameter matrix of the ith local model (rule),

$$\boldsymbol{\theta}_i = [\mathbf{A}_1^i, \dots, \mathbf{A}_{n_y}^i, \mathbf{B}_1^i, \dots, \mathbf{B}_{n_u}^i, \mathbf{c}^i]$$

and $\mathbf{e}(k)$ is a zero mean white noise sequence.

The output of this model is linear in the elements of the $\mathbf{A}_j^i$, $\mathbf{B}_j^i$ consequent matrices and the $\mathbf{c}^i$ offset vector. Therefore, these parameters can be estimated from input-output process data by linear least-squares techniques. The N identification data pairs and the truth values of the fuzzy rules are arranged in the following matrices.

$$\boldsymbol{\phi} = \left[\boldsymbol{\phi}^T(1)|\boldsymbol{\phi}^T(2)|\cdots|\boldsymbol{\phi}^N(N)\right]^T \tag{4.28}$$

$$\mathbf{Y} = [\mathbf{y}(2)|\mathbf{y}(3)|\cdots|\mathbf{y}(N+1)]^T \tag{4.29}$$

$$\mathcal{B}_i = \begin{bmatrix} \beta_i(1) & 0 & \cdots & 0 \\ 0 & \beta_i(2) & \cdots & 0 \\ \vdots & \vdots & \ddots & \vdots \\ 0 & 0 & \cdots & \beta_i(N) \end{bmatrix}. \tag{4.30}$$

By using this notation, the weighted least squares solution of $\boldsymbol{\theta}_i$ is:

$$\boldsymbol{\theta}_i = \left[\boldsymbol{\phi}^T \mathcal{B}_i \boldsymbol{\phi}\right]^{-1} \boldsymbol{\phi}^T \mathcal{B}_i \mathbf{Y}\,. \tag{4.31}$$

As this method forces the local linear models to fit the data locally, it does not give an optimal fuzzy model in terms of a minimal global prediction error, but it ensures that the fuzzy model is interpretable as a Linear Parameter Varying (LPV) system [6].

In the following, the extension of the method presented in Section 3.3 (Algorithm 3.3.1) is considered for MIMO model identification. Another main difference compared to Algorithm 3.3.1 is that so-called scheduling variables are also taken into account. Scheduling variables are most representative to handle the nonlinearity of the system. See also Section 3.3.4 and Section 3.4.2.

Problem Formulation

The available identification data, $\mathbf{R} = [\boldsymbol{\phi}, \mathbf{Y}]$ is arranged to form a regression data matrix $\boldsymbol{\phi}$ and a regression vector $\mathbf{Y}$. This means, each observation consists of $(n_y \times n_a + n_u \times n_b) + n_y$ measured variables, grouped into a row vector $\mathbf{r}_k = [\boldsymbol{\phi}(k)\, \mathbf{y}(k+1)^T]$, where the k subscript denotes the kth row of the $\mathbf{R}$ matrix. For sake of simplicity and space, in the following the $\mathbf{r}_k = [\boldsymbol{\phi}_k\, \mathbf{y}_k]$ notation is used according to the partition of $\mathbf{r}_k$ to regressors and regressand. Furthermore, the transpose of the $\mathbf{z}(k)$ vector containing the scheduling variables (that is the subset of $\boldsymbol{\phi}_k$) will be denoted by $\mathbf{z}_k$.

To form an easily interpretable model that does not use the transformed input variables, a new clustering algorithm has been presented based on the Expectation Maximization (EM) identification of Gaussian mixture models [18]. The aim of this section is to show how this EM based identification technique can be extended to the identification of MIMO fuzzy models.

The basic idea of the presented algorithm is to define the cluster prototype which contains three terms:

$$p(\mathbf{y}, \boldsymbol{\phi}, \mathbf{z}) = \sum_{i=1}^{c} p(\mathbf{y}|\boldsymbol{\phi}, \eta_i) p(\mathbf{z}|\eta_i) p(\eta_i) \tag{4.32}$$

where $\boldsymbol{\phi}$ and $\mathbf{z}$ may be identical, overlapping in some dimensions or completely distinct.

The $p(\mathbf{y}|\boldsymbol{\phi}, \eta_i)$ distribution of the output variables is taken to be

$$p(\mathbf{y}|\boldsymbol{\phi}, \eta_i) = \frac{\exp\left(-(\mathbf{y} - \boldsymbol{\phi}^* \boldsymbol{\theta}_i^T)^T (\mathbf{F}_i^{yy})^{-1} (\mathbf{y} - \boldsymbol{\phi}^* \boldsymbol{\theta}_i^T)\right)}{(2\pi)^{\frac{n_y}{2}} \sqrt{\det(\mathbf{F}_i^{yy})}} \tag{4.33}$$

where $\boldsymbol{\phi}^* = [\boldsymbol{\phi}, \mathbf{I}_{1\times n_y}]$ represents the extended regression vector, $\boldsymbol{\theta}_i$ the parameters of the ith local linear model, and the $\mathbf{F}_i^{yy}$ the weighted covariance matrix of the modelling error of this local model.

The $p(\mathbf{z}|\eta_i)$ distribution of the scheduling variables is parameterized as Gaussian distributions [95], and defines the domain of influence of a cluster

$$p(\mathbf{z}|\eta_i) = \frac{\exp\left(-\frac{1}{2}(\mathbf{z} - \mathbf{v}_i^z)^T (\mathbf{F}_i^{zz})^{-1} (\mathbf{z} - \mathbf{v}_i^z)\right)}{(2\pi)^{\frac{n_z}{2}} \sqrt{\det(\mathbf{F}_i^{xx})}}, \tag{4.34}$$

where the $\mathbf{F}_i^{zz}$ represent the covariance matrix and $\mathbf{v}_i^z$ the center of the ith cluster.

When the simplicity and the interpretability of the model is important, the $\mathbf{F}^{zz}$ cluster weighted covariance matrix is reduced to its diagonal elements, $\sigma_{i,j}^2$, which resembles to the simplified axis-parallel version of the Gath–Geva clustering algorithm [119]

$$p(\mathbf{z}|\eta_i) = \prod_{j=1}^{n_z} \frac{1}{\sqrt{2\pi\sigma_{i,j}^2}} \exp\left(-\frac{1}{2}\frac{(z_j - v_{i,j})^2}{\sigma_{i,j}^2}\right) \tag{4.35}$$

$$< \mathbf{y} \mid \mathbf{z} > = \int \mathbf{y} \mid \mathbf{z} d\mathbf{y} = \int \mathbf{y} \frac{p(\mathbf{y}, \boldsymbol{\phi}, \mathbf{z})}{p(\mathbf{z})} d\mathbf{y}. \tag{4.36}$$

The identification of the model means the determination of the $\eta_i = \{p(\eta_i), \mathbf{v}_i^z, \mathbf{F}_i^{zz}, \boldsymbol{\theta}_i, \mathbf{F}_i^{yy}\}$ parameters of the clusters. Below, the EM identification of the model is presented that is re-formulated in the form of Gath–Geva fuzzy clustering.

Estimation of the Model Parameters

The basics of EM are the followings. Suppose we know some observed values of a random variable $\mathbf{r}$ and we wish to model the density of $\mathbf{r}$ by using a model parameterized by η. The EM algorithm obtains an estimate $\hat{\eta}$ that maximizes the likelihood $\mathcal{L}(\eta) = p(\mathbf{r}|\eta)$.

E-step In the E-step the current cluster parameters η_i are assumed to be correct and the posterior probabilities that relate each data point in the conditional probability $p(\eta_i|\mathbf{z}, y)$ are evaluated. These posterior probabilities can be interpreted as the probability that a particular piece of data was generated by a particular cluster. By using the Bayes theorem,

$$p(\eta_i|\mathbf{z}, y) = \frac{p(\mathbf{z}, y|\eta_i)p(\eta_i)}{p(\mathbf{z}, y)} = \frac{p(\mathbf{z}, y|\eta_i)p(\eta_i)}{\sum_{i=1}^{c} p(\mathbf{z}, y|\eta_i)p(\eta_i)}. \tag{4.37}$$

M-step In the M-step the current data distribution is assumed to be correct and the parameters of the clusters that maximize the likelihood of the data are found. According to this, the new unconditional probabilities

$$p(\eta_i) = \frac{1}{N}\sum_{k=1}^{N} p(\eta_i|\mathbf{z}, y) \tag{4.38}$$

and the mean and the covariance of the weighted covariance matrices can be calculated:

$$\mathbf{v}_i^z = \frac{\sum_{k=1}^{N} \mathbf{z}_k p(\eta_i|\mathbf{z}_k, y_k)}{\sum_{k=1}^{N} p(\eta_i|\mathbf{z}_k, y_k)} \tag{4.39}$$

$$\mathbf{F}_i^{zz} = \frac{\sum_{k=1}^{N} (\mathbf{z}_k - \mathbf{v}_i^z)(\mathbf{z}_k - \mathbf{v}_i^z)^T p(\eta_i|\mathbf{z}_k, y_k)}{\sum_{k=1}^{N} p(\eta_i|\mathbf{z}_k, y_k)}. \tag{4.40}$$

In order to calculate the maximizing parameters of the local linear models, the derivative of the log-likelihood is calculated and set equal to zero:

$$0 = \frac{\partial}{\partial \theta_i} \ln \prod_{k=1}^{N} p(\mathbf{z}_k, y_k) = \sum_{k=1}^{N} \frac{\partial}{\partial \theta_i} \ln p(\mathbf{z}_k, y_k) \tag{4.41}$$

$$= \frac{1}{Np(\eta_i)} \sum_{k=1}^{N} p(\eta_i|\mathbf{z}, y)\,(y_k - f_i(\boldsymbol{\phi}_k, \theta_i))\,\frac{\partial f_i(\boldsymbol{\phi}_k, \theta_i)}{\partial \theta_i}.$$

Here, $f_i(\boldsymbol{\phi}_k, \theta_i)$ represents the local consequent models, $f_i(\boldsymbol{\phi}_k,\theta_i)=\mathbf{a}_i^T\boldsymbol{\phi}_k+b_i$.

The above equation results in weighted least-squares identification of the local linear models with the weighting matrix

$$\mathbf{\Phi}_j = \begin{bmatrix} p(\eta_i|\mathbf{z}_1, y_1) & 0 & \cdots & 0 \\ 0 & p(\eta_i|\mathbf{z}_2, y_2) & \cdots & 0 \\ \vdots & \vdots & \ddots & \vdots \\ 0 & 0 & \cdots & p(\eta_i|\mathbf{z}_N, y_N) \end{bmatrix}. \tag{4.42}$$

Finally, the standard deviations σ_i are calculated as

$$\sigma_i^2 = \frac{\sum_{k=1}^{N} (y_k - f_i(\boldsymbol{\phi}_k, \theta_i))^T (y_k - f_i(\boldsymbol{\phi}_k, \theta_i)) p(\eta_i|\mathbf{z}_k, y_k)}{Np(\eta_i)}. \tag{4.43}$$

The parameters of the above presented model can be derived by taking the derivative of the log-likelihood of the data with respect to the η_i parameters.

Clustering Algorithm

The clustering is based on the minimization of the sum of weighted squared distances between the data points, $\mathbf{r}_k$ and the cluster prototypes, η_i.

$$J(\mathbf{R}, \mathbf{U}, \eta) = \sum_{i=1}^{c} \sum_{k=1}^{N} (\mu_{i,k})^m \, d^2(\mathbf{r}_k, \eta_i) \tag{4.44}$$

where the distance measure consists of two terms and inversely proportional to the probability of the data. The first term is based on the $\mathbf{v}_i^z$ geometrical distance between the cluster center and the $\mathbf{z}$ scheduling vector, while the second distance term is based on the performance of the local linear models:

$$\frac{1}{d^2(\mathbf{r}_k, \eta_i)} = p(\eta_i)p(\mathbf{z}|\eta_i)p(\mathbf{y}|\boldsymbol{\phi}, \eta_i) \tag{4.45}$$

$$= w_i \prod_{j=1}^{n} \exp\left(-\frac{1}{2}\frac{(z_{j,k} - v_{i,j})^2}{\sigma_{i,j}^2}\right) \cdot \frac{\exp\left(-(\mathbf{y} - \boldsymbol{\phi}_k^* \boldsymbol{\theta}_i^T)^T (\mathbf{F}_i^{yy})^{-1} (\mathbf{y} - \boldsymbol{\phi}_k^* \boldsymbol{\theta}_i^T)\right)}{(2\pi)^{\frac{n_y}{2}} \sqrt{\det(\mathbf{F}_i^{yy})}}$$

Algorithm 4.1.1 (Clustering for MIMO Model Identification).

Initialization *Given a set of data* $\mathbf{R}$ *specify the number of clusters, c, choose a weighting exponent (usually* $m = 2$*) and a termination tolerance* $\epsilon > 0$*. Initialize the partition matrix (randomly),* $\mathbf{U} = [\mu_{i,k}]_{c \times N}$.

Repeat *for* $l = 1, 2, \ldots$

Step 1 *Calculate the parameters of the clusters*

- *Centers of the membership functions:*

$$\mathbf{v}_i^{z\,(l)} = \frac{\sum_{k=1}^{N} \mu_{i,k}^{(l-1)} \mathbf{z}_k}{\sum_{k=1}^{N} \mu_{i,k}^{(l-1)}}, \; 1 \leq i \leq c. \tag{4.46}$$

- *Standard deviation of the Gaussian membership function:*

$$\sigma_{i,j}^{2\,(l)} = \frac{\sum_{k=1}^{N} \mu_{i,k}^{(l-1)} (z_{j,k} - v_{i,j})^2}{\sum_{k=1}^{N} \mu_{i,k}^{(l-1)}}, \; 1 \leq i \leq c. \tag{4.47}$$

- *Parameters of the local models (see (4.31), where the weights in the* $\mathcal{B}_i$ *matrix are* $\beta_i(k) = \mu_{i,k}^{(l-1)}$*).*
- *Covariance of the modelling errors of the local models (4.85).*
- *A priori probability of the cluster*

$$p(\eta_i) = \frac{1}{N} \sum_{k=1}^{N} \mu_{i,k}. \tag{4.48}$$

- *Weight (impact) of the rules:*

$$w_i = p(\eta_i) \prod_{j=1}^{n_z} \frac{1}{\sqrt{2\pi\sigma_{i,j}^2}}. \tag{4.49}$$

Step 2 *Compute the distance measure* $d^2(\mathbf{r}_k, \eta_i)$ *by (4.45).*

Step 3 *Update the partition matrix*

$$\mu_{i,k}^{(l)} = \frac{1}{\sum_{j=1}^{c} \left(d(\mathbf{r}_k, \eta_i)/d(\mathbf{r}_k, \eta_j) \right)^{2/(m-1)}}, \quad 1 \leq i \leq c, \; 1 \leq k \leq N. \tag{4.50}$$

until $||\mathbf{U}^{(l)} - \mathbf{U}^{(l-1)}|| < \epsilon$.

and the $\mu_{i,k} = p(\eta_i|\phi)$ weight denotes the membership that the $\mathbf{r}_k$ input-output data is generated by the ith cluster

$$p(\eta_i|\phi) = \sum_{i=1}^{c} \frac{p(\phi|\eta_i)p(\eta_i)}{p(\phi)}. \tag{4.51}$$

As can be seen in the previous chapters, the alternating optimization method can be used to solve this minimization problem with respect to the conditions of the fuzzy partitioning space (1.12), (1.13) and (1.14).

The remainder of this section is concerned with the theoretical convergence properties of the presented algorithm. Since this algorithm is a member of the family of algorithms discussed in [113], the following discussion is based on the results of Hathaway and Bezdek [113]. Using Lagrange multiplier theory, it is easily shown that for $d(\mathbf{r}_k, \eta_i) \geq 0$, (4.50) defines $\mathbf{U}^{(\mathbf{l}+\mathbf{1})}$ to be a global minimizer of the restricted cost function (4.44). From this it follows that the presented iterative algorithm is a special case of grouped coordinate minimization, and the general convergence theory from [41] can be applied for reasonable choices of $d(\mathbf{r}_k, \eta_i)$ to show that any limit point of an iteration sequence will be a minimizer, or at worst a saddle point of the cost function J. The local convergence result in [41] states that if the distance measures $d(\mathbf{r}_k, \eta_i)$ are sufficiently smooth and a standard convexity holds at a minimizer $(\mathbf{U}^*, \eta^*)$ of J, then any iteration sequence started with $\mathbf{U}^{(\mathbf{0})}$ sufficiently close to $\mathbf{U}^*$ will converge to $(\mathbf{U}^*, \eta^*)$. Furthermore, the rate of convergence of the sequence will be w-linear. This means that there is a norm $\| * \|$and constants $0 < \gamma < 1$ and $l_0 > 0$, such that for all $l \geq l_0$, the sequence of errors $\{e^l\} = \{\|(\mathbf{U}^\mathbf{l}, \eta^\mathbf{l}) - (\mathbf{U}^*, \eta^*)\|\}$ satisfies the inequality $e^{l+1} < \gamma e^l$.

Example 4.2 (Identification of a distillation column). *The examined process is a first-principle model of a binary distillation column (see Figure 4.7). The column has 39 trays, a reboyler and a condenser. The modelling assumptions are equilibrium on all trays, total condenser, constant molar flows, no vapor holdup, linearised liquid dynamic. The simulated system covers the most important effects for the dynamic of a real distillation column. The studied column operates in LV configuration [249] with two manipulated variables (reflux and boilup rate, u_1 and u_2) and two controlled variables (top and bottom impurities, y_1 y_2). Further details of the simulation model are described in [249].*

A closed-loop identification experiment has been performed by two simple PI controllers. The sampling time is 2 minutes and 10000 samples are collected for the experiment. The identification data is divided into four sections of 2500 samples such that each section corresponds to data gathered around one operating point (see Figure 4.8). The first 2000 samples of each section are used to form a model estimation while the last 500 samples are used for validation. The order of the local models is chosen to be $n_a = n_b = 2$. The process was identified without time delay, $n_d = 0$.

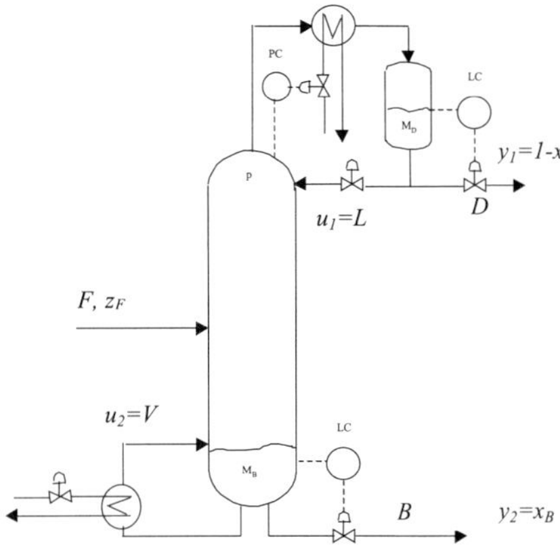

Figure 4.7: Schematic diagram of the distillation column.

Four different models were identified:

(1) *Linear: MIMO ARX model,*
(2) *FMID: combination of two MISO TS models obtained by Fuzzy Model Identification Toolbox [26] (two times four rules),*
(3) *FIX: grid-type MIMO TS model with fixed Gaussian membership functions (four rules),*
(4) *Proposed: MIMO fuzzy model where the Gaussian membership functions and the local models are identified by the presented method (four rules).*

As the process gain varies in direct proportion to the concentrations, the fuzzy sets – the operating regions of the local models – are defined on the domain of the product impurities $\mathbf{z}(k) = [y_1(k), y_2(k)]$. *This results in the following MIMO TS fuzzy model structure:*

$$\begin{aligned} R_i \quad &: \quad \textbf{If } y_1(k) \textit{ is } A_{i,1} \textbf{ and } y_2(k) \textit{ is } A_{i,2} \textbf{ then} \\ \mathbf{y}^i(k+1) &= \sum_{j=1}^{2} \mathbf{A}_j^i \mathbf{y}(k-j+1) + \sum_{j=1}^{2} \mathbf{B}_j^i \mathbf{u}(k-j+1) + \mathbf{c}^i \end{aligned} \tag{4.52}$$

Figure 4.8 shows the simulated output of the presented MIMO fuzzy model, whose membership functions are depicted in Figure 4.9. The results are obtained by the free run simulation of the model where the predicted outputs are fed back to the model as inputs. This free run simulation is a very rigorous test of the predictive power of the model, because in this way small errors can accumulate to major ones. Hence, the presented results are quite promising and indicate the usability of the model in model-based control.

Table 4.2 compares the performance of this model to other models. As this table shows, the presented model gives superior performance. Only the performance of

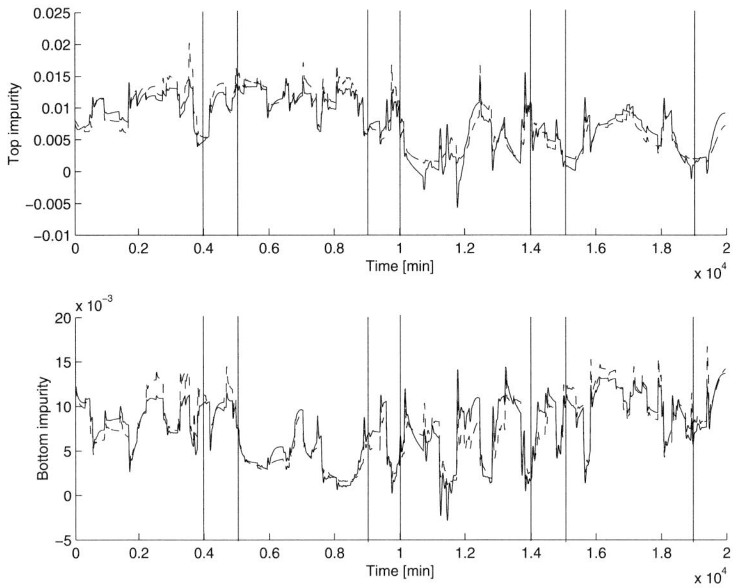

Figure 4.8: Measured (–) and predicted (- -) process outputs.

the model obtained by the FMID toolbox is comparable. However, this model is built up from two MISO fuzzy models, hence this model has two times four rules and $2 \times 4 \times 4$ *membership functions, which suggests that this model is unnecessarily more complex than the presented MIMO TS model.*

Table 4.2: Prediction errors obtained by different models (sum of square error (SSE)).

Model	SSEy_1	SSEy_2
Linear	$11.4e^{-3}$	$9.5e^{-3}$
FMID	$6.3e^{-3}$	$1.7e^{-3}$
Fixed	$10.3e^{-3}$	$9.1e^{-3}$
Proposed	$3.9e^{-3}$	$3.3e^{-3}$

The high-purity binary distillation column is a nonlinear process, because as the demanded product purities increase, the gains of the process are decreasing. As the arrangement of the membership functions depicted in Fig. 3 *shows, this phenomenon is correctly detected by the clustering algorithm.*

In this example, the number of rules was determined manually. However, the identification of the number of the clusters (rules) is an important task that is our current research, along with the selection of relevant input variables [11]. *Similarly*

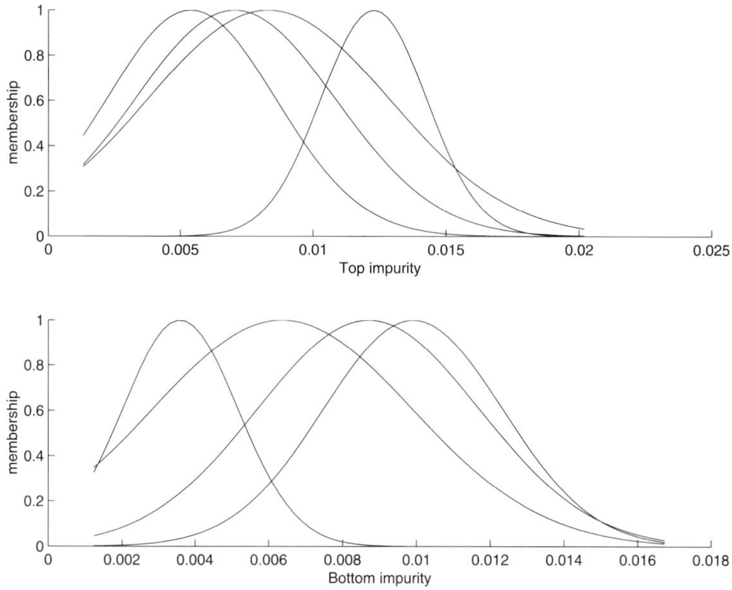

Figure 4.9: Operating regimes (fuzzy sets) obtained by the presented fuzzy clustering.

to the algorithm proposed by Gath and Geva [93] *and Bezdek* [38], *the application of cluster validity measures like fuzzy hypervolume and density, or cluster flatness* [146] *can be applied for this purpose.*

□

4.1.3 Conclusions

In this section the identification of nonlinear multiple-input, multiple-output systems is discussed. A fuzzy model structure has been presented, where the multivariable process is represented by a MIMO fuzzy model that consists of local linear MIMO ARX models. The local models and the antecedent part of the fuzzy model are identified by a new clustering algorithm, which clustering algorithm is an effective approximation of a nonlinear optimization problem related to the identification of fuzzy systems. The approach is demonstrated by means of the identification of a high-purity distillation column. The results show that the presented clustering based identification obtains compact and accurate models for MIMO processes.

4.2 Interpretable Semi-Mechanistic Fuzzy Models by Clustering, OLS and FIS Model Reduction

4.2.1 Introduction to Semi-Mechanistic Modelling

Fuzzy modelling and identification from process data proved to be effective for approximation of uncertain nonlinear processes [115], as can be seen from the previous examples. Different approaches have been proposed to obtain the most frequently applied Takagi–Sugeno models from data – one effective method can be found in the previous section. Most approaches, however, utilize only the function approximation capabilities of fuzzy systems, and little attention is paid to the qualitative aspects. This makes them less suited for applications in which emphasis is not only on accuracy, but also on interpretability, computational complexity and maintainability [241]. Furthermore, completely data-driven black-box identification techniques often yield unrealistic and non-interpretable models. This is typically due to an insufficient information content of the identification data and due to overparameterization of the models. Another disadvantage in process modelling is the non-scalability of black box models, i.e., one has to collect new training-data when the process is modified.

Due to the given drawbacks, combinations of *a priori* knowledge with black-box modelling techniques is gaining considerable interest. Two different approaches can be distinguished: grey-box modelling and semi-mechanistic modelling. In a grey-box model, *a priori* knowledge or information enters the black-box model as, e.g., constraints on the model parameters or variables, the smoothness of the system behavior, or the open-loop stability [48, 177, 268]. For example, Lindskog and Ljung [177] applied combinations or (nonlinear) transformations of the input signals corresponding to physical variables and used the resulting signals in a black-box model. A major drawback of this approach is that it may suffer from the same drawbacks as the black-box model, i.e., no extrapolation is possible and time-varying processes remain a problem.

On the other hand, transparency properties of fuzzy systems proved to be useful in the context of *grey-box* modelling, because it allows to effectively combine different types of information, namely linguistic knowledge, first-principle knowledge and information from data. For instance, if the model developer has *a priori* knowledge about steady-state or gain-independent dynamic behavior of the process, a Hybrid Fuzzy Convolution Model [14] can be used as a combination of a fuzzy model and *a priori* knowledge based impulse response model of the system [8]. For the incorporation of prior knowledge into data-driven identification of dynamic fuzzy models of the Takagi–Sugeno type, a constrained identification algorithm has been developed [5]. This approach is based on a transformation of the *a priori* knowledge about stability, bounds on the stationary gains, and the settling time of the process into linear inequalities on the parameter set of the fuzzy model, similar to [264, 268]. This constrained identification is useful, because the TS fuzzy

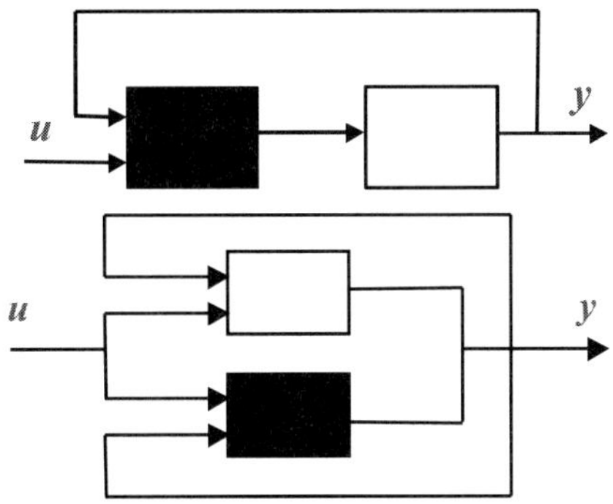

Figure 4.10: Serial and parallel combinations of white and black box models.

model is often over-parameterized, hence explicit regularization, like penalties on non-smooth behavior of the model and application of prior knowledge based parameter constraints can dramatically improve the robustness of the identification algorithm, eventually leading to more accurate parameter estimates [138].

One may also start by deriving a model based on first-principles and then include black-box elements as parts of the white-box model frame [137, 222, 245, 260, 263]. This modelling approach is usually denoted as *hybrid-modelling* or *semi-mechanistic modelling*. The latter term is used in the sequel because the first one is rather confusing with other methods. Johansen [137] describes various techniques to incorporate knowledge into black-box models. They use a local model structure and formulate the global identification as a nonlinear optimization problem. Thompson and Kramer [263] use the so-called parallel approach of semi-mechanistic modelling (Figure 4.10). Here, the error between the data and the white-box model is represented by a neural network. They also describe the serial approach where a neural network is used to model unknown parameters (Figure 4.10).

Several authors applied hybrid models for the modelling of biotechnological systems [222, 245, 262, 271]. Fuzzy models can also be incorporated into semi-physical models as linguistically interpretable black box elements [32]. A more complex semi-physical model was designed as a combination of a first-principles model, an artificial neural network and a fuzzy expert system for modelling a fed-batch cultivation [235]. The application of the fuzzy system was motivated by the fact that bioprocesses are often divided up into different phases (operating regimes) in which different mechanisms dominate. In many practical cases, the phase boundaries are not clear-cut boarders that can be described by crisp time instants. Hence, it is straightforward to use fuzzy techniques to determine in which phase the process is at a given time.

In this section, which is mainly based on [15], we also focus on semi-mechanistic models where a fuzzy model is used to represent, in a mechanistic sense, difficult-to-model parts of our system. It will be show that fuzzy models can be efficiently incorporated into the semi-mechanistic modelling environment and we show also that this approach provides interpretable and accurate submodels. The presented method is an extension of the method presented in the previous section. In this section we present a more advanced clustering method that pursues a further step in accomplishing the total parameter and structure identification of TS models. The clusters are represented by fuzzy sets and local semi-mechanistic models. The unknown parameters of the model are identified by expectation maximization (EM) [44] similarly to our new clustering algorithm: the modified Gath–Geva clustering [18].

Next, the obtained model is reduced by reducing the amount of antecedent variables and also the amount of consequent variables. Using too many antecedent variables may result in difficulties in the prediction and interpretability capabilities of the model due to redundancy, non-informative features and noise. Hence, selection of the scheduling variables is usually necessary. For this purpose, we modify our method that is based on the Fisher interclass separability method and have been developed a feature selection of fuzzy classifiers [231]. Others have focused on reducing the antecedent by similarity analysis of the fuzzy sets [230, 241], however this method is not very suitable for feature selection. Reduction of the consequent space is approached by an Orthogonal Least Squares (OLS) method. The application of orthogonal transforms for the reduction of the number of rules has received much attention in recent literature [230, 294]. These methods evaluate the output contribution of the rules to obtain an importance ordering. In 1999 Yen and Wang investigated various techniques such as orthogonal least-squares, eigenvalue decomposition, SVD-QR with column pivoting method, total least square method and direct SVD method [294]. SVD based fuzzy approximation technique was initiated by Yam in 1997 [289], which directly finds the minimal number of rules from sampled values. Shortly after, this technique was introduced as SVD reduction of the rules and structure decomposition in [289]. For modelling purposes, the OLS is the most appropriate tool [294]. In this section, OLS is applied for a different purpose; the selection of the most relevant input and consequent variables based on the OLS analysis of the local models of the clusters.

4.2.2 Structure of the Semi-Mechanistic Fuzzy Model

Semi-Mechanistic Modelling Approach

Generally, white box models of process systems are formulated by macroscopic balance equations, for instance, mass or energy balances. These balances are based on conservation principle that leads to differential equations written as

$$\begin{bmatrix}\text{accumulation}\\ \text{of } x_i\end{bmatrix} = \begin{bmatrix}\text{inflow}\\ \text{of } x_i\end{bmatrix} - \begin{bmatrix}\text{outflow}\\ \text{of } x_i\end{bmatrix} + \begin{bmatrix}\text{amount of}\\ x_i\text{generated}\end{bmatrix} + \begin{bmatrix}\text{amount of}\\ x_i\text{consumed}\end{bmatrix} \tag{4.53}$$

where x_i is a certain quantity, for example mass or energy.

Equation (4.53) can be simply formulated as commonly used state-space model of process systems given by

$$\begin{aligned} \dot{\mathbf{x}} &= f(\mathbf{x}, \mathbf{u}) \\ \mathbf{y} &= g(\mathbf{x}) \end{aligned} \tag{4.54}$$

where $\mathbf{x} = [x_1, \ldots, x_n]^T$ represents the state, and $\mathbf{y} = [y_1, \ldots, y_{n_y}]^T$ represents the output of the system, and f and g are nonlinear functions defined by the balance equations.

Besides the prediction properties of white box models, they also have the capability to explain the underlying mechanistic relationships of the process. These models are in general to a certain extent applicable independently of the process scale. Furthermore, these models are easy to modify or extend by changing parts of the model. White-box modelling is generally supported by experiments to get some of the parameters. In that case, even after scaling, new experiments are necessary because some of the parameters are difficult to express in a simple way as functions of process characteristics. In a practical situation, the construction of white-box models can be difficult because the underlying mechanisms may be not completely clear, experimental results obtained in the laboratory do not carry over to practice, or parts of the white-box models are in fact not known.

In general, not all of the terms in (4.53) are exactly or even partially known. Hence, when we do not have detailed first principles knowledge about the process, the control law should be determined using approximations of functions of f and g. In semi-mechanistic modelling black-box models, like neural networks are used to represent the otherwise difficult-to-obtain parts of the model. The semi-mechanistic modelling strategy is be combined quite naturally with the general structure of white-box models in process systems, since this structure is usually based on macroscopic balances (e.g., mass, energy or momentum). These balances specify the dynamics of the relevant state variables and contain different rate terms. Some of these terms are directly associated with manipulated or measured variables (e.g., in- and out-going flows) and do not have to be modeled any further. In contrast, some rate terms (e.g., reaction rate) have a static mathematical relation with one or more state variables which should be modeled in order to obtain a fully specified model. These terms are then considered as inaccurately known terms. They can be modeled in a white-box way if a static mathematical relation for the rate terms can be based on easy obtainable first principles. If this is not possible, they can be modeled in a black-box way with a nonlinear modelling technique. In the latter case, one obtains the semi-mechanistic model configuration. One of the advantages of the semi-mechanistic modelling strategy is that it seems to be more promising with respect to extrapolation properties [271]. The resulting semi-mechanistic, also called hybrid model can be formulated similarly to (4.54)

$$\begin{aligned} \dot{\mathbf{x}} &= f_{FP}(\mathbf{x}, \mathbf{u}, \boldsymbol{\theta}, f_{BB}(\mathbf{x}, \mathbf{u}, \theta)) \\ \mathbf{y} &= g(\mathbf{x}) \end{aligned} \tag{4.55}$$

where f_{FP} represents the first-principle part of the model and f_{BB} the black-box part parameterized by the $\boldsymbol{\theta}$ parameter vector.

Usually, the reaction rates (the amount of generated and consumed materials in chemical reactions) and the thermal effects (e.g., reaction heat) are especially difficult to model, but the first two transport terms (inlet and outlet flows) in (4.53) can be obtained easily and accurately. Hence, in the modelling phase it turns out which parts of the first principles model are easier and which are more laborious to obtain and often we can get the following hybrid model structure

$$\dot{\mathbf{x}} = f_{FP1}(\mathbf{x}, \mathbf{u}) + f_{FP2}(\mathbf{x}) \odot f_{BB}(\mathbf{x}, \boldsymbol{\theta}) \tag{4.56}$$

where $f_{FP2}(\mathbf{x}) \odot f_{BB}(\mathbf{x}, \theta)$ represents the reaction and thermal effects of the system that can be modelled as a function of the measured state variables, where $\odot$ denotes element-wise product.

Example 4.3 (Semi-mechanistic modelling of a biochemical process). *In this section, we use an example process from the biochemical industry. Here, most processes are carried out in a batch or fed-batch reactor, e.g., production of beer, penicillin and bakers yeast. These types of processes are characterized by their time-varying nature and their recipe-based processing. Furthermore, the measurement of all relevant state-variables is either impossible or very expensive. The on-line control of these processes is therefore usually limited to the setpoint-based control of several process-parameters like temperature, pH, and gasflows. The global process behavior is controlled by process operators that apply standard recipes, stating what to do at a certain time instant. These recipes are usually based on the available process experience and in some cases these are fine-tuned by neural-networks. This method highly depends on the gained expertise and large amounts of data. On the other hand, when accurate process models can be developed based on small amounts of data, one may use these to calculate model-based recipes or control strategies. Besides, one can apply these models to predict process behavior at different scales or optimize the process lay-out, e.g., bioreactor geometry, batch scheduling, etc.*

Modelling of batch and fed-batch processes is difficult due to the in general non-linear characteristics and time-varying dynamics. Moreover, often reaction mechanisms and orders are unknown and have to be postulated and experimentally verified. Fortunately, however, these processes can be partly described by mass-balances that are easily obtained from general process knowledge. These mass balances describe the mass flows and dilution effects. The unknown parts, are the reaction kinetics, which are the conversion rates in the process. Based on this structure, we develop a hybrid model structure for a simulated fed-batch bioprocess. We use a simple model for biomass growth on a single substrate:

$$\frac{dx_1}{dt} = \frac{u}{x_3}(x_{1,in} - x_1) - \sigma(\mathbf{x})x_2, \quad \frac{dx_2}{dt} = -\frac{u}{x_3}(x_2) + \mu(\mathbf{x})x_2, \quad \frac{dx_3}{dt} = u \tag{4.57}$$

where x_1 is the concentration of substrate and x_2 is the biomass concentration of substrate, x_3 the volume of the liquid phase, $x_{1,in}$ is the substrate concentration the volumetric feed rate u, $\sigma(\mathbf{x})$ the specific substrate consumption rate and $\mu(\mathbf{x})$ the overall specific growth rate. Generally, the kinetic submodels $\sigma(\mathbf{x})$ and $\mu(\mathbf{x})$, are unknown, and therefore the following hybrid model of the process can be obtained:

$$\dot{\mathbf{x}} = \underbrace{\begin{bmatrix} \frac{u}{x_3}(x_{1,in} - x_1) \\ -\frac{u}{x_3}(x_2) \\ u \end{bmatrix}}_{f_{FP1}(\mathbf{x},\mathbf{u})} + \underbrace{\begin{bmatrix} -x_2 \\ x_2 \\ 0 \end{bmatrix}}_{f_{FP2}(\mathbf{x})} \odot \underbrace{\begin{bmatrix} \sigma(\mathbf{x}) \\ \mu(\mathbf{x}) \\ 0 \end{bmatrix}}_{f_{BB}(\mathbf{x})} . \tag{4.58}$$

□

Operating Regime Based Semi-Mechanistic Model

In the given example, difficulties in modelling $\sigma(\mathbf{x})$ and $\mu(\mathbf{x})$ are due to lack of knowledge and understanding of relations between kinetic behavior and process states. To handle such uncertainties one can develop nonlinear data-based models. One specific class of nonlinear models with several special characteristics will be applied here, namely the *operating regime based model* [197]. Here, the global model is decomposed in several simple local models that are valid for predefined operating regions. For this, one should examine what are the important states that can be applied for specification of operating points of the submodels and what are the most relevant effectors. This type of local understanding, in fact, is a key to identification of reliable local models based on a limited amount of data.

In the operating regime based modelling approach, $f_{BB}(\mathbf{x})$ is defined by local linear models:

$$\mathbf{A}_i\mathbf{x} + \mathbf{b}_i, \tag{4.59}$$

where the $\mathbf{A}_i$-matrix and $\mathbf{b}_i$-vector define the parameters of the ith local linear model; $i = 1, \ldots, c$. The global black-box model f_{BB} is then given by:

$$f_{BB}(\mathbf{x}) = \sum_{i=1}^{c} \beta_i(\mathbf{z})\left(\mathbf{A}_i\mathbf{x} + \mathbf{b}_i\right), \tag{4.60}$$

where the $\beta_i(\mathbf{z})$-function describes the ith operating region defined by the "scheduling" vector $\mathbf{z} = [z_1, \ldots, z_{n_z}] \in \mathbf{x}$, that is usually a subset of the state variables; an example is given in Figure 4.11 where overlapping operating regions of four local models are defined by two state variables. Integration of the operating regime based model into hybrid a model (4.56) results in the following formulation:

$$\dot{\mathbf{x}} \;=\; \sum_{i=1}^{c} \beta_i(\mathbf{z})\left(f_{FP1}(\mathbf{x},\mathbf{u}) + (\mathbf{A}_i\mathbf{x} + \mathbf{b}_i) \odot f_{FP2}(\mathbf{x})\right). \tag{4.61}$$

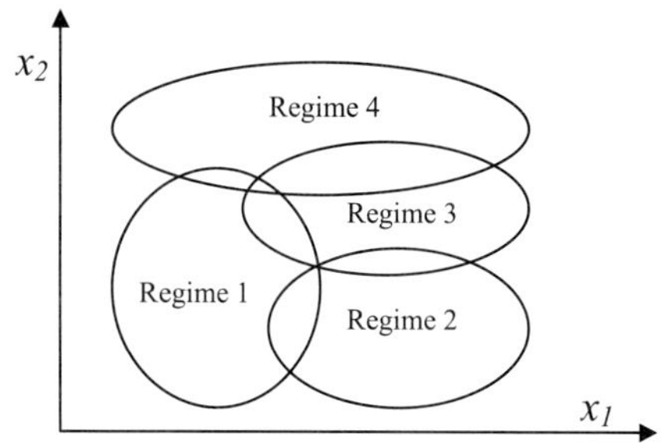

Figure 4.11: Example of an operating regime based model decomposed into four regimes.

Example 4.4 (Cont'd of Example 4.3). *We consider the process introduced in the previous section. The growth rate $\mu(\mathbf{x})$ and substrate consumption rate $\sigma(\mathbf{x})$ are probably defined by the biomass and substrate concentrations, however the relationship is unknown. Therefore, we can choose the scheduling vector as $\mathbf{z} = [x_1, x_2]^T$. This results in the semi-mechanistic:*

$$\begin{aligned}
\dot{\mathbf{x}} &= \sum_{i=1}^{c} \beta_i(\mathbf{z}) \left(\begin{bmatrix} \frac{u}{x_3}(x_{1,in} - x_1) \\ -\frac{u}{x_3}(x_2) \\ u \end{bmatrix} \right. \\
&+ \left. \left(\underbrace{\begin{bmatrix} a_i^{1,1} & a_i^{1,2} & 0 \\ a_i^{2,1} & a_i^{2,2} & 0 \\ 0 & 0 & 0 \end{bmatrix}}_{\mathbf{A}_i} \begin{bmatrix} x_1 \\ x_2 \\ x_3 \end{bmatrix} + \underbrace{\begin{bmatrix} b_i^1 \\ b_i^2 \\ 1 \end{bmatrix}}_{\mathbf{b}_i} \right) \odot \begin{bmatrix} -x_2 \\ x_2 \\ 0 \end{bmatrix} \right) \qquad (4.62)
\end{aligned}$$

□

Semi-Mechanistic Model Representation by Fuzzy Rules

The operating regime based modelling framework provides a method to accurately model nonlinear systems by an efficient interpolation of local linear models defined by operating regimes. In addition, it provides a transparent structure that can be represented by fuzzy rules, because the operating regimes can be given by fuzzy sets [31].

Such a fuzzy representation is appealing, since many systems show smooth changes in their dynamics as a function of the operating point, and the soft transition between the regimes introduced by the fuzzy set representation captures this feature in an elegant fashion. Hence, the entire global model can be conveniently represented by Takagi–Sugeno fuzzy rules [257]. Our previous hybrid model (4.61) can thus be given by MIMO Takagi–Sugeno fuzzy model rules:

$$\begin{aligned} R_i \quad : \quad & \textbf{If } z_1 \text{ is } A_{i,1} \textbf{ and } \ldots \textbf{ and } z_{n_z} \text{ is } A_{i,n_z} \textbf{ then} \\ & \dot{\mathbf{x}}_i = f_{FP1}(\mathbf{x}, \mathbf{u}) + (\mathbf{A}_i \mathbf{x} + \mathbf{b}_i) \odot f_{FP2}(\mathbf{x}), \quad [w_i], \end{aligned} \tag{4.63}$$

where $A_{i,j}(z_j)$ is the ith antecedent fuzzy set for the jth input and $w_i = [0, 1]$ is the rule-weight that represents the desired impact of the ith rule. The value of w_i is often chosen by the designer of the fuzzy system based on his or her belief in the goodness and accuracy of the ith rule. When such knowledge is not available w_i is set as $w_i = 1, \forall\, i$.

Gaussian membership function are used in this chapter as well to represent the $A_{i,j}(z_j)$ fuzzy sets:

$$A_{i,j}(z_j) = \exp\left(-\frac{1}{2} \frac{(z_j - v_{i,j})^2}{\sigma_{i,j}^2} \right), \tag{4.64}$$

where $v_{i,j}$ represents the center and $\sigma_{i,j}^2$ the variance of the Gaussian function. The overall output of the MIMO TS-fuzzy model is inferred by computing a weighted average of the outputs of the consequent multivariable models:

$$\dot{\mathbf{x}} = \sum_{i=1}^{c} \beta_i(\mathbf{z}) \dot{\mathbf{x}}_i, \tag{4.65}$$

where c is the number of rules, and β_i the degree of fulfilment of the ith rule, calculated by:

$$\beta_i(\mathbf{z}) = \frac{w_i \prod_{j=1}^{n_z} A_{i,j}(z_j)}{\sum_{i=1}^{c} w_i \prod_{j=1}^{n_z} A_{i,j}(z_j)}. \tag{4.66}$$

The fuzzy model can be formulated in the compact regression form:

$$\dot{\mathbf{x}}(k)^T = f_{FP1}(\mathbf{x}(k), \mathbf{u}(k))^T + \sum_{i=1}^{c} \beta_i(\mathbf{z}(k)) \boldsymbol{\phi}(k) \boldsymbol{\theta}_i^T, \tag{4.67}$$

where k denotes the index of the kth data-pair, $\boldsymbol{\phi}(k)$ the regressor matrix:

$$\boldsymbol{\phi}(k) \quad = \quad [\mathbf{x}(k) \odot f_{FP2}(\mathbf{x}(k)) \;\; f_{FP2}(\mathbf{x}(k))],$$

and $\boldsymbol{\theta}_i$ the parameter matrix of the ith local model:

$$\boldsymbol{\theta}_i = [\mathbf{A}_i, \mathbf{b}_i].$$

From (4.67) it is clear that the developed semi-mechanistic fuzzy model is linear in the consequent parameters of the fuzzy model. This allows for an effective identification of these parameters. It is interesting to see that the resulting model is a TS-fuzzy model with *semi-physical regressors*.

Next, we rewrite the MIMO fuzzy model as a set of MISO fuzzy models with same antecedents. This representation helps in the model analysis and interpretation because the input-output channels can be separately analyzed.

$$\begin{aligned} R_{i,j} \quad : \quad & \textbf{If } z_1 \text{ is } A_{i,1} \textbf{ and } \ldots \textbf{ and } z_{n_z} \text{ is } A_{i,n_z} \textbf{ then} \\ & \dot{x}_{i,j}(k) = f_{Hyb,j}(\mathbf{x}(k), \mathbf{u}(k)), [w_i], \end{aligned} \tag{4.68}$$

where $j = 1, \ldots, n$ denotes the index of the state variable whose predicted value is calculated and $f_{Hyb,j}$ represents the corresponding rule consequent function of the presented hybrid fuzzy model.

Note that this hybrid fuzzy model is much more parsimonious than the ordinary Takagi–Sugeno model of the process that is formulated as:

$$\begin{aligned} R_{i} \quad : \quad & \textbf{If } x_1 \text{ is } A_{i,1} \textbf{ and } \ldots \textbf{ and } x_{n} \text{ is } A_{i,n} \textbf{ then} \\ & \dot{\mathbf{x}}_i = \hat{\mathbf{A}}_i \mathbf{x} + \hat{\mathbf{B}}_i \mathbf{u} + \hat{\mathbf{c}}_i, \ [w_i]. \end{aligned} \tag{4.69}$$

The latter model has much more antacedent and consequent parameters because the sizes of $\hat{\mathbf{A}}_i$, $\hat{\mathbf{B}}_i$, and $\hat{\mathbf{c}}_i$ are bigger than those of $\mathbf{A}_i$ and $\mathbf{b}_i$. Furthermore, the fuzzy model can be independently interpreted and analyzed from the first-principle part of the model since *the output of the fuzzy model has physical meaning!*

Example 4.5 (Cont'd of Example 4.4). *The operating regime based model of our bioreactor* (4.62) *can now be formulated by fuzzy rules of the form:*

$$R_i : \textbf{If } x_1 \textit{ is } A_{i,1} \textbf{ and } x_2 \textit{ is } A_{i,2} \textbf{ then } \dot{\mathbf{x}} = \begin{bmatrix} \frac{u}{x_3}\left(x_{1,in} - x_1\right) \\ -\frac{u}{x_3}\left(x_2\right) \\ u \end{bmatrix}$$

$$+ \left(\underbrace{\begin{bmatrix} a_i^{1,1} & a_i^{1,2} & 0 \\ a_i^{2,1} & a_i^{2,2} & 0 \\ 0 & 0 & 0 \end{bmatrix}}_{\mathbf{A}_i} \begin{bmatrix} x_1 \\ x_2 \\ x_3 \end{bmatrix} + \underbrace{\begin{bmatrix} b_i^1 \\ b_i^2 \\ 1 \end{bmatrix}}_{\mathbf{b}_i} \right) \odot \begin{bmatrix} -x_2 \\ x_2 \\ 0 \end{bmatrix}, [w_i]. \tag{4.70}$$

Now, the input-output channels can be separately analyzed, e.g., the fuzzy rule that describes the change of the first state of the system (substrate concentration), can be represented by a fuzzy model with rules:

$$\begin{aligned} & R_i : \textbf{If } x_1 \textit{ is } A_{i,1} \textbf{ and } x_2 \textit{ is } A_{i,2} \textbf{ then } \dot{x}_{i,1}(k) = f_{Hyb,j}(\mathbf{x}(k), u(k)) = \\ & \frac{u(k)}{x_3(k)}\left(x_{1,in} - x_1(k)\right) + \left(a_i^{1,1} x_1(k) + a_i^{2,1} x_2(k) + b_i^1\right)\left(-x_2(k)\right). \end{aligned} \tag{4.71}$$

The fuzzy models can now be independently interpreted and analyzed from the first-principle part of the model. For instance, the first output channel of the fuzzy bioprocess model can be written as:

$$\begin{aligned} R_i : \textbf{If } x_1 \textit{ is } A_{i,1} \textbf{ and } x_2 \textit{ is } A_{i,2} \textbf{ then} \\ \sigma(\mathbf{x})_i = a_i^{1,1} x_1 + a_i^{1,2} x_2 + b_i^1, \; [w_i]. \end{aligned} \tag{4.72}$$

□

4.2.3 Clustering-based Identification of the Semi-Mechanistic Fuzzy Model

The remaining part of this chapter deals with the identification of the presented hybrid structures. For this purpose we developed a clustering based identification method, with subsequent reduction of antecendent and consequent variables.

Problem Formulation

In general, measured input-output data are used in data-driven model identification. In many industrial cases it may be difficult to obtain large datasets. Moreover, the measurements may contain a high noise level and often is asynchronous, which often is the case for bioprocesses. In this case, $\dot{\mathbf{x}}(k)$ cannot be calculated by using the simple forward Euler discretization:

$$\dot{\mathbf{x}}(k) = \frac{(\mathbf{x}(k) - \mathbf{x}(k-1))}{\Delta T}. \tag{4.73}$$

Generation of synchronous data by an interpolation is often a good solution to deal with this problem. We applied Hermite interpolation to interpolate between the measured data and estimate the corresponding derivatives. This method is described in Appendix 6.4.

After application of this spline-smoothing, and with the use of the first principle model parts, N semi-physical identification data pairs are obtained that can be arranged into the following matrices based on (4.67):

$$\boldsymbol{\phi} = [\boldsymbol{\phi}(1)|\boldsymbol{\phi}(2)|\cdots|\boldsymbol{\phi}(N)]^T, \tag{4.74}$$

$$\mathbf{f}_{FP1} = [f_{FP1}(1)|f_{FP1}(2)|\cdots|f_{FP1}(N)]^T, \tag{4.75}$$

$$\dot{\mathbf{X}} = [\dot{\mathbf{x}}(1)|\dot{\mathbf{x}}(2)|\cdots|\dot{\mathbf{x}}(N)]^T. \tag{4.76}$$

Identification of the Consequent Parameters

The output of the presented semi-mechanistic model is linear in the elements of the $\mathbf{A}_i$-consequent matrices and the $\mathbf{b}_i$-offset vector. Therefore, if the membership functions are given, these parameters can be estimated from input-output process data by a linear least-squares technique based on the calculated degree of fulfillments:

$$\mathcal{B}_i = \begin{bmatrix} \beta_i(1) & 0 & \cdots & 0 \\ 0 & \beta_i(2) & \cdots & 0 \\ \vdots & \vdots & \ddots & \vdots \\ 0 & 0 & \cdots & \beta_i(N) \end{bmatrix}, \tag{4.77}$$

similarly to (4.30). The local (weighted least squares for each rule) identification of the rule consequent parameters forces the local linear models to fit the systems separately and locally, i.e., resulting rule consequences are local linearizations of the nonlinear system [26]. Hence, in this chapter the parameters of the rule consequences are estimated separately by dividing the identification task into c weighted least-squares problems (the description of ordinary and total least squares techniques can be found in Section 1.6, and of the orthogonal least squares method in Section 3.3.4). By using this notation, the weighted least squares solution of θ_i is:

$$\theta_i = \left[\phi^T \mathcal{B}_i \phi\right]^{-1} \phi^T \mathcal{B}_i \left(\dot{\mathbf{X}} - \mathbf{f}_{FP1}\right). \tag{4.78}$$

Using too many consequent variables results in difficulties in the prediction and interpretability capabilities of the fuzzy model due to redundancy, non-informative features and noise. To avoid these problems in the following the application of orthogonal least squares based model reduction is proposed.

Reduction of the Number of Consequent Parameters

The least squares identification of θ_i, (4.78) can also be formulated as:

$$\theta_i = \mathbf{B}^+ \left(\dot{\mathbf{X}} - \mathbf{f}_{FP1}\right) \sqrt{\mathcal{B}_i} \tag{4.79}$$

where $\mathbf{B}^+$ denotes the Moore-Penrose pseudo inverse of $\phi^T \sqrt{\mathcal{B}_i}$.

The OLS method transforms the columns of $\mathbf{B}$ into a set of orthogonal basis vectors in order to inspect the individual contribution of each variable. To do this Gram-Schmidt orthogonalization of $\mathbf{B} = \mathbf{WA}$ is used, where $\mathbf{W}$ is an orthogonal matrix $\mathbf{W}^T\mathbf{W} = \mathbf{I}$ and $\mathbf{A}$ is an upper triangular matrix with unity diagonal elements. If $\mathbf{w}_i$ denotes the ith column of $\mathbf{W}$ and g_i is the corresponding element of the OLS solution vector $\mathbf{g} = \mathbf{A}\theta_i$, the output variance

$$\left((\dot{\mathbf{X}} - \mathbf{f}_{FP1})\sqrt{\mathcal{B}_i}\right)^T \left((\dot{\mathbf{X}} - \mathbf{f}_{FP1})\sqrt{\mathcal{B}_i}\right)/N$$

can be explained by the regressors $\sum_{i=1}^{n_r} g_i \mathbf{w}_i^T \mathbf{w}_i / N$. Thus, the error reduction ratio, ϱ, due to an individual rule i can be expressed as

$$\varrho^i = \frac{g_i^2 \mathbf{w}_i^T \mathbf{w}_i}{\left((\dot{\mathbf{X}} - \mathbf{f}_{FP1})\sqrt{\mathcal{B}_i}\right)^T \left((\dot{\mathbf{X}} - \mathbf{f}_{FP1})\sqrt{\mathcal{B}_i}\right)}. \tag{4.80}$$

This ratio offers a simple mean for ordering the consequent variables, and can be easily used to select a subset of the inputs in a forward-regression manner.

Semi-Mechanistic Model-based Clustering

In the previous section, we showed that the unknown consequent parameters of the semi-mechanistic fuzzy model are linear in the output. Therefore, a (weighted) least squares method can be applied for the identification of these parameters when the antecedent membership functions (rule-weights) are given. Determination of the antecedent part of the fuzzy model, however, is more difficult because it requires nonlinear optimization. Hence, often heuristic approaches, like fuzzy clustering are applied for this purpose.

In this chapter Gath–Geva algorithm is applied for this purpose. The main advantages have been mentioned in the previous sections. A main disadvantage of clustering tools for the identification of interpretable Takagi–Sugeno fuzzy models is that the fuzzy covariance matrix is defined among all of the model inputs and the correlation among all the input variables. This causes difficulties for the extraction of membership functions from the clusters. This can be done by membership projection methods [29] or a transformed input approach [151]. Generally, however, this decreases the interpretability and accuracy of the initial fuzzy model based on the multivariable cluster representation. Furthermore, the decomposed univariate membership functions are defined on the domains of all input variables of the model $\mathbf{x}$ instead of only the domain of the scheduling variables, $\mathbf{z}$. This makes the model unnecessarily complex due to a large number of fuzzy sets which decreases the model transparency.

To avoid these problems, we propose a new clustering algorithm that generates hybrid fuzzy models in a direct approach. This algorithm obtains a set of clusters, that all contain an input distribution (the distribution of the scheduling variables), $p(\mathbf{z}(k)|\eta_i)$, a local hybrid model $\dot{\mathbf{x}}(k) = f_{Hyb,i}(\mathbf{x}(k), \mathbf{u}(k))$, and an output distribution, $p(\dot{\mathbf{x}}(k)|\mathbf{x}(k), \mathbf{u}(k), \eta_i)$, where

$$\begin{aligned} p(\mathbf{x}(k), \mathbf{u}(k), \dot{\mathbf{x}}(k)) &= \sum_{i=1}^{c} p(\mathbf{x}(k), \mathbf{u}(k), \dot{\mathbf{x}}(k), \eta_i), \\ &= \sum_{i=1}^{c} p(\mathbf{x}(k), \mathbf{u}(k), \dot{\mathbf{x}}(k)|\eta_i) p(\eta_i), \\ &= \sum_{i=1}^{c} p(\dot{\mathbf{x}}(k)|\mathbf{x}(k), \mathbf{u}(k), \eta_i) p(\mathbf{z}(k)|\eta_i) p(\eta_i). \end{aligned} \tag{4.81}$$

The input distribution is parameterized as an unconditioned Gaussian [95] and defines the domain of influence of a cluster:

$$p(\mathbf{z}(k)|\eta_i) = \frac{\exp\left(-\frac{1}{2}(\mathbf{z}(k) - \mathbf{v}_i)^T (\mathbf{F}_i^{zz})^{-1} (\mathbf{z}(k) - \mathbf{v}_i)\right)}{(2\pi)^{\frac{n_z}{2}} \sqrt{\det(\mathbf{F}_i^{zz})}}, \tag{4.82}$$

while the output distribution is taken to be:

$$p(\dot{\mathbf{x}}(k)|\mathbf{x}(k),\mathbf{u}(k),\eta_i) \tag{4.83}$$
$$= \frac{\exp\left(-(\dot{\mathbf{x}}(k) - f_{Hyb,i}(\mathbf{x}(k),\mathbf{u}(k)))^T(\mathbf{F}_i^{xx})^{-1}(\dot{\mathbf{x}}(k) - f_{Hyb,i}(\mathbf{x}(k),\mathbf{u}(k)))\right)}{(2\pi)^{\frac{n}{2}}\sqrt{\det(\mathbf{F}_i^{xx})}},$$

where $\mathbf{F}_i^{zz}$ and $\mathbf{F}_i^{xx}$ are covariance matrices calculated as:

$$\mathbf{F}_i^{zz} = \frac{\sum\limits_{k=1}^{N}(\mathbf{z}(k)-\mathbf{v}_i)(\mathbf{z}(k)-\mathbf{v}_i)^T p(\eta_i|\mathbf{x}(k),\mathbf{u}(k),\dot{\mathbf{x}}(k))}{\sum\limits_{k=1}^{N} p(\eta_i|\mathbf{x}(k),\mathbf{u}(k),\dot{\mathbf{x}}(k))}, \tag{4.84}$$

$$\mathbf{F}_i^{xx} = \frac{\sum\limits_{k=1}^{N}(\dot{\mathbf{x}}(k) - f_{Hyb,i}(\mathbf{x}(k),\mathbf{u}(k)))}{\sum\limits_{k=1}^{N} p(\eta_i|\mathbf{x}(k),\mathbf{u}(k),\dot{\mathbf{x}}(k))} \cdot \frac{(\dot{\mathbf{x}}(k) - f_{Hyb,i}(\mathbf{x}(k),\mathbf{u}(k)))^T p(\eta_i|\mathbf{x}(k),\mathbf{u}(k),\dot{\mathbf{x}}(k))}{\sum\limits_{k=1}^{N} p(\eta_i|\mathbf{x}(k),\mathbf{u}(k),\dot{\mathbf{x}}(k))}, \tag{4.85}$$

in which $p(\eta_i|\mathbf{x}(k),\mathbf{u}(k),\dot{\mathbf{x}}(k))$ are posterior probabilities that can be interpreted as the probability that a particular piece of data has been generated by a particular cluster. By using Bayes Theorem,

$$\begin{aligned} p(\eta_i|\mathbf{x}(k),\mathbf{u}(k),\dot{\mathbf{x}}(k)) &= \frac{p(\mathbf{x}(k),\mathbf{u}(k),\dot{\mathbf{x}}(k)|\eta_i)p(\eta_i)}{p(\mathbf{x}(k),\mathbf{u}(k),\dot{\mathbf{x}}(k))} \\ &= \frac{p(\mathbf{x}(k),\mathbf{u}(k),\dot{\mathbf{x}}(k)|\eta_i)p(\eta_i)}{\sum_{i=1}^{c} p(\dot{\mathbf{x}}(k)|\mathbf{x}(k),\mathbf{u}(k),\eta_i)p(\mathbf{z}(k)|\eta_i)p(\eta_i)}. \end{aligned} \tag{4.86}$$

When simplicity and interpretability of the model are important, then the $\mathbf{F}_i^{xx}$-cluster weighted covariance matrix is reduced to its diagonal elements, resulting in the following density function:

$$p(\mathbf{z}(k)|\eta_i) = \prod_{j=1}^{n} \frac{1}{\sqrt{2\pi\sigma_{i,j}^2}} \exp\left(-\frac{1}{2}\frac{(z_j(k)-v_{i,j})^2}{\sigma_{i,j}^2}\right). \tag{4.87}$$

Identification of the model thus means determination of the

$$\eta_i = \{p(\eta_i), \mathbf{v}_i, \mathbf{F}_i^{zz}, \boldsymbol{\theta}_i, \mathbf{F}_i^{xx}\}\text{-parameters of the clusters},$$

where $\boldsymbol{\theta}_i$ represents the unknown parameters of the hybrid local models. In the following the Expectation Maximization identification of the model, which was re-formulated in the form of fuzzy clustering, is presented.

Clustering Algorithm for Hybrid Modelling

Clustering is based on minimization of the sum of weighted squared distances between the data pairs, $\mathbf{x}(k), \mathbf{u}(k), \dot{\mathbf{x}}(k)$ and the cluster prototypes, η_i. Here, the square of the $d^2(\mathbf{x}(k), \mathbf{u}(k), \dot{\mathbf{x}}(k), \eta_i)$ distances are weighted with the membership values $\mu_{i,k}$. Thus the objective function to be minimized by the clustering algorithm is formulated as:

$$J(\mathbf{Z}, \mathbf{U}, \eta) = \sum_{i=1}^{c} \sum_{j=1}^{N} (\mu_{i,k})\, d^2(\mathbf{x}(k), \mathbf{u}(k), \dot{\mathbf{x}}(k), \eta_i), \tag{4.88}$$

where $\mathbf{Z}$ denotes the available training data, $\mathbf{U}$ the matrix of membership values, and η the set of cluster parameters to be identified. The EM based algorithm with respect to the constrains of fuzzy partitioning (1.12), (1.13) and (1.14), can be formulated as shown in Algorithm 4.2.1.

Note that $\mu_{i,k} = p(\eta_i|\mathbf{x}(k), \mathbf{u}(k), \dot{\mathbf{x}}(k))$ denotes the membership (posterior probability) that the $\mathbf{x}(k), \mathbf{u}(k), \dot{\mathbf{x}}(k)$ input-output data is generated by the ith cluster. It can also be seen that the distance measure (4.95) consists of two terms. The first term is based on the geometrical distance between the cluster centers and the scheduling variable, $\mathbf{z}$, while the second is based on the performance of the local hybrid models.

Note that the application of (4.87) results in the direct determination of the parameters of the membership functions, as (4.87) is identical to (4.64), because the weight of the rule are determined as

$$w_i = \frac{p(\eta_i)}{(2\pi)^{n/2}\sqrt{\det(\mathbf{F}_i^{xx})}} \tag{4.89}$$

or equivalently,

$$w_i = \frac{p(\eta_i)}{\sqrt{2\pi\sigma_{i,j}^2}}. \tag{4.90}$$

Reduction of the Number of Fuzzy Sets

In general, application of too many features and fuzzy sets decreases the models' interpretability. Hence, feature selection is usually necessary. For this purpose, we modified Fisher's interclass separability method, as we described in Section 3.3.4 (see also in Example 3.4). As it was mentioned there, the method makes an importance ranking based on the interclass separability criterion which is obtained from statistical properties of the data. This criterion is based on the $\mathbf{F}_B$ between-class and the $\mathbf{F}_W$ within-class covariance matrices that sum up to the total covariance

Algorithm 4.2.1 (Clustering for Hybrid Modelling).

Initialization
Given a set of data $\mathbf{Z}$ *specify* c, m *and* ϵ. *Initialize the partition matrix,* $\mathbf{U} = [\mu_{i,k}]_{c\times N}$.

Repeat *for* $l = 1, 2, \ldots$

Step 1 *Calculate cluster parameters:*

- *Calculate centers of membership functions:*

$$\mathbf{v}_i^{(l)} = \sum_{k=1}^{N} \mu_{i,k}^{(l-1)} \mathbf{z}(k) / \sum_{k=1}^{N} \mu_{i,k}^{(l-1)}, \ 1 \le i \le c. \tag{4.91}$$

- *Calculate standard deviation of the Gaussian membership functions:*

$$\sigma_{i,j}^{2\,(l)} = \sum_{k=1}^{N} \mu_{i,k}^{(l-1)} (z_j(k) - v_{i,j})^2 / \sum_{k=1}^{N} \mu_{i,k}^{(l-1)}, \ 1 \le i \le c. \tag{4.92}$$

- *Calculate the unknown parameters of the local models by a weighted least squares algorithm, where the weights of the data pairs are:* $\mu_{i,k}^{(l-1)}$.
- *Calculate covariance of the modelling errors of the local models* (4.85).
- *Calculate* a priori *probability of the cluster:*

$$p(\eta_i) = \frac{1}{N} \sum_{k=1}^{N} \mu_{i,k}. \tag{4.93}$$

- *Calculate rule weights:*

$$w_i = p(\eta_i) \prod_{j=1}^{n} \frac{1}{\sqrt{2\pi\sigma_{i,j}^2}}. \tag{4.94}$$

Step 2 *Compute the distance measure* $d^2(\mathbf{x}(k), \mathbf{u}(k), \dot{\mathbf{x}}(k), \eta_i)$.

$$\begin{aligned} &1/d^2(\mathbf{x}(k), \mathbf{u}(k), \dot{\mathbf{x}}(k), \eta_i) \\ &= w_i \prod_{j=1}^{n_z} \exp\left(-\frac{1}{2}\frac{\left(z_{j,k}-v_{i,j}\right)^2}{\sigma_{i,j}^2}\right) \cdot \\ &\frac{\exp\left(-(\dot{\mathbf{x}}(k)-f_{Hyb,i}(\mathbf{x}(k),\mathbf{u}(k)))^T (\mathbf{F}_i^{xx})^{-1} (\dot{\mathbf{x}}(k)-f_{Hyb,i}(\mathbf{x}(k),\mathbf{u}(k)))\right)}{(2\pi)^{\frac{n}{2}} \sqrt{\det(\mathbf{F}_i^{xx})}}. \end{aligned} \tag{4.95}$$

Step 3 *Update the partition matrix:*

$$\mu_{i,k}^{(l)} = \frac{1}{\sum_{j=1}^{c} \left(d(\mathbf{x}(k), \mathbf{u}(k), \dot{\mathbf{x}}(k), \eta_i) / d(\mathbf{x}(k), \mathbf{u}(k), \dot{\mathbf{x}}(k), \eta_j)\right)^{2/(m-1)}}, \quad 1 \le i \le c, \ 1 \le k \le N. \tag{4.96}$$

until $||\mathbf{U}^{(l)} - \mathbf{U}^{(l-1)}|| < \epsilon$.

of the training data $\mathbf{F}_T$, with:

$$\begin{aligned}\mathbf{F}_W &= \sum_{i=1}^{c} p(\eta_i)\mathbf{F}_i, \\ \mathbf{F}_B &= \sum_{i=1}^{c} p(\eta_i)\left(\mathbf{v}_i - \mathbf{v}_0\right)^T \left(\mathbf{v}_i - \mathbf{v}_0\right), \\ \mathbf{v}_0 &= \sum_{i=1}^{c} p(\eta_i)\mathbf{v}_i. \end{aligned} \tag{4.97}$$

The feature interclass separability selection criterion is then given as a trade-off between $\mathbf{F}_W$ and $\mathbf{F}_B$:

$$J = \frac{\det \mathbf{F}_B}{\det \mathbf{F}_W}. \tag{4.98}$$

The importance of each feature is measured by leaving the feature out and calculating J for the reduced covariance matrices. Subsequent feature selection is done in an iterative way by leaving out the least needed feature in each iteration. Note that this method, in case of axis-parallel clusters (the fuzzy sets are directly obtained from such clusters), shows some similarities to similarity-based model reduction techniques [241].

Example 4.6 (Semi-mechanistic model identification of a bio-reactor based on GG clustering). Process Description. *We will demonstrate the presented approach for identification of the previously presented bio-reactor example. Here, we assume that the* true *nonlinear kinetics are given by a previously published nonlinear process model. In this model, the two specific rates $\sigma(\mathbf{x})$ and $\mu(\mathbf{x})$ are inter-related by the following law:*

$$\sigma(\mathbf{x}) = \frac{1}{Y_{x_2/x_1}}\mu(\mathbf{x}) + m, \tag{4.99}$$

where Y_{x_2/x_1} the biomass on substrate yield coefficient and m the specific maintenance demand, and the $\mu(\mathbf{x})$ specific growth rate was described by non-monotonic Haldane kinetics:

$$\mu(\mathbf{x}) = \mu_m \frac{x_1}{K_p + x_1 + \frac{x_1^2}{K_i}}, \tag{4.100}$$

where the K_p parameter relates the substrate level to the specific growth rate μ, together with the K_i inhibition parameter, which adds a negative substrate level effect. Nominal values of the model parameters are given in Table 4.3.

The identification data has been gained from normal operation of the process and five batches have been simulated to obtain process data. Note that these types of laboratory experiments are in reality expensive and time-consuming, which makes such a low number of experiments very realistic. Others, however, have used simulation studies with more than hundred simulated batches [86]*. Each batch experiment was run for max.* 16 *h, and all states were each measured for* 30 *min. For*

Table 4.3: Kinetic and operating parameter values

μ_m	2.1	h^-1
K_p	10	g/l
K_i	0.1	g/l
$Y_{x2/x1}$	0.47	$g\ DW/g$
m	0.29	$g\ DW/g\ h$
$x_{1,in}$	500	g/l
$x_3(0)$	7	l

every batch, a simple constant feeding strategy has been applied. In the five experiments we varied initial conditions $(x_1(0), x_2(0))$ *and the feeding rate u. To make the simulation experiments more realistic, gaussian white noise on the measurements of the substrate concentration* x_1 *and on the biomass concentration* x_2 *was considered; variances of the measurement errors are given by* $\sigma^2_{x_1} = 10^{-2}g^2/l^2$ *and* $\sigma^2_{x_2} = 6.25^{-4}g^2/l^2$*, respectively.*

In the next step, our spline-smoothing algorithm (see the Appendix) has been used to generate smoothed training data. Four knots were used to generate the splines; one at time zero and the others placed at 60%, 80% *and* 100% *of the time till substrate concentration became zero,* $x_1 = 0$. *These knots were determined such that they provide a good fit at small substrate concentrations realized at the end of the batches where the effect of the process nonlinearity is the biggest. From every batch,* 100 *data pairs were sampled from the spline-interpolation and the change of the state variables was estimated from the analytic derivatives of the splines. This procedure resulted in* 500 *data points from the five batches.*

Semi-Mechanistic Model of the Process. *Based on the collected identification data, we first identified a standard MIMO TS model with eight rules. The previously presented fuzzy clustering algorithm has been applied without prior knowledge. To illustrate the complexity of the obtained fuzzy model, we show the first rule:*

$$\begin{aligned} R_1 \quad &: \quad \textbf{If } x_1 \textit{ is } A_{1,1} \textbf{ and } x_2 \textit{ is } A_{1,2} \textbf{ then} \\ \dot{\mathbf{x}}_1 &= \underbrace{\begin{bmatrix} -0.0875 & -0.4927 & 0.0569 \\ 0.0381 & 0.0937 & -0.0013 \\ 0.0000 & 0.0000 & 0.0000 \end{bmatrix}}_{\hat{\mathbf{A}}_1} \begin{bmatrix} x_1 \\ x_2 \\ x_3 \end{bmatrix} \\ &+ \underbrace{\begin{bmatrix} 76.9633 \\ 0.9222 \\ 1.0000 \end{bmatrix}}_{\hat{\mathbf{b}}_1} u + \underbrace{\begin{bmatrix} -0.3093 \\ -0.0419 \\ 0 \end{bmatrix}}_{\hat{\mathbf{c}}_1} . \end{aligned} \tag{4.101}$$

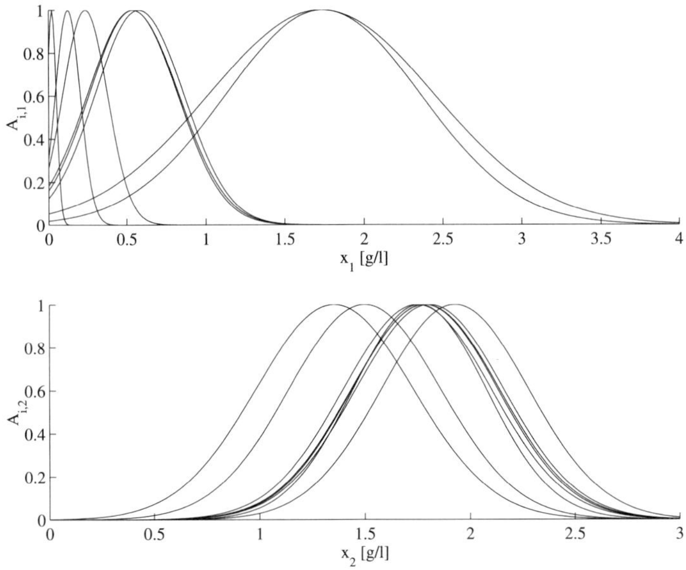

Figure 4.12: Membership functions obtained by standard TS fuzzy modelling approach.

The fuzzy sets $A_{1,i}$ *(Figure* 4.12*) were obtained from the clusters, i.e., the clusters could be analytically projected and decomposed to membership functions defined on the individual features, because the presented clustering method obtains axis-parallel clusters.*

It is clear that the fuzzy sets are more separated on the input x_1 *than on* x_2*. Application of Fisher Interclass separability therefore also showed that application of only the first input on antecedent part of the model was enough. This reflects our prior knowledge, since we know that the main source of the nonlinearity is given by the nonlinear relation between growth rate and substrate concentration* x_1*.*

Second, we identified a eight rule semi-mechanistic model based on the presented theory. The obtained rules are similar to (4.70) *and the membership functions are given by Figure* 4.13*.*

The main difference between the semi-mechanistic and the standard TS models is that the semi-mechanistic model is better interpretable, i.e., its outputs are not the derivatives of the state variables but the unknown growth rates. For example, the specific growth rate has physical meaning that can be interpreted by the bio-engineers and is given by Figure 4.14*.*

Moreover, the application of the presented clustering and model reduction tools resulted in a model where the growth rates are only the function of the substrate concentration, x_1 *even in case that additional prior knowledge was considered to*

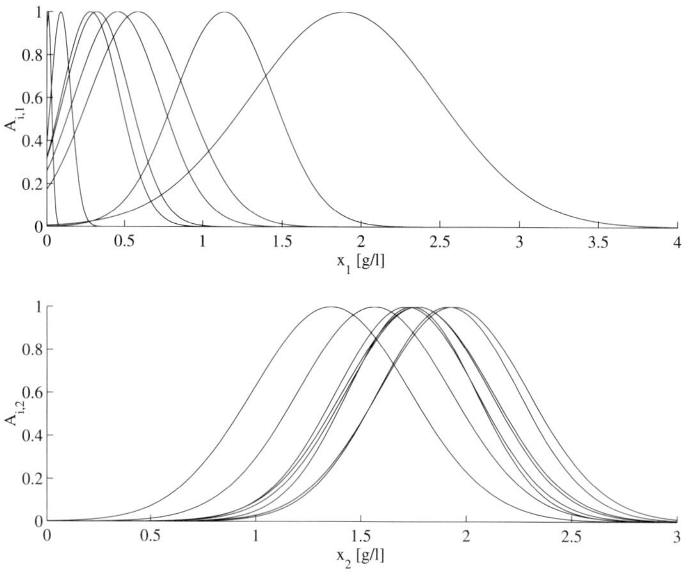

Figure 4.13: Membership functions obtained by semi-mechanistic fuzzy modelling approach.

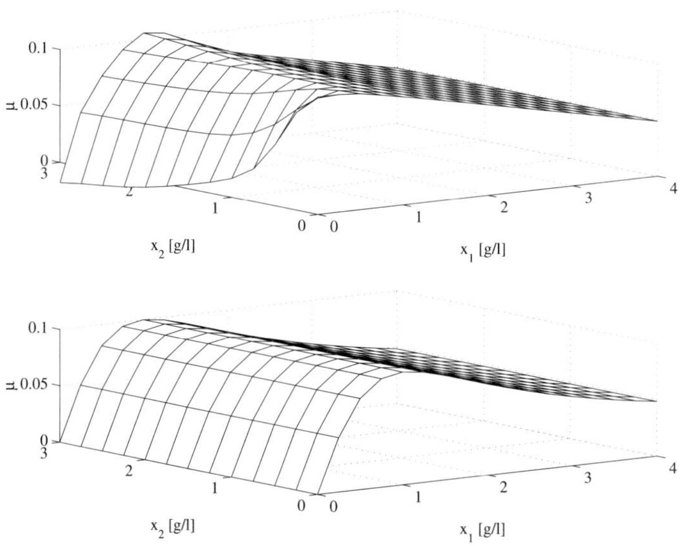

Figure 4.14: Estimated (top) and "real" (bottom) growth rate $\mu(\mathbf{x})$.

be unknown:

$$
\begin{aligned}
R_i \quad &: \quad \textbf{If } x_1 \textit{ is } A_{i,1} \textbf{ then} \\
\dot{\mathbf{x}}_i \quad &= \quad \begin{bmatrix} \frac{u}{x_3}(x_{1,in} - x_1) \\ -\frac{u}{x_3}(x_2) \\ u \end{bmatrix} \\
&+ \left(\underbrace{\begin{bmatrix} a_i^{1,1} & 0 & 0 \\ a_i^{2,1} & 0 & 0 \\ 0 & 0 & 0 \end{bmatrix}}_{\mathbf{A}_i} \begin{bmatrix} x_1 \\ x_2 \\ x_3 \end{bmatrix} + \underbrace{\begin{bmatrix} b_i^1 \\ b_i^2 \\ 1 \end{bmatrix}}_{\mathbf{b}_i} \right) \odot \begin{bmatrix} -x_2 \\ x_2 \\ 0 \end{bmatrix}. \qquad (4.102)
\end{aligned}
$$

The fuzzy part of the obtained semi-mechanistic fuzzy model can be inspected separately, e.g.,

$$
R_i \quad : \quad \textbf{If } x_1 \textit{ is } A_{i,1} \textbf{ then } \sigma(\mathbf{x})_i = a_i^{1,1} x_1 + b_i^1. \qquad (4.103)
$$

The predicted growth rates obtained by this simple model are shown in Figure 4.15.

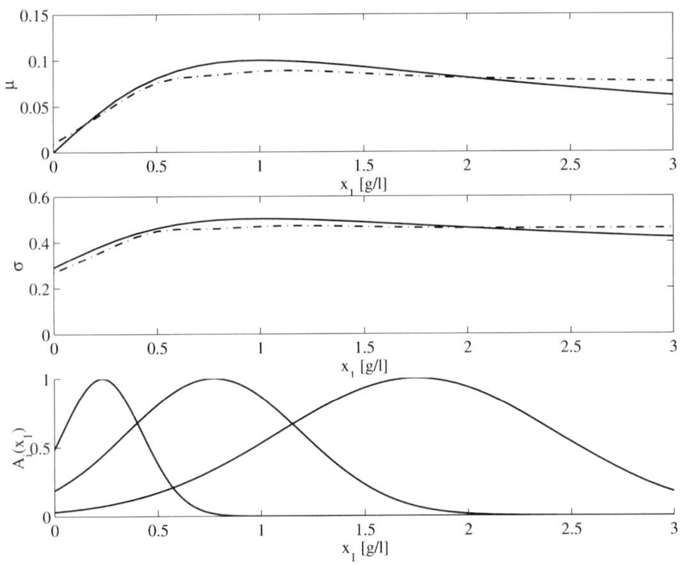

Figure 4.15: True and estimated kinetic model and the operating regions of the local models.

Finally, the obtained models have been validated on a number of additional batches that were not used for identification. A typical ballistic prediction is shown

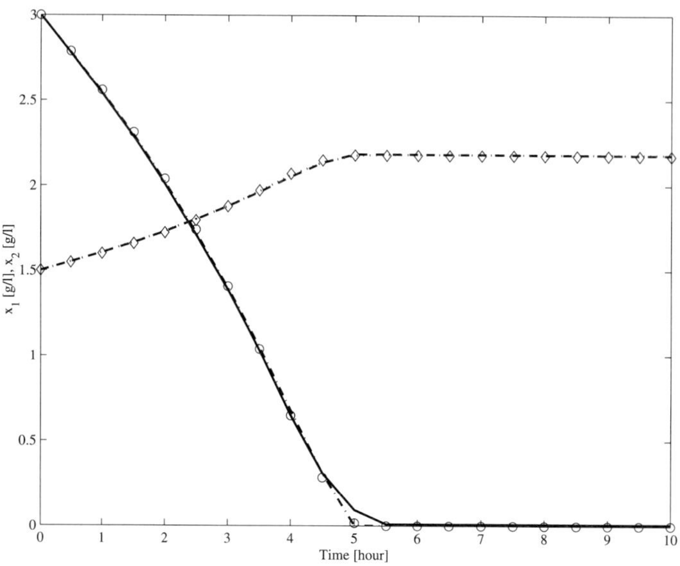

Figure 4.16: Simulation (ballistic prediction) from correct initial states for a typical batch and the "true" measured variables, marked with circles o.

in Figure 4.16, *and indicates that the prediction accuracy of the presented model is excellent. It is observed that the performance of the presented semi-mechanistic model is better than that of the complex TS fuzzy model which gives a worse prediction at low substrate concentrations where the effect of the nonlinearity is the biggest.*

Concluding this example, we have shown that interpretable TS-fuzzy and semi-mechanistic fuzzy models can be generated by using fuzzy clustering and rule-base reduction methods. Moreover, in case of structured semi-mechanistic fuzzy models, one can extract simple and interpretable fuzzy submodels that describe the unknown physical effects of the modeled system.

□

4.2.4 Conclusions

We have presented a semi-mechanistic modelling strategy to obtain compact and transparent models of process systems based on small datasets. The semi-mechanistic fuzzy identification method is based on white-box modelling combined with fuzzy model parts. This method is suitable to model systems where a priori knowledge is available and experimental data is insufficient to make a conventional fuzzy model. First, for identification of the semi-mechanistic fuzzy model a new fuzzy clustering method is presented. Here, clustering is achieved by the simultaneous identification of fuzzy sets defined on some of the scheduling variables and identification of the parameters of the local semi-mechanistic submodels. Second, application of orthogonal least squares and Fisher's interclass separability are presented to keep the models as compact as possible. Next, the overall modelling procedure is described by the development of a model for a biochemical process and we showed that the presented modelling approach was able to effectively identify compact and transparent models. The obtained semi-mechanistic fuzzy models are compared with classic TS fuzzy models and appeared to be more accurate and better interpretable.

4.3 Model Order Selection of Nonlinear Input-Output Models

4.3.1 Introduction

Most data-driven identification algorithms assume that the model structure is *a priori* known or that it is selected by a higher-level 'wrapper' structure-selection algorithm. Several information-theoretic criteria have been proposed for structure selection in linear dynamic input–output models. Examples of the classical criteria are the Final Prediction-Error (FPE) and the Akaike Information Criterion (AIC) [20]. Later, the Minimum Description Length (MDL) criterion developed by Schwartz and Rissanen was proven to produce consistent estimates of the structure of linear dynamic models [173]. With these tools, determining the structure of linear systems is a rather straightforward task.

However, relatively little research has been done into the structure selection for nonlinear models. In the paper of Aguirre and Billings [169], the concepts of term clusters and cluster coefficients are defined and used in the context of system identification. It is argued that if a certain type of term in a nonlinear model is spurious, the respective cluster coefficient is small compared with the coefficients of the other clusters represented in the model. In [168], this approach is used to the structure selection of polynomial models. In [82] an alternative solution to the model structure selection problem is introduced by conducting a forward search through the many possible candidate model terms initially and then performing an exhaustive all subset model selection on the resulting model. A backward search approach based on orthogonal parameter-estimation is also applied [1, 183].

As can be seen, these techniques are 'wrapped' around a particular model construction method. Hence, the result of the estimate can be biased due to the particular construction method used. To avoid this problem a 'model free' approach is followed where no particular model needs to be constructed in order to select the order of the model. The advantage of this approach is that this estimate is based on geometrical/embedding procedures and does not depend on the model representation that will be used a posteriori, i.e., the results would have a rather general character.

This is an important advantage, as the construction of a NARX model consists of the selection of many structural parameters which have significant effect on the performance of the designed model: e.g., the model order, type of the nonlinearity (Hammerstein or Wiener type system) [213], scheduling variables, number of neurons in a neural network, etc. The simultaneous selection of these structural parameters is a problematic task. The primary objective of this section is to decompose this complex problem by providing some useful guidance in selecting a tentative model order.

However, it should be borne in mind that there is no escape of performing a model-driven structure selection, once a certain model representation is chosen. For instance, suppose a model-free model order selection algorithm is used to determine the correct model order. If a neural network is used to model the process, the designer still needs to decide on the activation function, the number of nodes etc. Therefore, the model order selection method that will be presented definitely doesn't spare the user of having to go through some sort of structure selection. Indeed, if the orders of the nonlinear input-output model are well chosen, then structure selection will be significantly facilitated.

Deterministic suitability measures [49] and false nearest neighbor (FNN) algorithms [229] have already been proposed for data-based selection of the model order. These methods build upon similar methods developed for the analysis of chaotic time series [148]. The idea behind the FNN algorithm is geometric in nature. If there is enough information in the regression vector to predict the future output, then for any two regression vectors which are close in the regression space, the corresponding future outputs are also close in the output space. The structure is then selected by computing the percentage of false neighbors, i.e., vectors that violate the above assumption. A suitable threshold parameter must be specified by the user. For this purpose, heuristic rules have been proposed [49]. Unfortunately, for nonlinear systems the choice of this parameter will depend on the particular system under study [229]. The computational effort of this method also rapidly increases with the number of data samples and the dimension of the model.

To increase the efficiency of this algorithm, we present two clustering-based algorithms, which can be found in one of our previous works [85] as well. The main idea of these algorithms is the following. When the available input-output data set is clustered in the product space of the regressors and the model output, the obtained clusters will approximate the regression surface of the model. Although a clustering algorithm is utilized, the method does not assume that the data exhibit a cluster substructure. Clustering is used to detect clusters that are local linear approximations of the regression surface. In the first algorithm the threshold constant that is used to compute the percentage of the false neighbors is estimated from the shape of the obtained clusters. Departing from the theory behind the MDL algorithm, a new direct model structure selection algorithm is also developed. If the right input variables are used, because of the functional relationship between the regressors and the model output, the data are locally highly correlated and the obtained clusters are flat. In this way, the problem of determining the appropriate regressors is transformed into the problem of checking the flatness of the clusters, similarly to [26, 28]. The main advantage of the presented solution is that it is model-free. This means that no particular model needs to be constructed in order to select the model structure. There is also no need for finding the nearest neighbors for each data point, which is a computationally expensive task.

4.3.2 FNN Algorithm

Many non-linear static and dynamic processes can be represented by the following regression model

$$y_k = f(\boldsymbol{\phi}_k) \tag{4.104}$$

where $f(.)$ is a nonlinear function and $\boldsymbol{\phi}_k$ represents its input vector and $k = 1, \ldots, N$ represents the index of the kth available input-output data.

Among this class of models, the identification of discrete-time, Non-linear Auto-Regressive models with eXogenous inputs (NARX) is considered (see also Section 4.1). In the NARX model, the model regressors are past values of the process outputs y_k and the process inputs u_k.

$$\boldsymbol{\phi}_k = [y_{k-1}, \ldots, y_{k-n_a}, u_{k-1}, \ldots, u_{k-n_b}]^T \tag{4.105}$$

while the output of the model is the one-step ahead prediction of the process, y_k. The number of past outputs used to calculate y_k is n_a, and the number of past inputs is n_b. The values n_a and n_b are often referred to as model orders. The above SISO system representation can be assumed without a loss of generality since the extension to MISO and MIMO systems is straightforward.

The method of false nearest neighbors (FNN) was developed by Kennel [148] specifically for determining the minimum embedding dimension, the number of time-delayed observations necessary to model the dynamic behavior of chaotic systems. For determining the proper regression for input-output dynamic processes, the only change to the original FNN algorithm involves the regression vector itself [229].

The main idea of the FNN algorithm stems from the basic property of a function. If there is enough information in the regression vector to predict the future output, then any of two regression vectors which are close in the regression space will also have future outputs which are close in some sense. For all regression vectors embedded in the proper dimensions, for two regression vectors that are close in the regression space, their corresponding outputs are related in the following way:

$$y_k - y_j = df\left(\boldsymbol{\phi}_k^{n_a,n_b}\right)\left[\boldsymbol{\phi}_k^{n_a,n_b} - \boldsymbol{\phi}_j^{n_a,n_b}\right] + o\left(\left[\boldsymbol{\phi}_k^{n_a,n_b} - \boldsymbol{\phi}_j^{n_a,n_b}\right]\right)^2 \tag{4.106}$$

where $df\left(\boldsymbol{\phi}_k^{n_a,n_b}\right)$ is the Jacobian of the function $f(.)$ at $\boldsymbol{\phi}_k^{n_a,n_b}$.

Ignoring higher-order terms, and using the Cauchy-Schwarz inequality the following inequality can be obtained:

$$|y_k - y_j| \leq \|df\,(\phi_k^{n_a,n_b})\|_2 \left\|\phi_k^{n_a,n_b} - \phi_j^{n_a,n_b}\right\|_2 \tag{4.107}$$

$$\frac{|y_k - y_j|}{\left\|\phi_k^{n_a,n_b} - \phi_j^{n_a,n_b}\right\|_2} \leq \|df\,(\phi_k^{n_a,n_b})\|_2 \,. \tag{4.108}$$

If the above expression is true, then the neighbors are recorded as true neighbors. Otherwise, the neighbors are false neighbors.

Based on this theoretical background, the outline of the FNN algorithm is the following.

1. Identify the nearest neighbor to a given point in the regressor space. For a given regressor:

$$\phi_k^{n_a,n_b} = [y_{k-1}, \ldots, y_{k-n_a}, u_{k-1}, \ldots, u_{k-n_b}]^T$$

   find the nearest neighbor $\phi_j^{n_a,n_b}$ such that the distance d is minimized:

$$d = ||\phi_k^{n_a,n_b} - \phi_j^{n_a,n_b}||_2.$$

2. Determine if the following expression is true or false

$$\frac{|y_k - y_j|}{||\phi_k^{n_a,n_b} - \phi_j^{n_a,n_b}||_2} \leq \mathcal{R}$$

   where $\mathcal{R}$ is a previously chosen threshold value. If the above expression is true, then the neighbors are recorded as true neighbors. Otherwise, the neighbors are false neighbors.

3. Continue the algorithm for all times k in the data set.

4. Calculate the percentage of points in the data that have false nearest neighbors $J(n_a, n_b)$.

5. Continue the algorithm for increasing n_a and n_b using the percentage of false nearest neighbors dropping to some acceptably small number.

The FNN algorithm is sensitive to the choice of the $\mathcal{R}$ threshold. In [49] the threshold value was selected by trial and error method based on empirical rules of thumb, $10 \leq \mathcal{R} \leq 50$. However, choosing a single threshold that will work well for all data sets is an impossible task. In this case, it is advantageous to estimate $\mathcal{R}$ based on (4.108) using the maximum of the Jacobian, $\mathcal{R} = \max_k \|df\,(\phi_k^{n_a,n_b})\|$, as it was suggested by Rhodes and Morari [229].

Since this method uses the Jacobian of the identified models, the performance and the flexibility of these models can deteriorate the estimate of the model order. When the Jacobian is overestimated, the algorithm underestimates the order of the system. Contrary, if the model estimates smaller Jacobian than the real Jacobian of the system, the model order selection algorithm overestimates the order of the model. Hence, the modeler should be careful at the construction of the model used to estimate the Jacobian of the system (e.g., the model can be over or under parameterized, etc.). To increase the efficiency of the FNN based structure selection, a clustering-based algorithm will be introduced in the following section.

4.3.3 Fuzzy Clustering based FNN

The main idea of this section is the following. When the available input-output data set is clustered in the product space of the regressors and the model output and when the appropriate regressors are used, the collection of the obtained clusters will approximate the regression surface of the model as it was described in Section 1.6, see in Figure 1.17. In this case the clusters can be approximately regarded as local linearizations of the system and can be used to estimate $\mathcal{R}$.

Clusters of different shapes can be obtained by different clustering algorithms by using an appropriate definition of cluster prototypes (e.g., points vs. linear varieties) or by using different distance measures (see also Section 1.5 and Section 1.6 for more details). The Gustafson–Kessel clustering algorithm (Section 1.5.6) [108] has been often applied to identify Takagi–Sugeno fuzzy systems that are based on local linear models [26]. The main drawback of this algorithm is that only clusters with approximately equal volumes can be properly identified which constrain makes the application of this algorithm problematic for the task of this section. To circumvent this problem, in this section Gath–Geva algorithm (Section 1.5.7) is applied [18, 93].

The clustering algorithm has only one parameter: c, the number of the clusters. In general, the increase of c increases the accuracy of the model. However, to avoid overfitting and the excessive computational costs, it is recommended to determine the number of the clusters automatically. For this purpose various methods can be applied [26, 93] (see Section 1.7 for more details).

The applied validity measure is based on the hyper-volume index:

$$V_{fc} = \sum_{i=1}^{c} \det(\mathbf{F}_i) . \tag{4.109}$$

This index represents the volume of the clusters. When the nonlinear hyper-surface of the identification data is correctly approximated by the clusters, this volume should be small.

We scale this index by the volume of covariance matrix of the data

$$I = \frac{V_{fc}}{\det(\mathrm{cov}(\mathbf{X}))} . \tag{4.110}$$

Estimation of the $\mathcal{R}$ Threshold Coefficient

The collection of c clusters approximates the regression surface as it is illustrated in Figure 1.17. Hence, the clusters can be approximately regarded as local linear subspaces described by the cluster ellipsoids as shown in Figure 1.18 (Section 1.6). The smallest eigenvalues λ_{i,n_b+n_a+1} of the cluster covariance matrices $\mathbf{F}_i$ are typically in orders of magnitude smaller than the remaining eigenvalues [18, 26].

The eigenvector corresponding to this smallest eigenvalue, $\mathbf{t}^i_{n_b+n_a+1}$, determines the normal vector to the hyperplane spanned by the remaining eigenvectors of that cluster

$$(\mathbf{t}^i_{n_b+n_a+1})^T(\mathbf{x}_k - \mathbf{v}_i) = 0. \tag{4.111}$$

Similarly to the observation vector $\mathbf{x}_k = [\boldsymbol{\phi}_k^T\, y_k]^T$, the prototype vector is partitioned as $\mathbf{v}_i = \left[\left(\mathbf{v}_i^{\phi}\right)^T \; v_i^y\right]$ into a vector $\mathbf{v}^{\phi}$ corresponding to the regressor $\boldsymbol{\phi}_k$, and a scalar v_i^y corresponding to the output y_k. The smallest eigenvector is partitioned in the same way, $\mathbf{t}^i_{n_a+n_b+1} = \left[\left(\mathbf{t}^{i,\phi}_{n_a+n_b+1}\right)^T \; t^{i,y}_{n_a+n_b+1}\right]^T$. By using this partitioned vectors (4.111) can be written as

$$\left[\left(\mathbf{t}^{i,\phi}_{n_a+n_b+1}\right)^T \; t^{i,y}_{n_a+n_b+1}\right]^T \left([\boldsymbol{\phi}_k^T\, y_k]^T - \left[\left(\mathbf{v}_i^{\phi}\right)^T \; v_i^y\right]\right) = 0 \tag{4.112}$$

from which the parameters of the hyperplane defined by the cluster can be obtained:

$$y_k = \underbrace{\frac{-1}{t^{i,y}_{n_a+n_b+1}}\left(\mathbf{t}^{i,\phi}_{n_a+n_b+1}\right)^T}_{\mathbf{a}_i^T} \boldsymbol{\phi}_k + \underbrace{\frac{1}{t^{i,y}_{n_a+n_b+1}}\left(\mathbf{t}^{i}_{n_a+n_b+1}\right)^T \mathbf{v}_i}_{b_i} = \mathbf{a}_i^T\boldsymbol{\phi}_k + b_i. \tag{4.113}$$

Although the parameters have been derived from the geometrical interpretation of the clusters, it can be shown [26] that (4.113) is equivalent to the weighted total least-squares estimation of the consequent parameters, where each data point is weighted by the corresponding $\sqrt{\mu_{i,k}}$.

The main contribution of this section is that it suggests the application of an adaptive threshold function to FNN, which takes into account the nonlinearity of the system. This means, based on the result of the fuzzy clustering, for all input-output data pairs different $\mathcal{R}_k$ values are calculated. Therefore, the optimal value of $\mathcal{R}_k$ is $\mathcal{R}_k = \|df\,(\boldsymbol{\phi}_k^{n_a,n_b})\|$ and the $df\,(\boldsymbol{\phi}_k^{n_a,n_b})$ partial derivatives can be estimated based on the shape of the clusters from (4.113)

$$df\,(\boldsymbol{\phi}_k^{n_a,n_b}) \approx \sum_{i=1}^{c} \mu_{i,k} \frac{-1}{t^{i,y}_{n_a+n_b+1}}\left(\mathbf{t}^{i,\phi}_{n_a+n_b+1}\right)^T \tag{4.114}$$

the threshold can be calculated as

$$\mathcal{R}_k = \left\| \sum_{i=1}^{c} \mu_{i,k} \frac{-1}{t^{i,y}_{n_a+n_b+1}} \left(\mathbf{t}^{i,\phi}_{n_a+n_b+1} \right)^T \right\|_2 . \tag{4.115}$$

4.3.4 Cluster Analysis based Direct Model Order Estimation

In the previous section a new cluster analysis based approach to the adaptive choice of the $\mathcal{R}$ threshold value of the FNN algorithm has been presented. Based on the geometric idea behind this algorithm, in this section an alternative structure selection algorithm will be presented that does not require the time-consuming calculation of the nearest neighbors in the identification data.

The idea of this second algorithm is based on the well-known fact that in the absence of the observation noise, the covariance matrix of the identification data generated by a linear system has a zero eigenvalue with multiplicity s given by

$$s = 1 + \min(n_a - n_{a,l}, n_b - n_{b,l}) \tag{4.116}$$

when the selected model orders n_b and n_a are greater than or equal to the true orders of the linear system, i.e., $n_b \geq n_{b,l}$ and $n_a \geq n_{a,l}$ [56]. This relationship between the parameters n_a, n_b and the eigenvalues of the covariance matrix can be used to select the model order.

In [173] it has been shown that the widely applied Minimum Description Length (MDL) model order selection criterion can be expressed based on the smallest eigenvalue of the data covariance matrix:

$$J^{n_a,n_b}_{MDL} = \frac{N}{2} \log(\lambda_{\min}) + \frac{1}{2}(n_a + n_b) \log N. \tag{4.117}$$

Multiplying both sides by $2/N$ and combining the terms results in

$$\frac{2}{N} J^{n_a,n_b}_{MDL} = \log \left(\lambda_{\min} \left(N^{1/N} \right)^{n_a+n_b} \right). \tag{4.118}$$

As log(.) is monotonically increasing, in [173] a criterion containing exactly the same information as MDL has been proposed:

$$J^{n_a,n_b} = \lambda_{\min} \left(N^{1/N} \right)^{n_a+n_b} . \tag{4.119}$$

Since $N^{1/N} \approx 1$ for large N, one can see that the MDL criterion asymptotically provides the same information as the minimum eigenvalue of the covariance matrix. The advantage of the above formulation is that it can also be applied to noise-free data, in which λ_{min} is zero and where the logarithm in (4.117) thus cannot be calculated.

The utilized fuzzy clustering obtains local linear approximation of the nonlinear system, (4.119) can be easily modified for cluster-based model order estimation by weighting the values of this simplified cost functions calculated from the cluster covariance matrices with the a priori probability of the clusters:

$$J^{n_a,n_b} = \sum_{i=1}^{c} \alpha_i \lambda_{i,\min}. \tag{4.120}$$

Because the model order is determined by finding the number of past outputs n_a and past inputs n_b, the $J(n_a, n_b)$ indices form a table in these two dimensions. It is possible to find a 'global' solution (or solutions) for the model orders by computing the index over all values of n_a and n_b in a certain range and search for a rapid decrease of $J(n_a, n_b)$. The indices that have the smallest $J(n_a, n_b)$ relative to $J(n_a - 1, n_b)$ and $J(n_a, n_b - 1)$ represents the 'best' estimate of the model order. Hence, each row of the table is divided by the previous row to form a row ratio table, and each column is divided by the previous column to create a column ratio table. With the use of these ratios the model order can be determined:

$$[n_a, n_b] = \arg\min_{n_a,n_b} \left\{ \max \left(\frac{J(n_a, n_b)}{J(n_a-1, n_b)}, \frac{J(n_a, n_b)}{J(n_a, n_b-1)} \right) \right\}. \tag{4.121}$$

4.3.5 Application Examples

The presented cluster analysis based model order selection algorithms will be illustrated using three examples taken from the relevant literature [49] and [229]. In the first example the order of a linear system is estimated. This example shows that the utilized excitation signal has a big impact on the quality of the results. In the second example the order of a polymerization reactor model is estimated and the usage of a ratio table based on (4.121) is illustrated. The process considered in the third example is a third-order exothermic van der Vusse reaction in a cooled Continuously Stirred Tank Reactor (CSTR). This process is strongly nonlinear, so the difference between results highlights the necessity of the application of the presented methods. In this example neural networks with different model orders are also identified, and the performances of the models show good correlation with the cluster based indices of the model structures.

Example 4.7 (Direct model order selection based on GG clustering in case of a linear system). *The first system is a difference equation, where $n_a = 3$ and $n_b = 2$.*

$$y_k = 1.5y_{k-1} - 0.75y_{k-2} + 0.125y_{k-3} + u_{k-1} + u_{k-2}. \tag{4.122}$$

For this example algorithms based on Lipschitz numbers and false nearest neighbors provided different solutions in [49], and the model orders were chosen incorrectly. Based on our theoretical and experimental study related to the properties of the FNN algorithm, we have realized that this erroneous result is caused by the Pseudo Random Binary Signal (PRBS) used for the excitation of the system. As PRBS generates small amount of identification data with diverse control signal sequences, $u_{k-1}, \ldots, u_{k-n_b}$*, the ratio of the nearest neighbors will be less sensitive to the examined input order,* n_b*, of the model. Hence, as PRBS is unsuitable to generate identification data for nearest neighbors approaches, in this study the process has been excited by a signal that is random also in its amplitude.*

Table 4.4: FNN results for linear system data when $\mathcal{R}$ is obtained by fuzzy clustering

Input Delays (n_b)	*Output Delays* (n_a)				
% FNN	0	1	2	3	4
0	100.00	98.37	77.85	22.96	1.62
1	96.74	87.80	31.50	2.64	0.00
2	75.00	40.24	1.42	0.00	0.00
3	43.29	7.52	0.00	0.00	0.00
4	49.18	1.02	0.00	0.00	0.00

As Table 4.4 shows, the presented method gives correct estimate of the model order in the noise-free case. To obtain a more realistic identification problem, normally distributed noise with zero mean and different levels of standard deviation was added to the output of the process; and SNR defined to be the ratio of signal to noise variance. In this case, $J(n_a, n_b)$ *will not be zero if* n_a *and* n_b *are chosen as* $n_a \geq \hat{n}_a$ *and* $n_b \geq \hat{n}_b$*, but it will tend to remain relatively small and flat. In this point of view the presented approach also estimated the correct order of the model.*

□

Example 4.8 (Direct model order selection based on GG clustering in case of a continuous polymerization reactor). *In the second example, taken from [229], we use data generated by a simulation model of a continuous polymerization reactor. This model describes the free-radical polymerization of methyl methacrylate with azobisisobutyronitrile as an initiator and toluene as a solvent. The reaction takes place in a jacketed CSTR. Under some simplifying assumption, the first-principle model is given by:*

$$\begin{aligned}
\dot{x}_1 &= 10(6 - x_1) - 2.4568 x_1 \sqrt{x_2} \\
\dot{x}_2 &= 80u - 10.1022 x_2 \\
\dot{x}_3 &= 0.024121 x_1 \sqrt{x_2} + 0.112191 x_2 - 10 x_3 \\
\dot{x}_4 &= 245.978 x_1 \sqrt{x_2} - 10 x_4 \\
y &= \frac{x_4}{x_3}
\end{aligned}$$

The dimensionless state variable x_1 *is the monomer concentration, and* x_4/x_3 *is the number-average molecular weight (the output* y*). The process input* u *is the dimensionless volumetric flow rate of the initiator. For further information on this model and its derivation, see* [73]. *According to* [229], *we apply a uniformly distributed random input over the range* 0.007 *to* 0.015 *with the sampling time of* 0.2 s.

With four states, a sufficient condition for representing the dynamics is a regression vector that includes four delayed inputs and outputs. In this case, however, the system has two states that are weakly observable. This can be observed by linearizing the system and performing balanced realization, which shows that two of the Hankel singular values are larger than the remaining singular values. This week by observation leads to the system and can be approximated by a smaller input-output description [55]. *Obviously, the results may change depending on where the system is linearized. Although in this case such effects have not occurred, the local linear behavior of a nonlinear system can significantly vary, even if the system is analyzed around off-equilibrium operating points* [1, 196]. *The main advantage of the presented clustering based approach is that the clusters are the local linear approximations of the nonlinear system, so they can be directly used to estimate the operating regions and the orders of the local linear models* [18].

The presented clustering-based algorithm was applied to 960 *data points. The indices* $J(n_a, n_b)$ *obtained by using the direct model order estimation (see* (4.120)) *are given in Table* 4.5.

Table 4.5: Polymerization data: $J(n_a, n_b)$ values obtained based on the eigenvalues of 3 clusters.

Input lags (n_b)	*Output lags* (n_a)				
	0	1	2	3	4
0	-	5.55	4.84	4.80	4.81
1	4.28	1.28	0.54	0.43	0.42
2	1.23	<u>0.30</u>	0.44	0.41	0.37
3	0.37	0.34	0.31	0.33	0.34
4	0.29	0.35	0.30	0.32	0.32

With the use of (4.121), *the ratios of the* $J(n_a, n_b)$ *values were computed and tabulated in Table* 4.6. *One can see that the structure with* $n_a = 1$ *and* $n_b = 2$ *is clearly indicated. This result is in agreement with the analysis of Rhodes* [229] *who showed that a nonlinear model with these orders is appropriate.*

The clustering based false nearest neighbor (FNN) algorithm is also applied to the data, and the results are given in Table 4.7. *The model structure with* $n_a = 1$ *and* $n_b = 2$ *is indicated, which confirms the above results. The main drawback of the FNN algorithm, however, is that it requires demanding calculations of the nearest neighbors for each data point.*

Table 4.6: Polymerization data: Ratios obtained from Table 4.5.

Input lags (n_b)	*Output lags* (n_a) 1	2	3	4
1	0.30	0.42	0.80	0.98
2	<u>0.24</u>	1.47	0.95	0.90
3	1.13	0.91	1.06	1.03
4	1.21	0.97	1.07	1.00

Table 4.7: Polymerization data: results obtained with the FNN method.

Input lags (n_b)	*Output lags* (n_a) 0	1	2	3	4
0	100.00	99.59	92.32	53.64	0.40
1	99.46	69.54	10.24	0.94	0.27
2	73.18	<u>3.10</u>	2.69	0.40	0.00
3	8.76	0.81	0.13	0.00	0.00
4	0.54	0.00	0.00	0.00	0.00

Table 4.8 *shows results obtained for the linear ARX model structure. Note that for this particular process, the linear method also works satisfactorily, although the decrease is less sharp.*

Table 4.8: Polymerization data: results obtained with a linear model (smallest eigenvalue of the covariance matrix of the data).

Input lags (n_b)	*Output lags* (n_a) 0	1	2	3	4
0	-	19.65	16.62	14.98	14.04
1	10.02	3.33	2.14	2.00	1.99
2	2.91	<u>1.94</u>	1.93	1.91	1.87
3	1.93	1.91	1.82	1.81	1.80
4	1.88	1.82	1.81	1.75	1.75

We can allocate that this linear model-based method does not give conspicuously incorrect results, as it behaves similarly to the method presented in [229]*. The only difference is that the linear model-based approach applies the "average" gain of the system, while the method of Rhodes and Morari utilizes the maximal gain of the nonlinear system* [229]*. For highly nonlinear systems both approaches can induce large model order estimation error, as the linear model-based approach can over-, while the maximum gain-based approach can under-estimate the order of the system.*

In the following example the order of a strongly nonlinear system will be estimated, so the difference between results obtained by linear and clustering based approaches will highlight the necessity of the application of the presented methods.

□

Example 4.9 (Direct model order selection based on GG clustering in case of the Van der Vusse Reactor). *The process considered in this section is a third-order exothermic van der Vusse reaction in a cooled Continuously Stirred Tank Reactor (CSTR). It is a strongly nonlinear process with a non-minimum-phase behavior and input multiplicity. The model of the system is given by*

$$
\begin{aligned}
\dot{x}_1 &= -x_1 k_1 e^{-\frac{E_1}{x_3}} - x_1^2 k_3 e^{-\frac{E_3}{x_3}} + (x_{10} - x_1)\, u_1 \\
\dot{x}_2 &= x_1 k_1 e^{-\frac{E_1}{x_3}} - x_2 k_2 e^{-\frac{E_2}{x_3}} - x_2 u_1 \\
\dot{x}_3 &= -\frac{1}{\varrho c_p}\left[\Delta H_1 x_1 k_1 e^{-\frac{E_1}{x_3}} + \Delta H_2 x_2 k_2 e^{-\frac{E_2}{x_3}}\right. \\
&+ \left. \Delta H_3 x_1^2 k_3 e^{-\frac{E_3}{x_3}}\right] + (x_{30} - x_3)\, u_1 + \frac{u_2}{\varrho c_p V} \\
y &= x_2
\end{aligned}
$$

where x_1, x_{10} and x_2 are concentration of the components involved in the reaction, x_3 is the temperature in the reactor, u_1 is the dilution rate of the reactor, u_2 is the heat exchanged between the CSTR and the environment, and x_{30} is the temperature of the inlet stream. The parameters are given in Table 4.9.

Table 4.9: Parameter values for the Van der Vusse reactor.

V	$= 0.01\,\mathrm{l}$	ρ	$= 0.9342\,\mathrm{kg\,l^{-1}}$
k_1	$= 1.287{\cdot}10^{12}\,\mathrm{s^{-1}}$	ΔH_1	$= 4.2\,\mathrm{J\,mol^{-1}\,s^{-1}}$
k_2	$= 1.287{\cdot}10^{12}\,\mathrm{s^{-1}}$	ΔH_2	$=$ -11.0 $\mathrm{J\,mol^{-1}\,s^{-1}}$
k_3	$= 9.043{\cdot}10^{9}\,\mathrm{s^{-1}}$	ΔH_3	$=$ -41.85 $\mathrm{J\,mol^{-1}\,s^{-1}}$
E_1	$= 9758.3\,\mathrm{K}$	x_{10}	$= 5.10\,\mathrm{mol\,l^{-1}}$
E_2	$= 9758.3\,\mathrm{K}$	x_{30}	$= 378.05\,\mathrm{K}$
E_3	$= 8560.0\,\mathrm{K}$	c_p	$= 8560.0\,\mathrm{K}$

The input flow rate u_1 is chosen as the system's input while u_2 is kept constant at $u_2 = -1.1258\ J/s$ [261]. To estimate the model orders, 960 data points were used (see Figure 4.17).

The eigenvalue based results obtained for a linear model structure are given in Table 4.10. No sharp decrease of the tabulated values $J(n_a, n_b)$ can be observed, which makes it difficult to choose the model order.

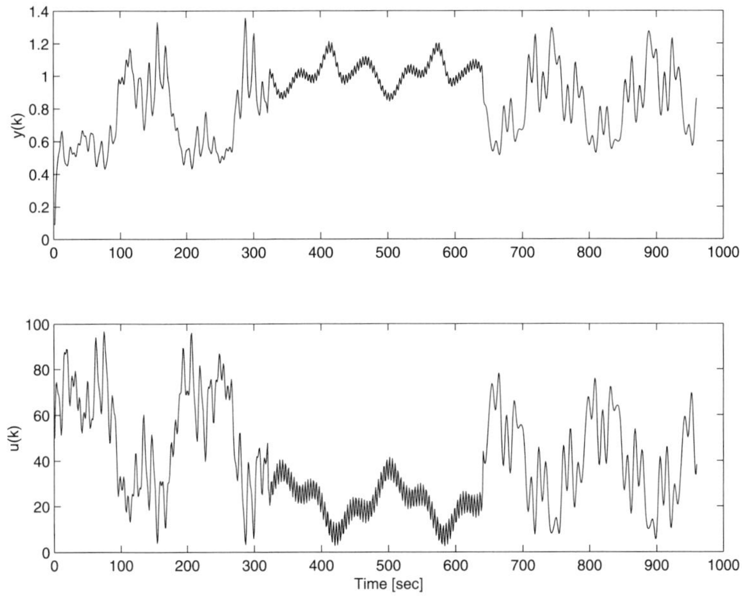

Figure 4.17: Input-output data of the van der Vusse reactor.

Table 4.10: Van der Vusse reactor: results obtained for the analysis of the smallest eigenvalue of the covariance matrix of the data (linear model).

Input lags (n_b)	*Output lags* (n_a)				
	0	1	2	3	4
0	–	4.19	1.79	0.98	0.58
1	5.85	2.80	1.53	0.97	0.56
2	4.94	1.88	0.60	0.54	0.43
3	2.75	1.45	0.53	0.25	0.22
4	1.61	1.28	0.43	0.25	0.12

The fuzzy clustering-based method, however, results in a sharp decrease of the cost function for the structure given by $n_b = 2$ *and* $n_a = 2$ *(see Table* 4.11*). A similar model structure was used in earlier papers on model based control of this process* [261].

This result is in a good agreement with the result of the FNN method (see Table 4.12*). The computational load of the FNN method, however, is much larger (three times in the case of this relatively small data set). As the computational complexity of the FNN method increases exponentially with the number of data samples, for larger data sets, this difference will become more pronounced.*

Table 4.11: Van der Vusse reactor: results obtained for the analysis of the smallest eigenvalues of 3 clusters.

Input lags (n_b)	*Output lags* (n_a)				
	0	1	2	3	4
0	–	1.36	0.56	0.26	0.14
1	1.43	0.52	0.35	0.22	0.13
2	0.98	0.41	<u>0.07</u>	0.07	0.07
3	0.66	0.30	0.06	0.04	0.03
4	0.42	0.19	0.07	0.04	0.02

Table 4.12: Van der Vusse reactor: results obtained with the false nearest neighbor method, where $\mathcal{R}$ has been determined based on cluster analysis.

Input lags (n_b)	*Output lags* (n_a)				
	0	1	2	3	4
0	1.00	0.99	0.67	0.18	0.006
1	0.99	0.54	0.18	0.025	0
2	0.70	0.23	<u>0.00</u>	0	0
3	0.33	0.08	0	0	0
4	0.05	0.01	0	0	0

In all of the examples the algorithm estimated the model order based on the shape of three clusters. In these cases the quality of the clustering was measured by the proposed hyper volume performance index (4.110). *When the number of clusters was increased, the* $I(n_a, n_b)$ *values did not decrease significantly. This shows that three clusters are enough to approximate the nonlinearity of the data. The fact that the same results were obtained when more clusters are used confirms the validity of this assumption.*

In order to verify the results of the presented algorithms, a three-layered neural network with six hidden neurons was identified and two types of error analysis were performed. The first determines the one-step ahead prediction error, where the actual past outputs and inputs are used to predict the next output of the system. The second analysis involves a simulated model that utilizes the inputs and only the initial conditions of the experimental time series.

In examining these results, it appears that large $J(n_a, n_b)$ *values correspond to larger prediction errors for both one-step predication and simulation (see Table* 4.13 *and Table* 4.14*). For the one-step-ahead prediction, the model error should always decrease when more terms are used. However, for the simulation results the error is not guaranteed to decrease as the model errors can highly accumulate during the simulation.*

Table 4.13: Van der Vusse reactor: mean square one-step ahead prediction errors obtained with neural networks.

Input lags (n_b)	*Output lags* (n_a)				
	0	1	2	3	4
0	-	0.0362	0.0141	0.0066	0.0051
1	0.0589	0.0079	0.0037	0.0025	0.0012
2	0.0366	0.0035	<u>0.0001</u>	0.0001	0.0001
3	0.0308	0.0018	0.0001	0.0001	0.0001
4	0.0272	0.0006	0.0001	0.0000	0.0001

Table 4.14: Van der Vusse reactor: mean square simulation prediction errors obtained with neural networks.

Input lags (n_b)	*Output lags* (n_a)				
	0	1	2	3	4
0	-	0.0362	0.0141	0.0066	0.0051
1	0.0589	0.0766	0.0359	0.0799	0.0335
2	0.0366	0.1033	<u>0.0029</u>	0.0214	12.9797
3	0.0308	0.0601	0.0124	0.0071	0.0047
4	0.0272	0.0073	0.0022	0.0007	0.0074

Without the presented methods it would be necessary to build many models with different model orders. After these models were built, the results would have to be analyzed. By utilizing the presented algorithms, time can be saved when performing the entire nonlinear identification process, and the results of the model order selection are not biased by the other structural parameters of the applied nonlinear model and the utilized parameter identification algorithm.

Although in this experiment a relatively small number of data (960 *data points*) *have been analyzed, there was a significant difference among the computational time of the identification of the neural networks, the FNN based and the direct cluster analysis based methods. The identification of the neural networks took approximately* 600 *seconds compared to the* 20 *seconds of the presented clustering based direct model order selection algorithm. Contrary, when the nearest neighbors were also calculated, the computational time increased to* 60 *seconds. As the computational complexity of FNN increases exponentially with the number of data* (*for* 2500 *data points the FNN took five times more time than the presented method* (500 *seconds*), *this difference is significant and shows the benefit of the presented cluster analysis based approach.*

□

4.3.6 Conclusions

A new approach for selecting the model order for nonlinear ARX models has been presented. Cluster analysis is first applied to the product space input-output data. The model orders are then estimated on the basis of the cluster covariance matrix eigenvalues. In the first approach the clusters are used to calculate the threshold parameter of the false nearest neighbors algorithm (FNN). The second approach directly estimates the order of the model based on the eigenvalues of the covariance matrices of the clusters. The presented eigenvalue based algorithm is several times faster than the false nearest neighbor method, as it does not require the time-consuming calculation of the nearest neighbors in the identification data set.

The main advantage of the presented approaches is that there is no need to extensively apply nonlinear model construction tools for the selection of the proper model order. However, it should be borne in mind that there is no escape of performing a model-driven structure selection, once a certain model representation is chosen. Indeed, if the orders of the nonlinear input-output model are well chosen, then structure selection will be facilitated a lot. Hence, the presented model order selection approach decreases the computational effort of model identification.

Numerical examples were given to illustrate the performance of the new technique and to compare it to other methods from the literature. These examples showed that for nonlinear systems linear model-based approaches can overestimate the order of the system, or that the result is not so clear as with the use of the presented cluster analysis based approach.

4.4 State-Space Reconstruction and Prediction of Chaotic Time Series

4.4.1 Introduction

Most of the developments in the field of nonlinear dynamics over the past century assume that one has a complete description of the dynamic system under consideration. The practical application of these results in principle requires the simultaneous measurement of all the state variables. Unfortunately, in many real applications one only has rough information about what these variables are, and there is certainly no hope of observing them all. Instead, one typically has a time series of one or more observables of the system, whose relationship to the state variables is at best uncertain. Fortunately, the remarkable result due to Takens [258] shows that the dynamics of an unknown deterministic finite-dimensional system can be reconstructed from a scalar time series generated by that system.

The state-space of the dynamic system is assumed to be a finite-dimensional compact manifold $\mathbf{M}$. It is assumed that the state of the system at time k, denoted $\mathbf{x}_k \in \mathbf{M}$, evolves according to $\mathbf{x}_{k+1} = \mathbf{f}(\mathbf{x}_k)$. The system is observed using a smooth measurement function $g : \mathbf{M} \rightarrow \mathbb{R}$ giving a scalar time series $y_k = \mathbf{g}(\mathbf{x}_k)$. The aim

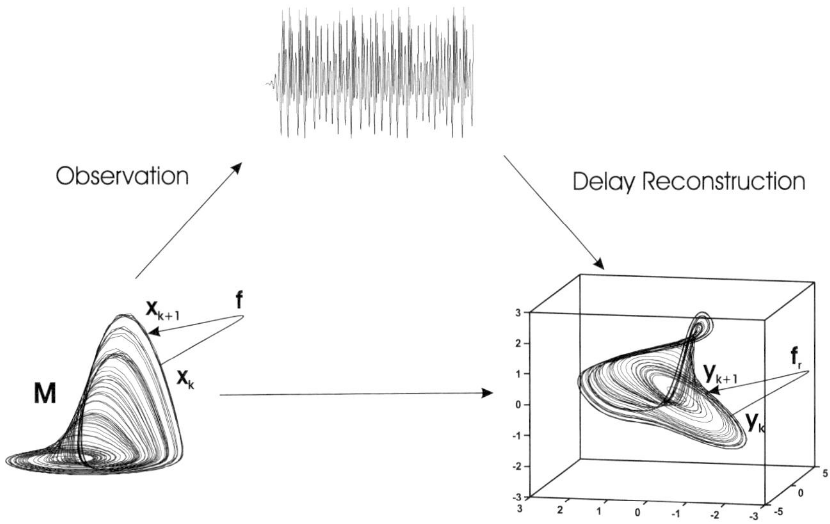

Figure 4.18: Scheme of delay embedding for deterministic systems.

of the so-called method of delays [258] is to reconstruct the state-space $\mathbf{M}$ and the dynamics $\mathbf{f}$ from the time series y_k. In Figure 4.18 the reconstructed version of the original system $\mathbf{f}$ is denoted by $\mathbf{f}_r$. Because $\mathbf{M}$ is high-dimensional and each component of y_k is only one-dimensional, it is clear that in order to obtain a suitable state-space, one has to group different elements of the time series. The most natural (but not the only) way of doing this is to put several successive values of y_k in a vector:

$$\mathbf{y}_k = \left[y_k, y_{k-\tau}, \ldots, y_{k-\tau(d_e-1)}\right]^T \tag{4.123}$$

where τ correspond to the interval on the time series that creates the reconstructed state-space (the lag time).

The number of components d_e is usually referred to as the *embedding dimension.* Although, the data samples are embedded in a d_e-dimensional space, they do not necessarily fill that space. The system defines a nonlinear hyper-surface in which the state variables reside. The dimension of this hyper-surface is referred to as *intrinsic*, *topological* or *local dimension*, d_l. The embedding and the local dimensions are not independent of each other. Takens proved that for an infinite number of data samples without noise the inequality $d_e \geq 2d_l + 1$ holds [258]. According to Sauer et al. [234], in practical cases $d_e \geq d_l + 1$ can be enough to reconstruct the original state-space.

The selection of the d_e embedding, and d_l local dimensions is a key step toward the analysis and prediction of nonlinear and chaotic time series. For the estimation of the embedding dimension Ataei et al. proposed a method based on a general polynomial autoregressive model [24]. Cao improved the False Neighbor method for this purpose [58]. Walker extended the so-called Wayland method to determine the minimum embedding dimension and compared the results with the Cao algorithm [278]. Ashkenazy proposed a method to measure local (fractal) dimension called generalized information dimension and applied it to chaotic time series [23]. Leok studied the effect of the noise level on the estimation of the local dimension in case of equatorial weather systems [54]. Camastra gave a survey of data dimensionality estimation methods in [57]. Jiang et al. [135] proposed a clustering-based approach to determining the embedding dimension. In their work a fuzzy c-means algorithm was used for clustering in the reconstruction space. The correct embedding dimension is estimated by comparing the number of iterations that are needed to achieve the termination tolerance of the clustering algorithm used to cluster the data in the reconstructed space of different dimensions. This method is based on a heuristic assumption, which assumes that the convergence curve of the clustering algorithm indicates the optimum embedding dimension when the number of clusters is identical to the discrete lag time.

A new clustering-based algorithm is introduced for the estimation of the dimensions of chaotic systems in this section, which is mainly based on [9]. The insights obtained by clustering can be well exploited in this case, namely the local dimension can be estimated based on cluster shape analysis. At the same time, the presented general Multiple Input-Multiple Output (MIMO) fuzzy modelling approach gives a model. The main advantage of the presented solution is that three tasks are simultaneously solved during clustering: selection of the embedding dimension, estimation of the intrinsic dimension, and identification of a model that can be used for prediction.

4.4.2 Clustering-based Approach to State-space Reconstruction

Overview of the Proposed Method

The idea of the presented method is illustrated in Figure 4.19. Clustering is applied in the reconstructed space defined by the lagged measured variables y_k. First, the lag time τ is chosen by using the average mutual information (**Step 1**).

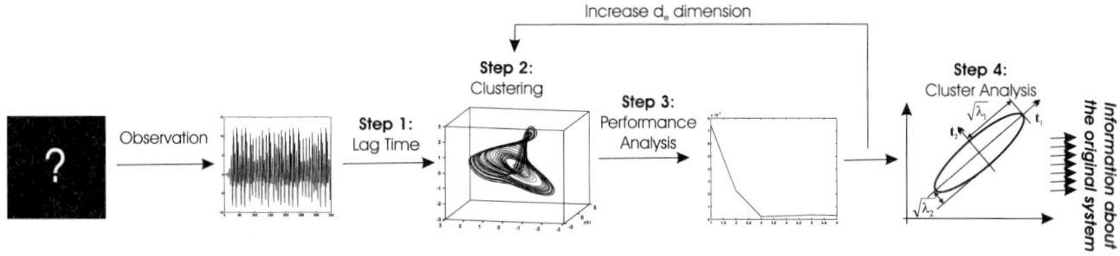

Figure 4.19: A scheme of the presented method.

The key step of the approach is the clustering of the data (**Step 2**). A model-based clustering algorithm has been developed for the identification of operating regimes and parameters of Gaussian mixture models using the Expectation Maximization (EM) method. The local models of the clusters are linear multi-input multi-output (MIMO) or multi-input single-output (MISO) models and the corresponding operating regions are represented by multivariate Gaussian functions. The embedding dimension is inferred from the one-step ahead prediction performance of the local models (**Step 3**). The intrinsic dimension of the reconstructed space is estimated by analyzing the eigenvalues of the fuzzy cluster covariance matrices (**Step 4**).

Selection of the Lag Time

The mutual information between two measurements is used to measure the generally nonlinear dependence of two variables. The mutual information for two measurements y_n and $y_{n+\tau}$ from the same time series is expressed by

$$I_n(\tau) = \log_2 \left[\frac{P(y_n, y_{n+\tau})}{P(y_n)P(y_{n+\tau})} \right] \tag{4.124}$$

where the individual probability densities, $P(y_n)$ and $P(y_{n+\tau})$, are equal to the frequency with which the data points y_n and $y_{n+\tau}$ appear in the time series, respectively. The frequency can be obtained directly through tracing the data points in the entire time series. The joint probability density, $P(y_n, y_{n+\tau})$, is obtained by counting the number of times the values of the pair are observed in the series. Average mutual information is computed for all data points in the following manner:

$$I(\tau) = \sum_{n=1}^{N-\tau} P(y_n, y_{n+\tau}) \log_2 \left[\frac{P(y_n, y_{n+\tau})}{P(y_n)P(y_{n+\tau})} \right]. \tag{4.125}$$

When $P(y_n, y_{n+\tau}) = P(y_n)P(y_{n+\tau})$ such that $I(\tau)$ approaches zero, the data points y_n and $y_{n+\tau}$ are completely independent. In practice the first minimum of the average mutual information function is chosen as the lag time [135].

Estimation of the Embedding Dimension based on Local Modelling

If the embedding dimension is properly chosen, the behavior of the time series in the reconstruction space can be described by a smooth nonlinear function, $\mathbf{f}_r$:

$$\mathbf{y}_{k+1} = \mathbf{f}_r(\mathbf{y}_k). \tag{4.126}$$

If this mapping is similar to the original dynamic system $\mathbf{f}$, it is required that $\mathbf{y}_{k+1} = \mathbf{y}_{p+1}$ whenever $\mathbf{y}_k = \mathbf{y}_p$. It means that the trajectories must not cross each other. These systems are deterministic, therefore the state-space would not be uniquely determined if the trajectories crossed each other. In terms of the time series this condition amounts to $\mathbf{y}_{n+d_e} = \mathbf{y}_{p+d_e}$ whenever $\mathbf{y}_k = \mathbf{y}_p, \ldots, \mathbf{y}_{k+d_e} = \mathbf{y}_{p+d_e}$ and is thus equivalent to the time series being predictable. The embedding dimension is determined by increasing the number of lagged outputs, d_e, and the analysis of the one-step ahead prediction error of the identified model.

The basic idea of the method can be described as follows: when the input vector of the model, the so-called regressor vector (4.123) contains enough information, i.e., it has the right dimension, the output of the model can be predicted with acceptably small error, i.e., the function $\mathbf{f}_r$ can be approximated accurately (4.126). The question is what the terms ‘acceptably small error’ and ‘being approximated accurately’ mean. They mean that the model error is due to the noise/measurement error present in the dataset and the imperfection of the modelling method rather than the lack of enough information. To satisfy this condition, the measurements should be accurate and an adequate modelling method should be chosen. If the dimension of the regressor vector is less than necessary, the output can be predicted with large error, but the model error is reduced with the increase of the information content/dimension of the regressor vector. When the reconstructed space with the proper embedding dimension is analyzed, the model error is reduced to an acceptable small value, and it will not be decreased significantly even with the increase of the dimension of the regressor vector, since extra terms in (4.123) do not increase the information content of the regressor. Consequently, a model should be identified by each dimension, and their accuracy should be stored and the changes should be analyzed. The accuracy of a model can be measured by the one-step ahead prediction error by chaotic systems because more than one step ahead prediction is not possible according to the nature of that kind of systems. The dimension showing a ‘knee’ in the model error, i.e., a drastic decrease, with no significant decrease by bigger dimensions, is the proper embedding dimension d_e. As can be seen, the model identification stands in the center of the method.

To identify a model that can be used for prediction of a time series global, local and semi-local methods can be used [176]. One of the main disadvantages of global methods is that a new sample pair may change the approximation function everywhere. Local interpolation overcomes this drawback by utilizing only a limited number of neighboring samples. There are two major classes of local methods, those applying neighbor samples directly in the prediction, and those fitting a function locally to the neighbors. The latter method fits a set of surfaces to the measurement points. These local surfaces can be hyperplanes but polynomials of higher degrees may also be used. For chaotic time series, this local approach was first used in [84]. Other works applying local methods are, e.g., [60] and [142]. The modelling framework that is based on combining a number of local models, where each local model has a predefined operating region in which it is valid, is called *operating regime based model* [197]. This approach is advantageous in the modelling of complex nonlinear systems, since it may not be possible to find a model that is universally applicable to describe the unknown MIMO system $\mathbf{f}_r(.)$, while with the application of the divide and conquer paradigm it is possible to decompose the complex problem into a set of smaller identification problems where standard linear models give satisfactory performance.

This operating-regime based model is formulated as follows:

$$\mathbf{y}_{k+1} = \sum_{i=1}^{c} \beta_i(\mathbf{y}_k) \underbrace{(\mathbf{A}_i \mathbf{y}_k + \mathbf{b}_i)}_{f_i} = \sum_{i=1}^{c} \beta_i(\mathbf{y}_k) \left[\mathbf{y}_k^T \; 1\right] \boldsymbol{\theta}_i^T = \sum_{i=1}^{c} \beta_i(\mathbf{y}_k) f_i(\mathbf{y}_k, \boldsymbol{\theta}_i) \tag{4.127}$$

where the function $\beta_i(\mathbf{y}_k)$ describes the operating region, and $\boldsymbol{\theta}_i = [\mathbf{A}_i \, \mathbf{b}_i]$ is the parameter matrix of the ith local model (rule). The output of the ith local model is denoted by $\mathbf{y}_{k+1}^i = f_i(\mathbf{y}_k, \boldsymbol{\theta}_i)$. In this section, it is assumed that β_i is known, in Section 4.4.2 this function will be computed through fuzzy clustering.

The main advantage of this framework is its transparency. The operating regions of the local models can be represented by fuzzy sets [31]. This representation is appealing, since many systems change their behavior smoothly as a function of the operating point, and the soft transition between the regimes introduced by the fuzzy set representation captures this feature in a natural way.

The output of this MIMO model is the vector $\mathbf{y}_{k+1} = [y_{k+1}, y_{k-\tau+1}, \ldots, y_{k-\tau(d_e-1)+1}]^T$. As only the first element y_{k+1} is unknown, another possible method for the state-space reconstruction problem is the Multiple Input-Single Output (MISO) modelling approach. In this case, only y_{k+1} is estimated. Both methods are used in the examples in Section 4.4.3 because MIMO approach gives not only a prediction error value but also takes into account the error covariance matrices of each local model, and therefore gives an insight into the reconstructed space.

The output of the MIMO and MISO model is linear in the parameters. Therefore, these parameters can be estimated from the data by linear least-squares techniques (see also Section 1.6). The N identification data pairs are arranged in the following regressor $\boldsymbol{\phi}$ and regressand $\mathbf{Y}$ matrices where they contain the same terms, $\mathbf{y}_k$, just shifted by one sample time.

$$\boldsymbol{\phi} = [\mathbf{y}_1, \mathbf{y}_2, \ldots, \mathbf{y}_N]^T \tag{4.128}$$

$$\mathbf{Y} = [\mathbf{y}_2, \mathbf{y}_3, \ldots, \mathbf{y}_{N+1}]^T \tag{4.129}$$

$$\mathcal{B}_i = \begin{bmatrix} \beta_i(\mathbf{y}_1) & 0 & \cdots & 0 \\ 0 & \beta_i(\mathbf{y}_2) & \cdots & 0 \\ \vdots & \vdots & \ddots & \vdots \\ 0 & 0 & \cdots & \beta_i(\mathbf{y}_N) \end{bmatrix}. \tag{4.130}$$

By using this notation, the weighted least squares solution of $\boldsymbol{\theta}_i$ is:

$$\boldsymbol{\theta}_i = \left[\boldsymbol{\phi}^T \mathcal{B}_i \boldsymbol{\phi}\right]^{-1} \boldsymbol{\phi}^T \mathcal{B}_i \mathbf{Y}. \tag{4.131}$$

As this method forces the local linear models to fit the data locally, it does not give an optimal model in terms of a minimal global prediction error, but it ensures that the fuzzy model is interpretable as a Linear Parameter Varying (LPV) system [6].

Cluster Analysis in the Reconstructed Space

The bottleneck of the data-driven identification of operating regime based models is the identification of the functions that can represent the operating regimes of the models (membership functions in the terminology of fuzzy models), which requires nonlinear optimization. In this section Gaussian functions are used to represent the operating regimes of the linear models:

$$\beta_i(\mathbf{y}_k) = \frac{\exp\left(-\frac{1}{2}(\mathbf{y}_k - \mathbf{v}_i)^T \mathbf{F}_i^{-1}(\mathbf{y}_k - \mathbf{v}_i)\right)}{\sum_{j=1}^{c} \exp\left(-\frac{1}{2}(\mathbf{y}_k - \mathbf{v}_j)^T \mathbf{F}_j^{-1}(\mathbf{y}_k - \mathbf{v}_j)\right)} \tag{4.132}$$

where $\mathbf{v}_i$ is the center and $\mathbf{F}_i$ is the covariance matrix of the multivariate Gaussian function.

A new clustering-based technique for the identification of these parameters is presented. The objective is to partition the identification data $\mathbf{Y} = [\mathbf{y}_1, \ldots, \mathbf{y}_N]^T$ into c clusters to reveal the underlying structure of the data. The patterns belong to clusters with degrees that are in inverse proportion to the distance from the respective cluster prototypes. The basic idea of the presented algorithm is to define the cluster prototype such that it locally approximates the MIMO function $\mathbf{y}_{k+1} = \mathbf{f}_r(\mathbf{y}_k)$. In this way, the algorithm simultaneously partitions the data (i.e., identifies the operating regimes of the local models) and determines the cluster prototypes (i.e., identifies the local model parameters). The fuzzy partition is represented by the $\mathbf{U} = [\mu_{i,k}]_{c\times N}$ matrix, where the element $\mu_{i,k}$ represents the membership degree of $\mathbf{y}_k$ in cluster i.

The clustering is based on minimizing the following cost function:

$$J(\mathbf{Y}, \mathbf{U}, \eta) = \sum_{i=1}^{c}\sum_{k=1}^{N} (\mu_{i,k})^m \, d^2(\mathbf{y}_{k+1}, \mathbf{y}_k, \eta_i). \tag{4.133}$$

where the squared distance $d^2(\mathbf{y}_{k+1}, \mathbf{y}_k, \eta_i)$ is given by the probability that the data belong to a given cluster:

$$\frac{1}{d^2(\mathbf{y}_{k+1}, \mathbf{y}_k, \eta_i)} = p(\eta_i) p(\mathbf{y}_k|\eta_i) p(\mathbf{y}_{k+1}|\mathbf{y}_k, \eta_i) \tag{4.134}$$

$$= p(\eta_i) \underbrace{\frac{1}{(2\pi)^{\frac{d_e}{2}}\sqrt{\det(\mathbf{F}_i)}} \exp\left(-\frac{1}{2}(\mathbf{y}_k - \mathbf{v}_i)^T \mathbf{F}_i^{-1}(\mathbf{y}_k - \mathbf{v}_i)\right)}_{p(\mathbf{y}_k|\eta_i)}$$

$$\underbrace{\frac{1}{(2\pi)^{\frac{d_e}{2}}\sqrt{\det(\mathbf{P}_i)}} \exp\left(-\frac{1}{2}(\mathbf{y}_{k+1} - \mathbf{y}^i_{k+1})^T \mathbf{P}_i^{-1}(\mathbf{y}_{k+1} - \mathbf{y}^i_{k+1})\right)}_{p(\mathbf{y}_{k+1}|\mathbf{y}_k, \eta_i)}.$$

The term $p(\eta_i)$ represents the *a priori* probability of the ith cluster defined by the η_i set of parameters (see equation (4.139) below). The Gaussian distribution $p(\mathbf{y}_k|\eta_i)$ defines the domain of influence of a cluster (see (4.132)). The third term is based on the performance of the local linear models where $\mathbf{P}_i$ is the weighted covariance matrix of the modelling error of the ith local model.

The above distance function can be seen as the geometric mean of the distance measure of Gath–Geva [93] (the first two terms) and the Fuzzy c-Regression (FCR) [113] (the last term) algorithms. Gath–Geva algorithm takes into account the distribution of the data, it is able to reveal clusters with arbitrary ellipsoidal shape and size, but it is not able to fit models to data. That is why the third term is introduced. However, the FCR distance measure is not sufficient in itself, either, because the same linear model (the same hyperspace in an n-dimensional space) happens to be suited for different regions of the input domain, therefore the local linear model would not be 'compact' and its operating regime could not be described as a (Gaussian) function. To avoid this, the Gath–Geva distance measure should also be taken into account as can be seen above.

The minimization of the functional (4.133) represents a non-linear optimization problem which is subject to the constrains (1.12), (1.13) and (1.14). It can be solved by using a variety of methods. The most popular one is the alternating optimization (AO), which is formulated as shown in Algorithm 4.4.1.

The main advantage of the presented clustering-based approach is that the local dimension can be estimated by the analysis of the shape of the clusters, as presented in the following section.

Estimation of the Local Dimension

Local (or topological) dimension of the reconstructed state-space is the basis dimension of the local linear approximation of the hypersurface on which the data resides, i.e., the tangent space [57]. The most popular algorithms to estimate the local dimension are Fukunaga-Olsen's algorithm [91], Near-Neighbor algorithms [217], Topological Representing Networks [189], different projection techniques, and fractal-based methods. A method is presented that exploits the shape of the clusters, and our results will be compared with the so-called Correlation dimension [102].

In the literature on nonlinear dynamics, many definitions of fractal dimensions have been proposed. Box-Counting (d_{BC}) and Correlation dimensions (d_C) are the most popular ones. The d_{BC} of a set $\mathbf{Y}$ is defined as follows. If $\nu(r)$ is the number of the boxes of size r to cover $\mathbf{Y}$, then d_{BC} is

$$d_{BC} = \lim_{r \to 0} \frac{\ln(\nu(r))}{\ln(1/r)}. \tag{4.135}$$

The Box-Counting dimension can only be computed for low-dimensional spaces because the algorithmic complexity grows exponentially with the set dimensionality. A good alternative to the Box-Counting dimension is the Correlation dimension. Due to its computational simplicity, the Correlation dimension has been successfully used to estimate the dimension of attractors of dynamical systems.

Algorithm 4.4.1 (Modified Gath–Geva Clustering for MIMO Modelling).

Initialization

Given a set of data $\mathbf{Y}$ *specify the number of clusters, c, choose a weighting exponent* $1 < m < \infty$ *(usually* $m = 2$*) and a termination tolerance* $\epsilon > 0$*. Initialize the partition matrix* $\mathbf{U} = [\mu_{i,k}]_{c\times N}$ *randomly in such a way that the constrains hold or use the partition matrix given by another clustering method such as fuzzy c-means or Gustafson–Kessel algorithm.*

Repeat *for* $l = 1, 2, \ldots$

Calculate the parameters of the clusters

- *Centers of the membership functions*

$$\mathbf{v}_i^{(l)} = \frac{\sum_{k=1}^{N} \left(\mu_{i,k}^{(l-1)}\right)^m \mathbf{y}_k}{\sum_{k=1}^{N} \left(\mu_{i,k}^{(l-1)}\right)^m}, \ 1 \le i \le c \tag{4.136}$$

- *Fuzzy covariance matrices of the Gaussian membership functions:*

$$\mathbf{F}_i = \frac{\sum_{k=1}^{N} \left(\mu_{i,k}^{(l-1)}\right)^m (\mathbf{y}_k - \mathbf{v}_i)(\mathbf{y}_k - \mathbf{v}_i)^T}{\sum_{k=1}^{N} \left(\mu_{i,k}^{(l-1)}\right)^m} \quad 1 \le i \le c. \tag{4.137}$$

- *Parameters of the local models are computed by (4.131) where the elements of the* $\mathcal{B}_i$ *matrix are equal to the membership:* $\beta_i(\mathbf{y}_k) = \mu_{i,k}$.
- *Covariance of the modelling errors of the local models:*

$$\mathbf{P}_i = \frac{\sum_{k=1}^{N} (\mathbf{y}_{k+1} - f_i(\mathbf{y}_k, \boldsymbol{\theta}_i))(\mathbf{y}_{k+1} - f_i(\mathbf{y}_k, \boldsymbol{\theta}_i))^T \left(\mu_{i,k}^{(l-1)}\right)^m}{\sum_{k=1}^{N} \left(\mu_{i,k}^{(l-1)}\right)^m}. \tag{4.138}$$

- A priori *probability of the cluster*

$$p(\eta_i) = \frac{1}{N} \sum_{k=1}^{N} \mu_{i,k}^{(l-1)}. \tag{4.139}$$

Compute the distance $d^2(\mathbf{y}_{k+1}, \mathbf{y}_k, \eta_i)$ **by** (4.134) **and update the partition matrix**

$$\mu_{i,k}^{(l)} = \frac{1}{\sum_{j=1}^{c} \left(d(\mathbf{y}_{k+1}, \mathbf{y}_k, \eta_i)/d(\mathbf{y}_{k+1}, \mathbf{y}_k, \eta_j)\right)^{2/(m-1)}}, \quad 1 \le i \le c, \ 1 \le k \le N. \tag{4.140}$$

until $||\mathbf{U}^{(l)} - \mathbf{U}^{(l-1)}|| < \epsilon$.

The Correlation dimension is defined as follows. Let $\mathbf{Y} = [\mathbf{y}_1, \mathbf{y}_2, \ldots, \mathbf{y}_N]^T$ be a set of points of cardinality N. Define the correlation integral $C_m(r)$ as

$$C_m(r) = \lim_{N\to\infty} \frac{2}{N(N-1)} \sum_{i=1}^{N} \sum_{j=i+1}^{N} I(\|\mathbf{x}_j - \mathbf{x}_i\| \leq r) \tag{4.141}$$

where $I(\cdot)$ is the indicator function ($I(\cdot)$ is 1 if condition $(\cdot)$ holds, 0 otherwise). The Correlation dimension d_C is

$$d_C = \lim_{r\to 0} \frac{\ln(C_m(r))}{\ln(r)}. \tag{4.142}$$

It can be proven that both dimensions are special cases of the generalized Renyi dimension [57].

The method presented in this section is able to determine simultaneously the embedding and the local dimension based on the result of the clustering. The local dimension can be estimated based on the shape of the clusters because the collection of c clusters approximate the tangent space. Hence, the clusters can be approximately regarded as local linear subspaces described by the cluster ellipsoids (covariance matrices).

To detect linear subspaces Principal Component Analysis (PCA), i.e., eigenanalysis, of the fuzzy covariance matrices has been used. The eigenvector associated with the largest eigenvalue has the same direction as the first principal component. The eigenvector associated with the second largest eigenvalue determines the direction of the second principal component. The first principal component accounts for as much of the variability in the data as possible, and each succeeding component accounts for as much of the remaining variability as possible. Hence, the first few biggest eigenvalues are much bigger than the remaining ones, and they cover the greatest part of the variance of the data set (see also Section 1.6 and Section 2.1.1). In general, the smallest eigenvalues of the cluster covariance matrices, $\mathbf{F}_i$, are typically in orders of magnitude smaller than the remaining ones [18, 26].

The local dimension is estimated based on the sum of the eigenvalues weighted by the *a priori* probability of the clusters:

$$J^{d_e,j} = \sum_{i=1}^{c} p(\eta_i)\lambda_{i,j} \tag{4.143}$$

where $\lambda_{i,j}$ is the jth biggest eigenvalue of the ith covariance matrix. The local dimension is determined by computing the index over all values of j, $j = 1, \ldots, d_e$ in case of the proper embedding dimension chosen beforehand based on the one-step-ahead prediction errors. The local dimension is selected as a cutoff point on a sharp change in the slope of the graph or such a way that

$$\frac{\sum_{j=1}^{k} J^{d_e,j}}{\sum_{j=1}^{d_e} J^{d_e,j}} > \text{threshold}. \tag{4.144}$$

This method is often referred to as screeplot analysis, which is commonly used to estimate the proper dimension of Principal Component Analysis models in the literature, e.g., in [251]. Srinivasan et al. used 0.93 as the threshold, but it can be replaced by another number, based on domain knowledge and the expected noise levels.

4.4.3 Application Examples and Discussion

The presented approach has been tested on several higher-dimensional chaotic time series and gave convincing results. In each example, the states of the original system were reconstructed from 15000 samples of the first state variable, $y = x_1$.

According to our experience, the presented approach is quite robust with respect to the choice of the clustering algorithm parameters. In all examples, the number of clusters is $c = 20$, the termination tolerance $\epsilon = 10^{-4}$ and the weighting exponent $m = 2$. The applied threshold is equal to 0.93 according to [251] (see (4.144)).

Example 4.10 (State-space reconstruction of the Rössler attractor). *The following three differential equations define this system:*

$$\dot{x_1} = -(x_2 + x_3) \tag{4.145}$$

$$\dot{x_2} = x_1 + a x_2 \tag{4.146}$$

$$\dot{x_3} = b + x_3(x_1 - c). \tag{4.147}$$

The initial conditions are $x_1(0) = 0$, $x_2(0) = 0$ *and* $x_3(0) = 0$ *and the parameters used are* $a = 0.38$, $b = 0.3$ *and* $c = 4.5$, *according to* [135]. *Under these conditions, the trajectories are depicted as in Figure* 4.20. *The variable* $y = x_1$ *is shown in the time domain in Figure* 4.21. *The lag time was* 20τ *in this case where* τ *is the sampling time* $\tau = 0.05$. *The lag time was chosen as the first minimum of the average mutual information function* [87, 135].

Figure 4.22 *shows that the proposed index based on the one-step-ahead prediction error correctly reflects the embedding dimension of the original three-dimensional system, for both the MIMO and MISO method. After the embedding dimension has been chosen, in this case it is* $d_e = 3$, *the local dimension,* d_l, *can be estimated based on the screeplot (see Section* 4.4.2). *In Figure* 4.22 *the subfigures* (c) *and* (d) *show that the local dimension is equal to* 2 *because the first two eigenvalues weighted by the* a priori *probability of the clusters contain* 99.8% *of the total variance. The Correlation dimension of the analyzed* 15000 *data is* $d_C = 1.9838$ *based on the result of the TSTOOL Toolbox, which is a software package for signal processing with emphasis on nonlinear time-series analysis* [191]. *Since the correlation dimension is a global measure of the local dimension (the calculation of the correlation dimension does not use the local approximation of the reconstruction*

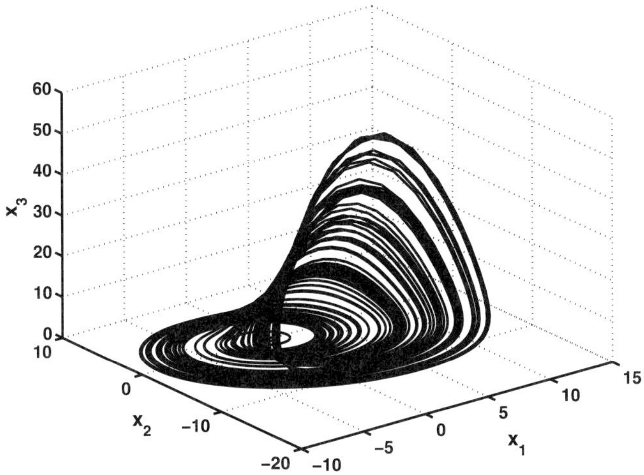

Figure 4.20: Trajectories in the original state-space in case of the Rössler system.

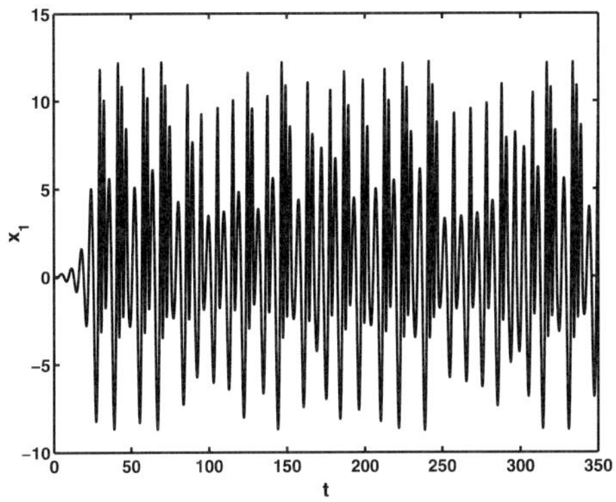

Figure 4.21: The variable $y = x_1$ in time in case of the Rössler system.

space), the agreement of the results of the two approaches with the visual inspection of phase space (see Figure 4.23) *indicate that the presented method gives good results in terms of the estimation of the dimensionality of the system.*

Furthermore, the identified MIMO and MISO models give excellent prediction performance (see Figure 4.23), *the original and the predicted data take the same subspace. The predicted data were generated by a free run simulation of* 5000 *samples based on the identified MIMO model. It can be seen that although the system is chaotic, the trajectories are similar, so the approximation is very good.*

□

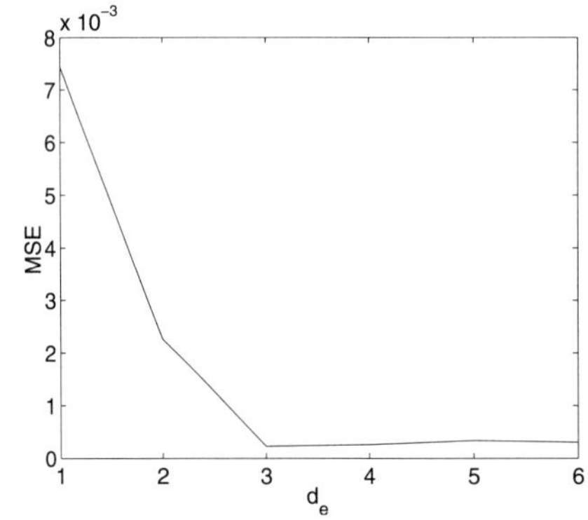

(a) Estimation of the embedding dimension d_e based on MIMO approach.

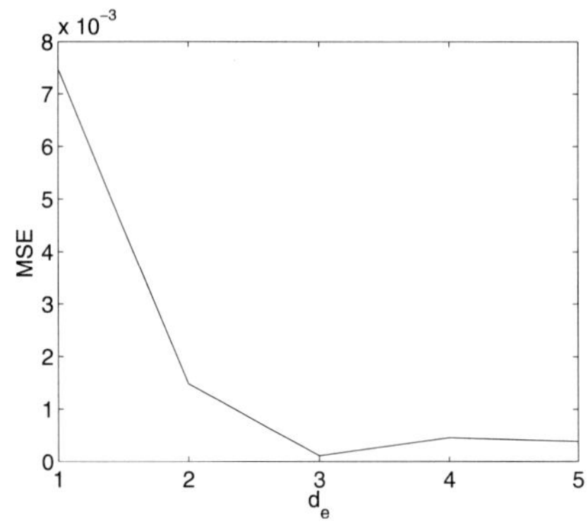

(b) Estimation of the embedding dimension d_e based on MISO approach.

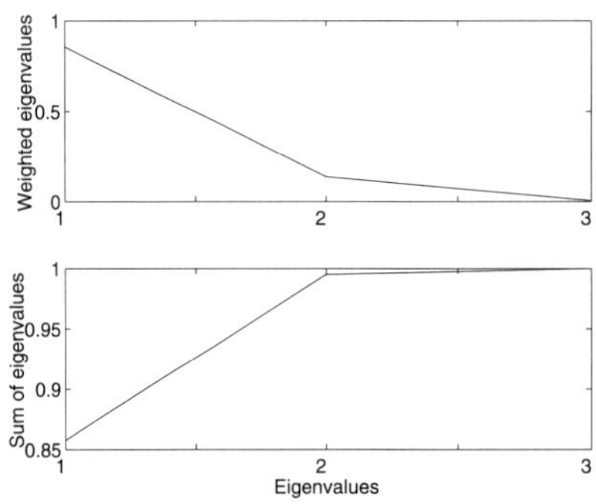

(c) Estimation of the local dimension d_l based on MIMO approach for $d_e = 3$.

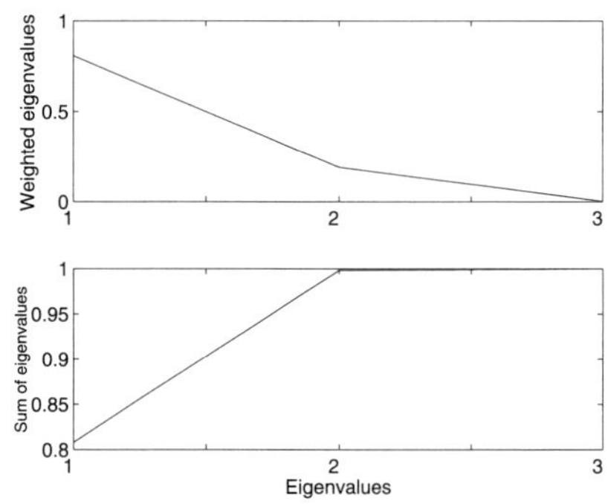

(d) Estimation of the local dimension d_l based on MISO approach for $d_e = 3$.

Figure 4.22: Estimation of d_e and d_l of the reconstruction space for the Rössler system.

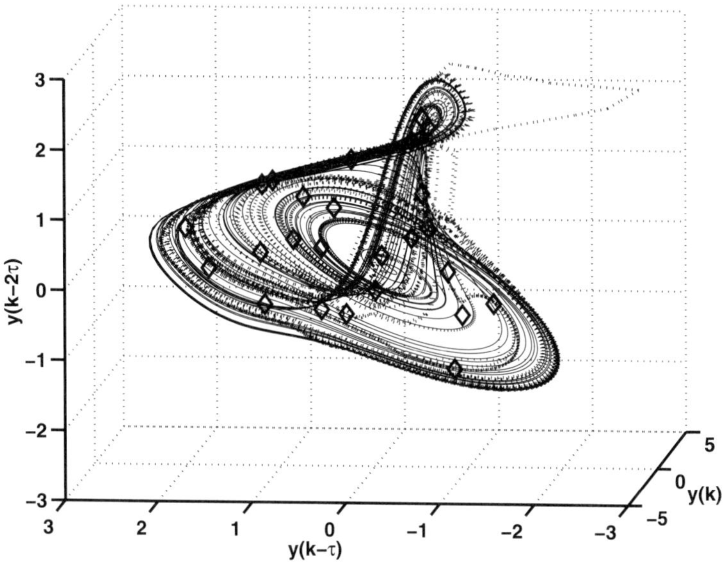

Figure 4.23: Prediction performance for the Rössler system. Cluster centers are denoted by diamonds.

Example 4.11 (State-space reconstruction of the Lorenz system). *The Lorenz system, describing the thermal driving of convection in the lower atmosphere, is defined by the following three differential equations:*

$$\dot{x}_1 = \sigma(x_2 - x_1) \tag{4.148}$$

$$\dot{x}_2 = x_1(r - x_3) - x_2 \tag{4.149}$$

$$\dot{x}_3 = x_1 x_2 - b x_3 \tag{4.150}$$

where the variables x_1, x_2 and x_3 are proportional to the intensity of the convection rolls, the horizontal temperature variation, and the vertical temperature variation, respectively, and the parameters σ, b and r are constants representing the properties of the system. The initial conditions are $x_1(0) = 0$, $x_2(0) = 1$ and $x_3(0) = 1$ and the parameter values used are $r = 28$, $b = \frac{8}{3}$ and $\sigma = 10$, following [278].

As shown in Figure 4.24, *the results are as good as in the case of the Rössler system, and the same statements can be given. The Correlation dimension of the data set is $d_C = 2.0001$.*

□

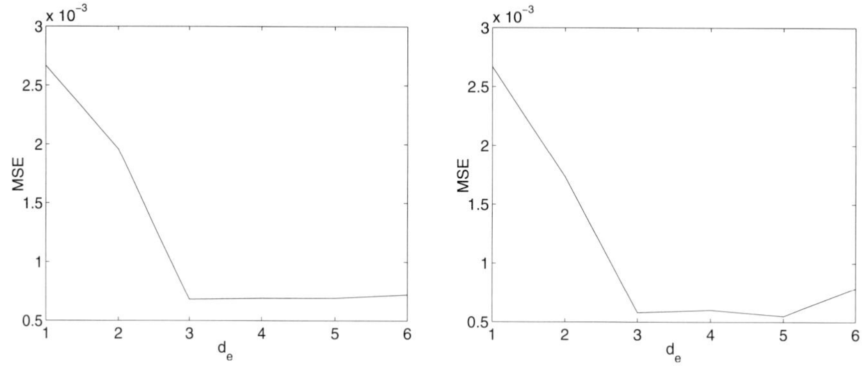

(a) Estimation of the embedding dimension d_e based on MIMO approach.

(b) Estimation of the embedding dimension d_e based on MISO approach.

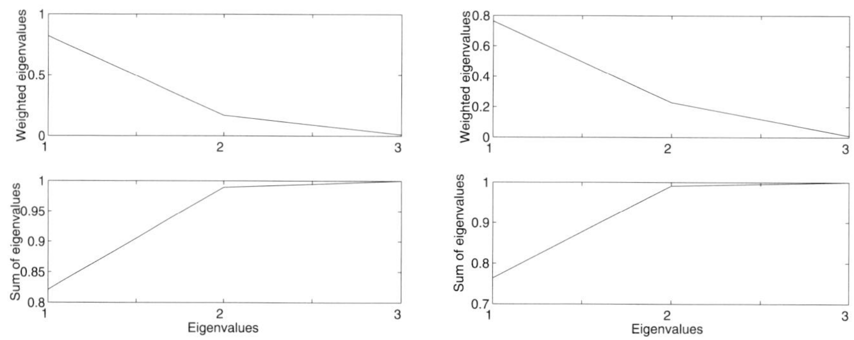

(c) Estimation of the local dimension d_l based on MIMO approach for $d_e = 3$.

(d) Estimation of the local dimension d_l based on MISO approach for $d_e = 3$.

Figure 4.24: Estimation of d_e and d_l dimensions of the reconstruction space for the Lorenz system.

Example 4.12 (State-space reconstruction of a four-dimensional system). *In this example let us consider the four-dimensional system published by Yao [292]. The system equations are*

$$\dot{x}_1 = x_3 \quad (4.151)$$

$$\dot{x}_2 = x_4 \quad (4.152)$$

$$\dot{x}_3 = -(\alpha + x_2^2)x_1 + x_2 \quad (4.153)$$

$$\dot{x}_4 = -(\beta + x_1^2)x_2 + x_1. \quad (4.154)$$

With $\alpha = 0.1$, $\beta = 0.101$ *and the initial condition* $x_1(0) = 0.1$, $x_2(0) = -0.1$, $x_3(0) = 0.1$, $x_4(0) = -0.1$, *this system is highly chaotic [292]. The sampling rate was 0.05, and the lag time is* $\tau = 50$ *estimated by the average mutual information function [135].*

In Figure 4.25 one can see that the estimated embedding dimension is four. The correct embedding dimension can be found by comparing the ratio of the neighboring MSE *values. While* $MSE(d_e = 3)/MSE(d_e = 4) = 2.031$, $MSE(d_e = 4)/MSE(d_e = 5) = 1.2329$ *in subfigure (a) of Figure 4.25 (MIMO approach). The values in case of MISO method are similar. The screeplots for* $d_e = 4$ *can be seen in Figure 4.25. The local dimension is equal to 2 according to the applied threshold value, 0.93. The Correlation dimension of the data set is* $d_C = 2.0445$.

In Figure 4.26 (a) the first three dimensions, in Figure 4.26 (b) the second three dimensions of the original (solid line) and the predicted (dashed line) four-dimensional data can be seen given by a free run simulation of 5000 data. The prediction performance is as good as in the previous examples.

□

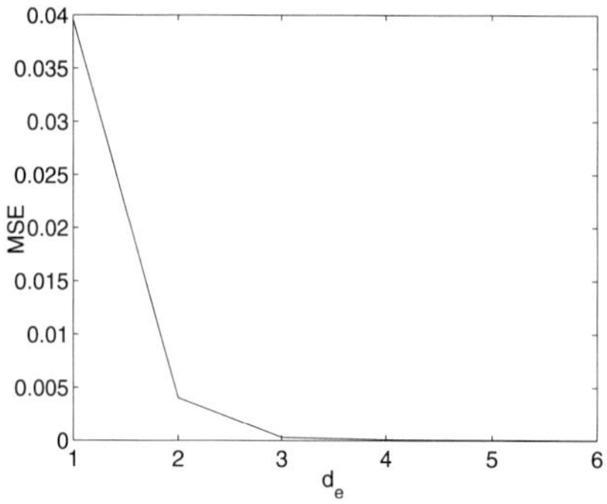

(a) Estimation of the embedding dimension d_e based on MIMO approach

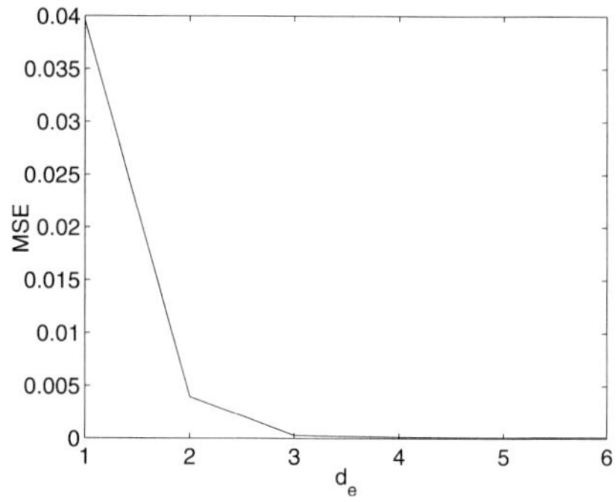

(b) Estimation of the embedding dimension d_e based on MISO approach

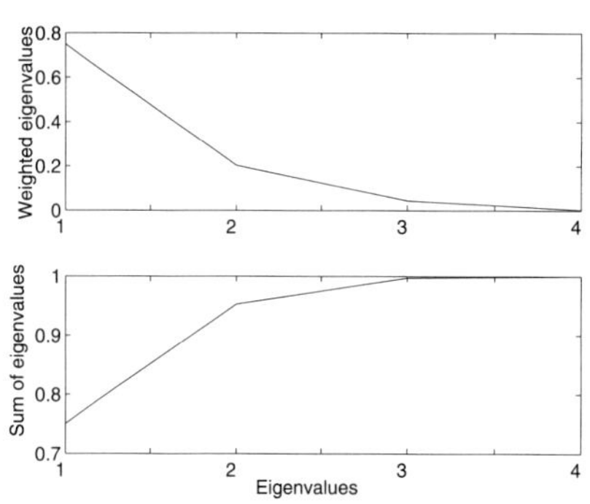

(c) Estimation of the local dimension d_l based on MIMO approach for $d_e = 4$.

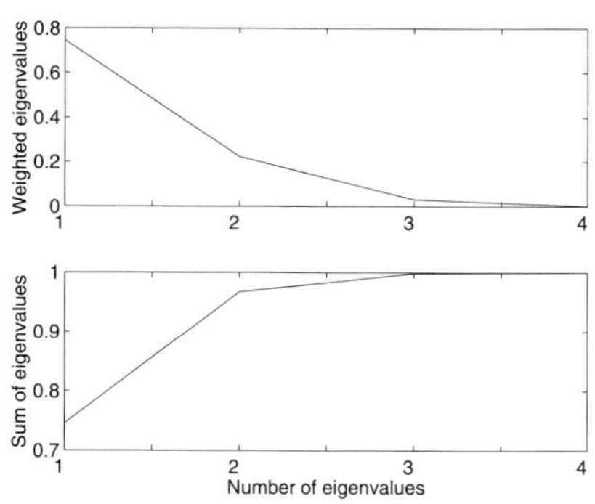

(d) Estimation of the local dimension d_l based on MISO approach for $d_e = 4$.

Figure 4.25: Estimation of the d_e and d_l dimensions of the reconstruction space for the four-dimensional system.

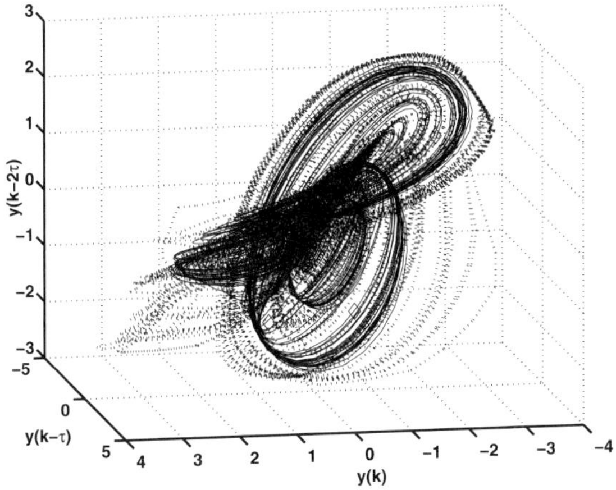

(a)

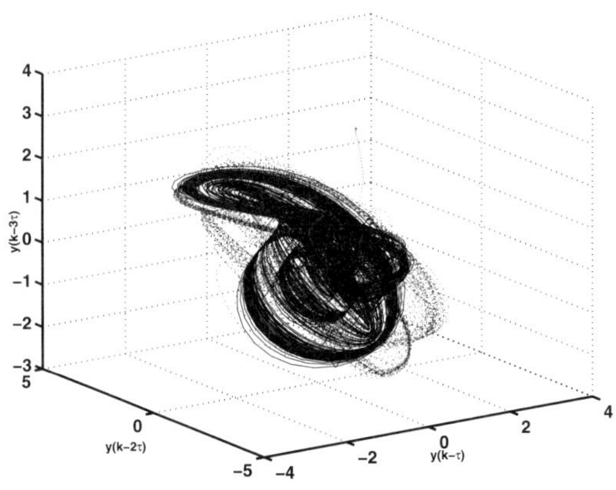

(b)

Figure 4.26: Prediction performance for the four-dimensional chaotic system.

4.4.4 Case Study

In Chapter 2 it can be seen how effective and useful the fuzzy Sammon mapping method can be to visualize the results of (fuzzy) clustering and based on this it enables the human inspector to validate the clustering procedure, e.g., cluster prototypes, number of clusters etc.

The advantages of the visualization are exploited in this example: the phase space trajectories of chaotic crystallizers are analyzed by clustering algorithms and the results are visualized.

Model of the Crystallizers

Most of the developments in the field of nonlinear dynamics assume that one has a complete description of the dynamic system under consideration. The practical application of these results, in principle, requires the simultaneous measurement of all state variables. In the case of crystallizers, however, in-line measurement of crystal size distribution, that is one of the most important properties of crystallization processes, is a difficult task so that application of the finite-dimensional moment equation model, computed from the crystal size distribution, often appears to be troublesome. Instead, observing a time series of one or more observables of the system the dynamics of the unknown deterministic finite-dimensional system can be reconstructed from this scalar time series as it was shown by Takens [258] and Sauer [234]. A number of algorithms have been proposed to handle this problem numerically [23], [24], [58], [135], but the method presented above based on fuzzy clustering of chaotic time series [9], allows solving three tasks simultaneously: selection of the embedding dimension, estimation of the intrinsic dimension, and identification of a model that can be used for prediction of chaotic time series.

Consider two continuous isothermal MSMPR crystallizers connected in cascade series where the first crystallizer is forced by sinusoidally varied solute input, while the second crystallizer is forced with the output solute signal of the first one. Let the crystallizers be identical in the sense that all kinetic and process parameters are of the same value. Further, let us assume that the working volumes are constant during the course of the operation, all new crystals are formed at a nominal size $L_n \approx 0$, crystal breakage and agglomeration are negligible, no growth rate fluctuations occur, the overall linear growth rate of crystals is size-independent and has the form of the power law expression of supersaturation, and the nucleation rate is described by Volmer's model.

The population balance model is an adequate mathematical description of crystallization processes. This model consists of a mixed set of ordinary and partial integral-differential equations even in the simplest case of MSMPR (mixed suspension, mixed product removal) crystallizers, and the state-space of a crystallizer is given by the Descartes product $\mathbb{R}^k \times \mathbb{N}$, where k is a positive natural number, of some vector space $\mathbb{R}^k$ of concentrations and temperatures, and the function space $\mathbb{N}$ of population density functions. Consideration of dynamical problems of crystallizers in this product space, however, seems to be quite complex, so that we usually concentrate on a reduced case, approximating the distributed parameter system by a finite-dimensional state-space model based on the moments of crystal size. Taking into account only the first four leading moments, the resulted space even in the simplest case of isothermal MSMPR crystallizers becomes 6-D. It is suitable for studying the dynamic phenomena of crystallizers [12], but it often appears to be too complex to apply for their model based control [12]. Consequently, it is reasonable to generate an appropriate reduced dimensional model, e.g., for control purposes. Then, the moment equation model of this system of crystallizers is a closed set of 6 ordinary differential equations by each crystallizer, and their connection can be described by algebraic equations. There are ranges of parameters in which the system behavior is chaotic and it can also be observed by real crystallization processes.

In the next section it is presented how the fuzzy clustering-based algorithm can be applied for reduction of the moment equation model of continuous isothermal crystallizers using the chaotic time series generated under some operation conditions, and how the visualization can be used in this case.

Results and Discussion

In this analysis, the zero-order moment of the second crystallizer is used to predict the embedding and local dimension of the system with simultaneous identification of a model that can be used for prediction. Using only this single time-series, the embedding and local dimension are to be estimated by the method described above. By the simulation, the sampling time was equal to 0.01 times of the dimensionless time unit ξ. By the state-space reconstruction, the sampling time was 10 times greater, so as to reduce the time and memory demand. The other parameters of the method: number of points: 1500; number of clusters: $c = 10$; termination tolerance: $\epsilon = 10^{-4}$ and the weighting exponent: $m = 2$. According to our experience, the presented approach is quite robust with respect to the choice of the clustering algorithm parameters.

The lag time was chosen as the first minimum of the average mutual information, and in this case it is equal to 0.6ξ (ξ is the dimensionless time unit). The embedding dimension was run from 1 to 10, and by each value the model identification was evaluated to get the one-step-ahead prediction error. These values can be seen on the left part of Figure 4.27. As can be seen, there is a 'knot' by $d_e = 4$, and mean square error reduction rate here also is greater. After the correct embedding dimension is chosen, it is possible to estimate the local dimension by the analysis of the rate of the cumulative and the total sum of eigenvalues weighted by the a priori possibility of the clusters. These values are depicted on the right part of Figure 4.27. It can be determined that a 2-dimensional subspace of the reconstructed 4-dimensional state-space contains the trajectories with high, 95% possibility. The disadvantage of this method is that it is only able to estimate integer local dimensions. To avoid this, fractal dimension of the trajectories can

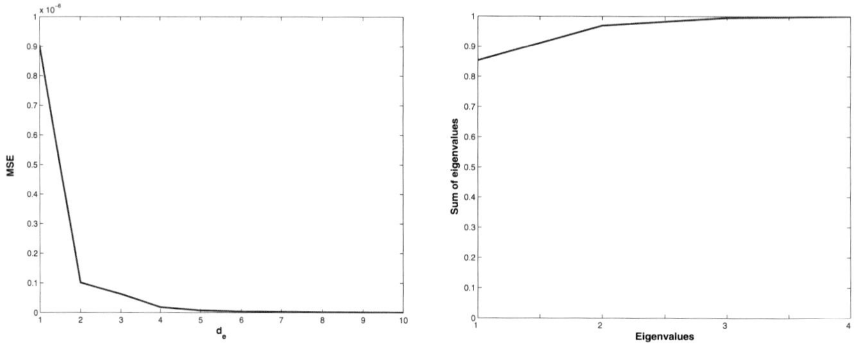

(a) (b)

Figure 4.27: Estimation of the embedding dimension and of the local dimension by $d_e = 4$.

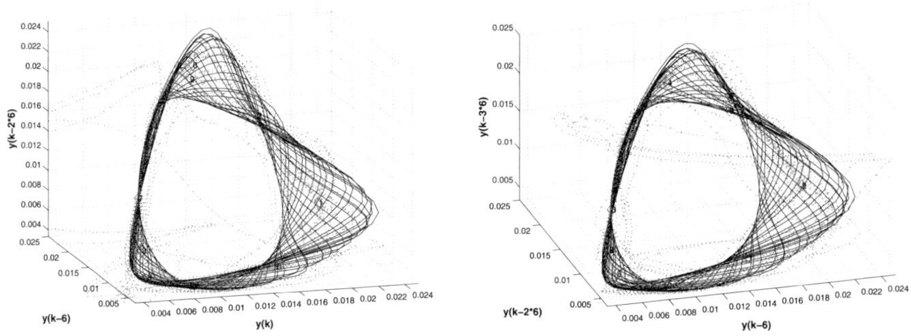

(a) (b)

Figure 4.28: Free run simulation in the state-space.

also be computed. In this case the so-called Correlation Dimension [103] is equal to 2.0009 which confirms the result of the clustering based method.

In Figure 4.28, the measured data are depicted with solid lines embedded in the 4-dimensional reconstructed state-space. To validate the model in some sense, a free-run simulation was evaluated whose results can also be seen on Figure 4.28 plotted with dotted lines. On the left side of Figure 4.28 the first, on the right the second three dimensions are depicted. (To visualize all four dimensions a scatterplot matrix can be used, which is a 4×4 matrix in this case.) It can be seen that although the system is chaotic, the trajectories are similar and take the same

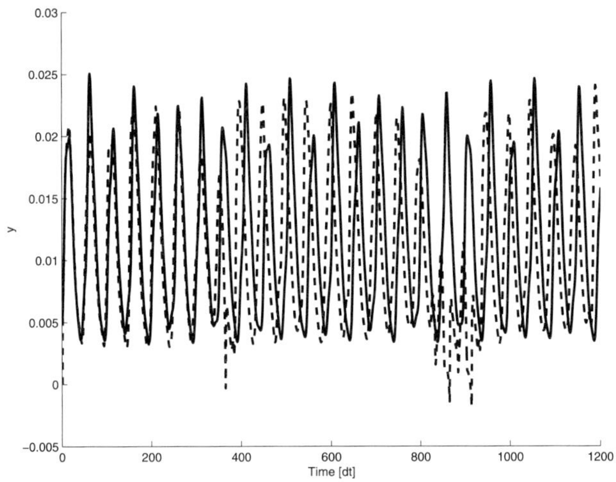

Figure 4.29: Free run simulation.

subspace, so the approximation is very good. The predicted one-dimensional data are depicted in Figure 4.29. It can be seen that there are ranges in the space that cannot be modeled as well as other ranges, it causes that the predicted trajectories draw away from the measured ones.

It is possible to depict the 'relationship' and 'order' of clusters in the 4-dimensional space in some sense. In the analyzed case there is a dynamic system so it can be computed how the data points are 'wandering' in the state-space from one cluster to another. This enables us to order clusters. The procedure is the following: the points with high probability $\{k\colon \mu_{i,k} > 0.99$ by the current cluster $i\}$ should be found and determined by which cluster the membership value $\mu_{j,k+m}, j \neq i$ will be first high again 'in the future', $m > 0$. In this way it can be computed which percentage of the points in the current cluster wanders to which clusters. In Figure 4.30 (a) this 'path' can be seen. In this figure only the paths with ratio greater than 40% are depicted because of transparency (the width of arrows is proportional to the ratio). It has to be mentioned that the size of the clusters is also very important from the viewpoint of the path because relatively small clusters have the same effect as large ones. However, in this case this effect is negligible because the size of the clusters (the *a priori* probability $p(\eta_i)$) are nearly the same as it can be seen in Table 4.15. This kind of figure can be useful to analyze the relationship of clusters in the original multidimensional space but it cannot be used to visualize the clustering results because it does not contain any information about that. For this purpose visualization techniques have to be used. At first, consider the result of PCA that is depicted in Figure 4.30 (b). It can be determined that the order of the cluster is completely the same as in Figure 4.30 (a) and the 'torsion' of the trajectories (see in Figure 4.28) can also be observed.

Table 4.15: The *a priori* probability of clusters.

Cluster	Probability (%)
1	12.87
2	10.40
3	11.14
4	12.54
5	10.32
6	6.95
7	11.91
8	8.30
9	7.08
10	8.45

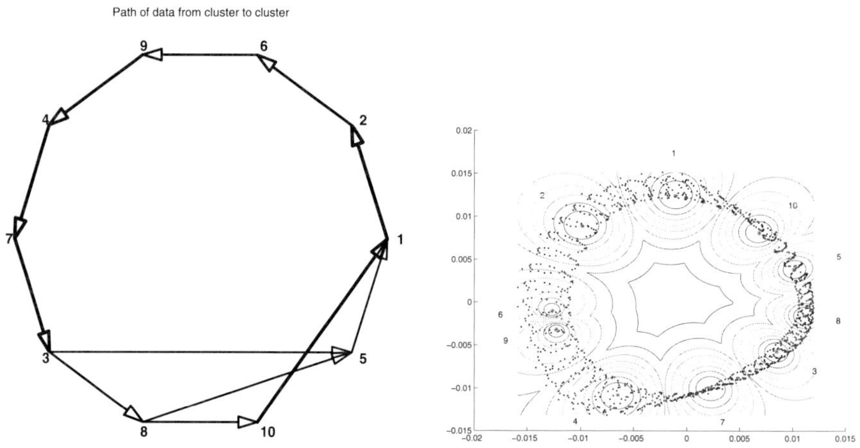

(a) (b)

Figure 4.30: Projection by PCA.

PCA is a linear method. It means that it finds the 2-dimensional linear subspace of this 4-dimensional space that fits the data best. If the data do not lie close to a linear 2-dimensional subspace, the mapping gives bad results. In this case, it is advisable to apply nonlinear techniques. Results given by Sammon mapping can be seen in Figure 4.31: on the left initialized by PCA and on the right with random initialization. It can be determined that these results are nearly the same and they are very similar to PCA results, only the data points and clusters have turned round in some directions or have been mirrored. It is caused by the

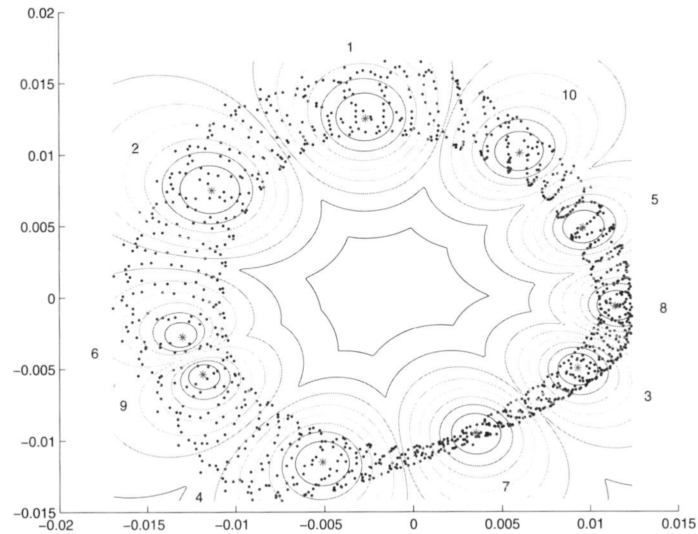

(a) Initialized by PCA

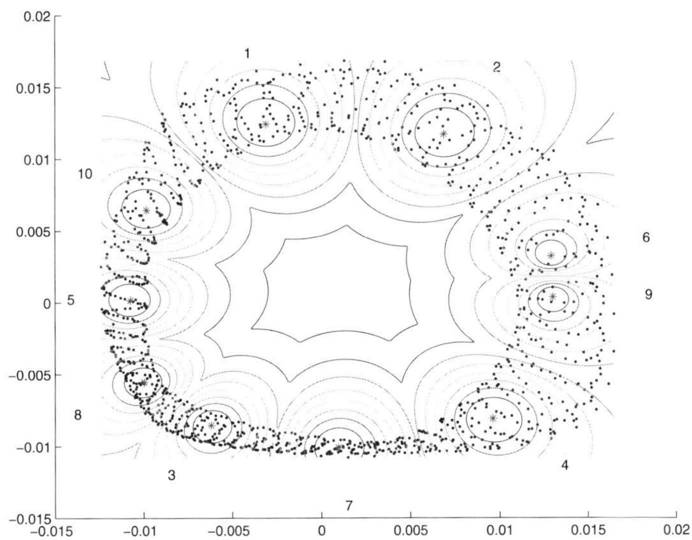

(b) Random initialization

Figure 4.31: Projection based on Sammon mapping.

procedure of Sammon mapping itself because it tries to preserve the distance between each data pair, so there is no 'fix point' in the projected space. Because of the similarity of results from linear PCA and nonlinear Sammon mapping, it can be determined that not only the local dimension of the trajectories is equal to 2, but also this subspace is nearly a linear one. It can also be represented by index-numbers (Table 4.16) in a way similar to the one presented in Section 2.2 (see, e.g., Example 2.4). It shows that PCA and Sammon mapping give nearly the same result regarding the visualization of clustering results.

Table 4.16: Comparison of the performance of the mappings (line of *FUZZSAM (rand)*) contains the average of results by 20 random initialization.

Method	P	F	F^*	E
PCA	0.0859	0.9911	0.5920	0.0035
SAMMON	0.0813	0.9911	0.5717	0.0018
FUZZSAM (PCA)	0.0625	0.9911	0.7004	2.6737
FUZZSAM (rand)	0.0408	0.9911	0.7727	10.1837

The presented fuzzy Sammon mapping gives completely different results as it can be seen in Figure 4.32. As it was the case in Figure 4.31, the result on the left is initialized by PCA and the ones on the right given by random initialization. Figure 4.32 (a) shows similar structure of clusters to PCA results (Figure 4.30). The order of clusters is similar as well (see, e.g., 1-2-6-9-4-7), but in the latter figure it can be seen which clusters are less compact, so less accurate: mainly cluster 8, but there are data points far from cluster 3, 5 and 7 as well. It has to be mentioned that also Figure 4.30 (a) shows cluster 8 as a 'critical point'. Fuzzy Sammon mapping with random initialization gives similar results but no such structure of clusters as in the previous examples. It can be seen that there are data points scattered in the range of cluster 3, 7 and 8, since the other clusters are very compact with similar *a priori* probability (Table 4.15). The presented FUZZSAM tool outperforms the linear method and the classical Sammon projection tools. It is interesting that FUZZSAMM with random initialization gives better results than with PCA initialization, because in the latter case PCA 'forces' FUZZSAMM algorithms to preserve the original structure of clusters in some sense.

4.4.5 Conclusions

An operating regime based model can be effectively used to approximate MIMO dynamic systems. For the identification of this model, fuzzy clustering is applied in the reconstruction space defined by the lagged output variables. The main advantage of the presented solution is that three tasks are simultaneously solved during clustering: selection of the embedding dimension, estimation of the intrinsic (local) dimension, and identification of a model that can be used for predic-

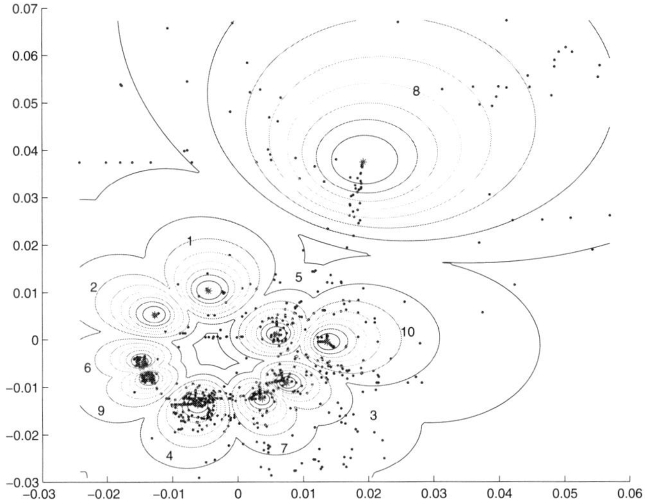

(a) Initialized by PCA

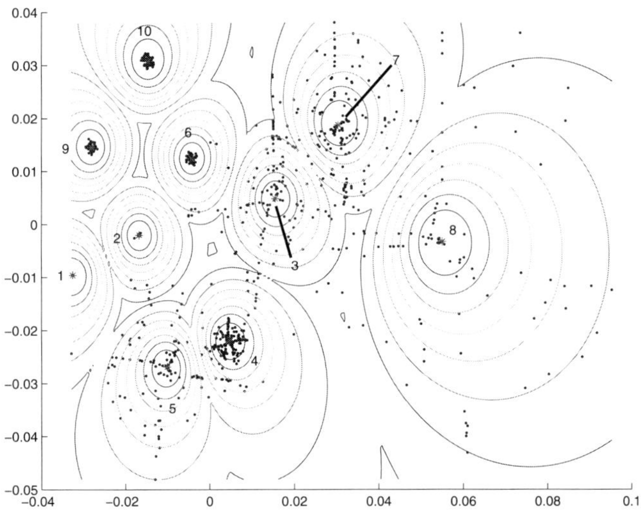

(b) Random initialization

Figure 4.32: Projection based on fuzzy Sammon mapping.

tion. In case of the analyzed three and four-dimensional systems the algorithm gave good estimations of the local and embedding dimensions and the model extracted from the clusters gave satisfactory prediction performance. The programs and data sets used to generate the results are available from the web page http://www.fmt.vein.hu/softcomp/timeseries.

Chapter 5

Fuzzy Model based Classifiers

Two forms of the data-driven modelling are *regression* and *classification*. Based on some measured variables, both of them predict the value of one or more variables we are interested in. In case of regression there are continuous or ordered variables, in case of classification there are discrete or nominal variables needed to be predicted. Classification is also called supervised learning because the labels of the samples are known beforehand. This is the main difference between classification and clustering. The later is unsupervised learning since clusters want to be determined and the labels of the data points are not known.

An example to training data can be found in Table 5.1 [110]. Data in each row represent a sample, and also a label belongs to each one, in this case it means whether a potential customer buys a computer or not.

Table 5.1: Data for classification (from [110]).

Age	Income	Student	CreditRating	Class: BuysComputer
≤ 30	high	no	fair	no
≤ 30	high	no	excellent	no
$31 \ldots 40$	high	no	fair	yes
> 40	medium	no	fair	yes
> 40	low	yes	fair	yes
> 40	low	yes	excellent	no
$31 \ldots 40$	low	yes	excellent	yes
≤ 30	medium	no	fair	no
≤ 30	low	yes	fair	yes
> 40	medium	yes	fair	yes
≤ 30	medium	yes	excellent	yes
$31 \ldots 40$	medium	no	excellent	yes
$31 \ldots 40$	high	yes	fair	yes
> 40	medium	no	excellent	no

Based on known training data a classifier model can be identified, and it can be predicted using that model whether a client would buy a computer, and the cost of advertisement can be reduced in this way. The structure of classifiers can be various, but a well-comprehensible and interpretable way is to generate (classification) rules:

If *Age* is ≤ 30 **and** *Student* is "*no*" **then** *BuysComputer* is "*no*"
If *Age* is ≤ 30 **and** *Student* is "*yes*" **then** *BuysComputer* is "*yes*"
If *Age* is $31\ldots40$ **and** *CreditRating* is "*excellent*" **then** *BuysComputer* is "*yes*"
If *Age* is > 40 **and** *CreditRating* is "*excellent*" **then** *BuysComputer* is "*yes*"
If *Age* is > 40 **and** *CreditRating* is "*fair*" **then** *BuysComputer* is "*no*"

These rules speak for themselves, it is unnecessary to explain them. The part of a rule after **If** is called antecedent, and after **then** is called consequent part. In the antecedent part one or more tests are defined on the measured variables, since the consequent part contains the class label. It can be seen as well that each potential customer can fulfil the antecedent part of exactly one rule, hence, the rules divide the "space of possible cases" into disjunct subspaces.

However, in practice there are numerous cases where the transitions are not as sharp as the classifier predicts. E.g., if someone has his or her 31st birthday soon, is not a student and his or her credit rating is excellent, it is not likely that this person will change his or her decision from day to day. These kinds of transitions cannot be handled by classifiers on the basis of "crisp" rules. Therefore it is advisable to use "fuzzy" classifiers, i.e., rules without sharp boundaries. In case of classical fuzzy classifiers each sample can fulfil more than one rule simultaneously, but with different weights. The consequent part contains "crisp" class labels, and the estimated (predicted) label can be computed from the weighted labels.

In order to get to know which classifier is better than others, these classifiers have to be compared. The following criteria can or have to be considered [110]:

- **Predictive accuracy.** The ability of the model to correctly predict the class label of previously unseen data.
- **Interpretability.** The level of understanding and insight that is provided by the model.
- **Speed.** The computation demand to generate and use the model.
- **Robustness.** The ability to make correct predictions given noisy data.
- **Scalability.** The ability to construct the model efficiently given large amounts of data.

The presented methods described in this section consider mainly the first two points of view. For that purpose, an iterative learning method, a supervised clustering algorithm and a fuzzy classification tree induction method are introduced.

5.1 Fuzzy Model Structures for Classification

5.1.1 Classical Bayes Classifier

The identification of a classifier system means the construction of a model that predicts the class $y_k = \{c_1, \ldots, c_C\}$ to which pattern $\mathbf{x}_k = [x_{1,k}, \ldots, x_{n,k}]^T$ should be assigned. The classic approach for this problem with C classes is based on Bayes rule. The probability of making an error when classifying an example $\mathbf{x}$ is minimized by Bayes decision rule of assigning it to the class with the largest *a posteriori* probability:

$$\mathbf{x} \text{ is assigned to } c_i \Longleftrightarrow p(c_i|\mathbf{x}) \geq p(c_j|\mathbf{x}) \, \forall j \neq i. \tag{5.1}$$

The *a posteriori* probability of each class given a pattern $\mathbf{x}$ can be calculated based on the $p(\mathbf{x}|c_i)$ class conditional distribution, which models the density of the data belonging to the class c_i, and the $P(c_i)$ class prior, which represents the probability that an arbitrary example out of data belongs to class c_i

$$p(c_i|\mathbf{x}) = \frac{p(\mathbf{x}|c_i)P(c_i)}{p(\mathbf{x})} = \frac{p(\mathbf{x}|c_i)P(c_i)}{\sum_{j=1}^{C} p(\mathbf{x}|c_j)P(c_j)}. \tag{5.2}$$

As (5.1) can be rewritten using the numerator of (5.2)

$$\mathbf{x} \text{ is assigned to } c_i \Longleftrightarrow p(\mathbf{x}|c_i)P(c_i) \geq p(\mathbf{x}|c_j)P(c_j) \, \forall j \neq i\,, \tag{5.3}$$

we would have an optimal classifier if we would perfectly estimate the class priors and the class conditional densities.

In practice one needs to find approximate estimates of these quantities on a finite set of training data $\{\mathbf{x}_k, y_k\}$, $k = 1, \ldots, N$. Priors $P(c_i)$ are often estimated on the basis of the training set as the proportion of samples of class c_i or using prior knowledge. The $p(\mathbf{x}|c_i)$ class conditional densities can be modeled with non-parametric methods like histograms, nearest-neighbors or parametric methods such as mixture models.

A special case of Bayes classifiers is the quadratic classifier, where the $p(\mathbf{x}|c_i)$ distribution generated by the class c_i is represented by a Gaussian function

$$p(\mathbf{x}|c_i) = \frac{1}{(2\pi)^{n/2}\sqrt{\det(\mathbf{F}_i)}} \exp\left(-\frac{1}{2}(\mathbf{x} - \mathbf{v}_i)^T \mathbf{F}_i^{-1} (\mathbf{x} - \mathbf{v}_i)\right) \tag{5.4}$$

where $\mathbf{v}_i = [v_{1,i}, \ldots, v_{n,i}]^T$ denotes the center of the ith multivariate Gaussian and $\mathbf{F}_i$ stands for a covariance matrix of the data of the class c_i. In this case, the (5.3) classification rule can be reformulated based on a distance measure. The sample $\mathbf{x}_k$ is classified to the class that minimizes the $d^2(\mathbf{x}_k, \mathbf{v}_i)$ distance, where the distance measure is inversely proportional to the probability of the data:

$$d^2(\mathbf{x}_k, \mathbf{v}_i) = \left(\frac{P(c_i)}{(2\pi)^{n/2}\sqrt{\det(\mathbf{F}_i)}} \exp\left(-\frac{1}{2}(\mathbf{x}_k - \mathbf{v}_i)^T \mathbf{F}_i^{-1} (\mathbf{x}_k - \mathbf{v}_i)\right)\right)^{-1}. \tag{5.5}$$

5.1.2 Classical Fuzzy Classifier

The classical fuzzy rule-based classifier consists of fuzzy rules each one describing one of the C classes. The rule antecedent defines the operating region of the rule in the n-dimensional feature space and the rule consequent is a crisp (non-fuzzy) class label from the $\{c_1, \ldots, c_C\}$ label set:

$$r_i : \quad \textbf{If } x_1 \text{ is } A_{i,1}(x_{1,k}) \textbf{ and } \ldots x_n \text{ is } A_{i,n}(x_{n,k}) \textbf{ then } \hat{y} = c_i, \ [w_i] \tag{5.6}$$

where $A_{i,1}, \ldots, A_{i,n}$ are the antecedent fuzzy sets and w_i is a certainty factor that represents the desired impact of the rule. The value of w_i is usually chosen by the designer of the fuzzy system according to his or her belief in the accuracy of the rule. When such knowledge is not available, w_i is fixed to value 1 for any i.

The **and** connective is modeled by the product operator allowing for interaction between the propositions in the antecedent. Hence, the degree of activation of the ith rule is calculated as:

$$\beta_i(\mathbf{x}_k) = w_i \prod_{j=1}^{n} A_{i,j}(x_{j,k}). \tag{5.7}$$

The output of the classical fuzzy classifier is determined by the *winner takes all* strategy, i.e., the output is the class related to the consequent of the rule that gets the highest degree of activation:

$$\hat{y}_k = c_{i^*}\ ,\ i^* = \arg \max_{1 \le i \le C} \beta_i(\mathbf{x}_k). \tag{5.8}$$

To represent the $A_{i,j}(x_{j,k})$ fuzzy set, we use Gaussian membership functions

$$A_{i,j}(x_{j,k}) = \exp\left(-\frac{1}{2}\frac{(x_{j,k} - v_{i,j})^2}{\sigma_{i,j}^2}\right) \tag{5.9}$$

where $v_{i,j}$ represents the center and $\sigma_{i,j}^2$ stands for the variance of the Gaussian function. The use of the Gaussian membership function allows for the compact formulation of (5.7):

$$\beta_i(\mathbf{x}_k) = w_i \mathbf{A}_i(\mathbf{x}_k) = w_i \exp\left(-\frac{1}{2}(\mathbf{x}_k - \mathbf{v}_i)^T \mathbf{F}_i^{-1} (\mathbf{x}_k - \mathbf{v}_i)\right) \tag{5.10}$$

where $\mathbf{v}_i = [v_{1,i}, \ldots, v_{n,i}]^T$ denotes the center of the ith multivariate Gaussian and $\mathbf{F}_i$ stands for a diagonal matrix that contains the $\sigma_{i,j}^2$ variances.

The fuzzy classifier defined by the previous equations is in fact a quadratic Bayes classifier when $\mathbf{F}_i$ in (5.4) contains only diagonal elements (variances).

In this case, the $\mathbf{A}_i(\mathbf{x})$ membership functions and the w_i certainty factors can be calculated from the parameters of the Bayes classifier following equations (5.4) and (5.10) as

$$\mathbf{A}_i(\mathbf{x}) = p(\mathbf{x}|c_i)(2\pi)^{n/2}\sqrt{\det(\mathbf{F}_i)},\ w_i = \frac{P(c_i)}{(2\pi)^{n/2}\sqrt{\det(\mathbf{F}_i)}}. \tag{5.11}$$

5.1.3 Bayes Classifier based on Mixture of Density Models

One of the possible extensions of the classical quadratic Bayes classifier is to use a mixture of models for estimating the class-conditional densities. The usage of mixture models in Bayes classifiers is not so widespread [144]. In these solutions each conditional density is modeled by a separate mixture of models. A possible criticism of such Bayes classifiers is that in a sense they are modelling too much: for each class many aspects of the data are modeled which may or may not play a role in discriminating between the classes.

In this section a new approach is presented. The $p(c_i|\mathbf{x})$ posteriori densities are modeled by $R > C$ mixture of models (clusters)

$$p(c_i|\mathbf{x}) = \sum_{l=1}^{R} p(r_l|\mathbf{x})P(c_i|r_l) \tag{5.12}$$

where $p(r_l|\mathbf{x})$, representing the *a posteriori* probability of $\mathbf{x}$, has been generated by the r_lth local model and $P(c_i|r_l)$ denotes the *prior* probability of this model representing the class c_i.

Similarly to (5.2) $p(r_l|\mathbf{x})$ can be written as

$$p(r_i|\mathbf{x}) = \frac{p(\mathbf{x}|r_i)P(r_i)}{p(\mathbf{x})} = \frac{p(\mathbf{x}|r_i)P(r_i)}{\sum_{j=1}^{R} p(\mathbf{x}|r_j)P(r_j)}. \tag{5.13}$$

By using this mixture of density models the posteriori class probability can be expressed following equations (5.2), (5.12) and (5.13) as

$$p(c_i|\mathbf{x}) = \frac{p(\mathbf{x}|c_i)P(c_i)}{p(\mathbf{x})} = \sum_{l=1}^{R} \frac{p(\mathbf{x}|r_i)P(r_i)}{\sum\limits_{j=1}^{R} p(\mathbf{x}|r_j)P(r_j)} P(c_i|r_l) = \frac{\sum\limits_{l=1}^{R} p(\mathbf{x}|r_i)P(r_i)P(c_i|r_l)}{p(\mathbf{x})} \tag{5.14}$$

The Bayes decision rule can be thus formulated similarly to (5.3) as

$$\begin{aligned} &\mathbf{x} \text{ is assigned to } c_i \\ \iff &\textstyle\sum_{l=1}^{R} p(\mathbf{x}|r_l)P(r_l)P(c_i|r_l) \geq \sum_{l=1}^{R} p(\mathbf{x}|r_l)P(r_l)P(c_j|r_l)\, \forall j \neq i \end{aligned} \tag{5.15}$$

where the $p(\mathbf{x}|r_l)$ distribution is represented by Gaussians similarly to (5.4).

5.1.4 Extended Fuzzy Classifier

A new fuzzy model that is able to represent Bayes classifier defined by (5.15) can be obtained. The idea is to define the consequent of the fuzzy rule as the probabilities of the given rule represents the $c_1, \ldots, c_C$ classes:

$$\begin{aligned} r_i: \quad &\textbf{If } x_1 \text{ is } A_{i,1}(x_{1,k}) \textbf{ and } \ldots x_n \text{ is } A_{i,n}(x_{n,k}) \textbf{ then} \\ &\hat{y}_k = c_1 \text{ with } P(c_1|r_i) \ldots, \hat{y}_k = c_C \text{ with } P(c_C|r_i)\ , [w_i] \end{aligned} \tag{5.16}$$

Similarly to Takagi–Sugeno fuzzy models [257], the rules of the fuzzy model are aggregated using the normalized fuzzy mean formula and the output of the classifier is determined by the label of the class that has the highest activation:

$$\hat{y}_k = c_{i^*} \ , \ i^* = \arg\max_{1 \le i \le C} \frac{\sum_{l=1}^{R} \beta_l(\mathbf{x}_k) P(c_i|r_l)}{\sum_{i=1}^{R} \beta_l(\mathbf{x}_k)} \tag{5.17}$$

where $\beta_l(\mathbf{x}_k)$ has the meaning expressed by (5.7).

As the previous equation can be rewritten using only its numerator, the obtained expression is identical to the Gaussian mixtures of Bayes classifiers (5.15) when similarly to (5.11) the parameters of the fuzzy model are calculated as

$$\mathbf{A}_i(\mathbf{x}) = p(\mathbf{x}|r_i)(2\pi \det(\mathbf{F}_i))^{n/2}, \ w_i = \frac{P(r_i)}{(2\pi \det(\mathbf{F}_i))^{n/2}} \,. \tag{5.18}$$

The main advantage of the previously presented classifier is that the fuzzy model can consist of more rules than classes and every rule can describe more than one class. Hence, as a given class will be described by a set of rules, it should not be a compact geometrical object (hyper-ellipsoid).

5.1.5 Fuzzy Decision Tree for Classification

Using not only crisp but also fuzzy predicates, decision trees can be used to model vague decisions (see Section 3.4 for Fuzzy Regression Trees). The basic idea of fuzzy decision trees is to combine example based learning in decision trees with approximative reasoning of fuzzy logic [133]. This hybridization integrates the advantages of both methodologies compact knowledge representation of decision trees with the ability of fuzzy systems to process uncertain and imprecise information. Viewing fuzzy decision trees as a compressed representation of a (fuzzy) rule set, enables us to use decision trees not only for classification, but also for approximation of continuous output functions.

An example of how a fuzzy decision tree can be used for the compressed representation of a fuzzy rule base is given in Figure 5.1, where the rule defined by the dashed path of the tree is the following:

If x_3 is *large* **and** x_2 is *medium* **and** x_1 is *small* **and** x_5 is *medium* **then** C_1 (5.19)

ID3 and its fuzzy variant (FID) assume discrete and fuzzy domains with small cardinalities. This is a great advantage as it increases comprehensibility of the induced knowledge, but may require an *a priori* partitioning of the numerical attributes (see the bottom of Figure 5.1 for the illustration of such fuzzy partitioning). Since this partitioning has significant effect to the performance of the generated model, recently some research has been done in the area of domain partitioning while

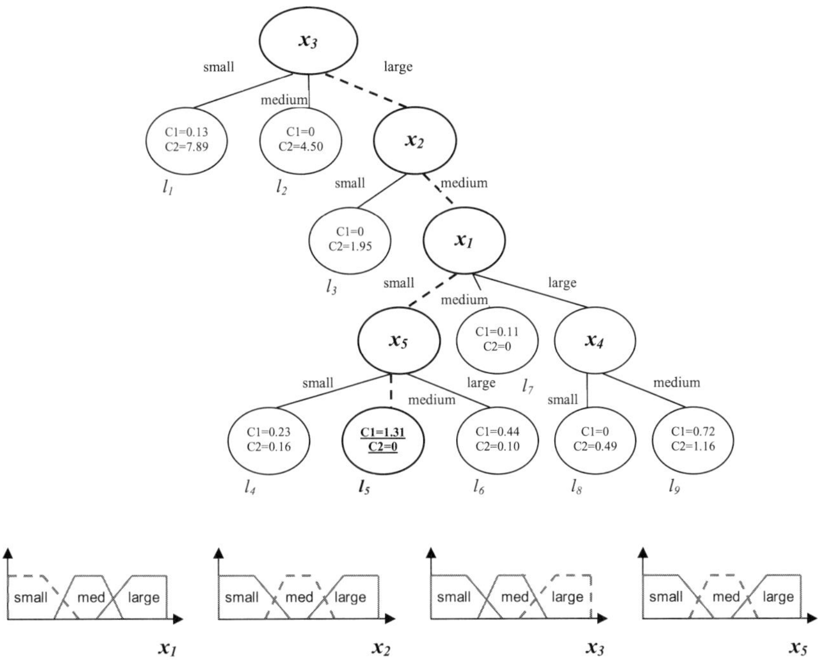

Figure 5.1: Example of a fuzzy decision tree and a fuzzy partitioning.

constructing a symbolic decision tree. For example, Dynamic-ID3 [92] clusters multivalued ordered domains, and Assistant [159] produces binary trees by clustering domain values (limited to domains of small cardinality). However, most research has concentrated on *a priori* partitioning techniques [171].

An example for a fuzzy decision tree is given in Figure 5.1. As can be seen in this figure each internal node is associated with a decision function (represented by a fuzzy membership function) to indicate which nodes to visit next. Each terminal node represents the output of a given input that leads to this node. In classification problems each terminal node contains the conditional probabilities $P(c_1|r_i), \ldots, P(c_C|r_i)$ of the predicted classes.

As a result of the increasing complexity and dimensionality of classification problems, it becomes necessary to deal with structural issues of the identification of classifier systems. Important aspects are the selection of the relevant features and the determination effective initial partition of the input domain [65]. Moreover, when the classifier is identified as part of an expert system, the linguistic interpretability is also an important aspect which must be taken into account.

In the following, clustering based identification techniques of classifier systems will be presented. It will be shown that the aspects mentioned above can be considered using clustering based methods.

5.2 Iterative Learning of Fuzzy Classifiers

To improve the classification capability of the rule base, we apply a genetic algorithm (GA) based optimization method [231, 243]. Also other model properties can be optimized by applying multi-objective functions, like, e.g., search for redundancy [230]. When an initial fuzzy model has been obtained from data, it is successively reduced, simplified and finally optimized in an iterative fashion. Combinations of the GA with the model reduction tools described above can lead to different modelling schemes. Three different approaches are shown in Figure 5.2.

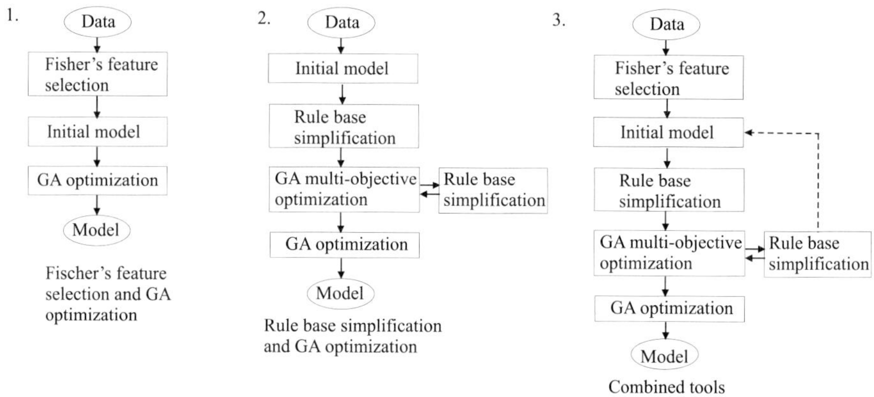

Figure 5.2: Modelling schemes resulting from combination of tools.

The model accuracy is measured in terms of the number of misclassifications. To further reduce the model complexity, the misclassification rate is combined with a similarity measure in the GA objective function. Similarity is rewarded during the iterative process, that is, the GA tries to emphasize the redundancy in the model. This redundancy is then used to remove unnecessary fuzzy sets in the next iteration. In the final step, fine tuning is combined with a penalized similarity among fuzzy sets to obtain a distinguishable term set for linguistic interpretation.

The GAs is subject to minimize the following multi-objective function:

$$J = (1 + \lambda S^*) \cdot MSE\,, \tag{5.20}$$

where $S^* \in [0, 1]$ is the average of the maximum pairwise similarity that is present in each input, i.e., S^* is an aggregated similarity measure for the total model. The weighting function $\lambda \in [-1, 1]$ determines whether similarity is rewarded ($\lambda < 0$) or penalized ($\lambda > 0$).

From the K available input-output data pairs $\{\mathbf{x}_k, y_k\}$ we construct the n-dimensional pattern matrix $\mathbf{X}^T = [\mathbf{x}_1, \ldots, \mathbf{x}_K]$ and the corresponding label vector $\mathbf{y}^T = [y_1, \ldots, y_K]$. The fuzzy antecedents $A_{i,j}(x_j)$ in the initial rule base are now determined by a three step algorithm. In the first step, M multivariable membership functions are defined in the product space of the features. Each describe a region where the system can be approximated by a single fuzzy rule. This partitioning is often realized by iterative methods such as clustering [230]. Here, given the labeled data, a one-step approach is presented. This assumes that each class is described by a single, compact construct in the feature space. (If this is not the case, other methods such as, e.g, relational classification [239], must be applied.) Similar to Gustafson and Kessel's clustering algorithm (Section 1.5.6), the approach presented here assumes that the shape of the partitions can be approximated by ellipsoids. Hence, each class prototype is represented by a center $\mathbf{v}$ and its covariance matrix $\mathbf{Q}$:

$$\mathbf{v}_i = \frac{1}{N_i} \sum_{k|y_k=i} \mathbf{x}_k \,, \tag{5.21}$$

$$\mathbf{F}_i = \frac{1}{N_i} \sum_{k|y_k=i} (\mathbf{x}_k - \mathbf{v}_i)(\mathbf{x}_k - \mathbf{v}_i)^T \,, \tag{5.22}$$

where i denotes the index of the classes, $i = 1, \ldots, N_c$, and K_i represents the number of samples that belong to the ith class. In the second step, the algorithm computes the fuzzy partition matrix $\mathbf{U}$ whose i,kth element $u_{i,k} \in [0,1]$ is the membership degree of the data object $\mathbf{x}_k$ in class i. This membership is based on the distance between the observation and the class center:

$$D^2_{i,k} = (\mathbf{x}_k - \mathbf{v}_i)\mathbf{F}_i^{-1}(\mathbf{x}_k - \mathbf{v}_i)^T \,. \tag{5.23}$$

Using this distance, the membership becomes:

$$\mu_{i,k} = 1 \Bigg/ \sum_{j=1}^{N} \left(\frac{D_{i,k}}{D_{j,k}} \right)^{2/(m-1)} \,, \tag{5.24}$$

where m denotes a weighting exponent that determines the fuzziness of the obtained partition ($m = 1.8$ is applied).

The rows of $\mathbf{U}$ now contain pointwise representations of the multidimensional fuzzy sets describing the classes in the feature space. In the third step, the univariate fuzzy sets $A_{i,j}$ for the classification rules (5.6) are obtained by projecting the rows of $\mathbf{U}$ onto the input variables x_j and subsequently approximated the projections by parametric functions [26]. In this section we apply triangular fuzzy sets.

5.2.1 Ensuring Transparency and Accuracy

Using too many input variables may result in difficulties in the prediction and interpretability capabilities of the classifier. Hence, the selection of the relevant features is usually necessary. Generally, there is a very large set of possible features to compose feature vectors of classifiers. As ideally the training set size should increase exponentially with the feature vector size, it is desired to choose a minimal subset among it. Some generic tips to choose a good feature set include the facts that they should discriminate as much as possible the pattern classes and they should not be correlated/redundant. There are two basic feature-selection approaches: The *closed-loop* algorithms are based on the classification results, while the *open-loop* algorithms are based on a distance between clusters. In the former, each possible feature subset is used to train and to test a classifier, and the recognition rates are used as a decision criterion: the higher the recognition rate, the better is the feature subset. The main disadvantage of this approach is that choosing a classifier is a critical problem on its own, and that the final selected subset clearly depends on the classifier. On the other hand, the latter depends on defining a distance between the clusters, and some possibilities are Mahalanobis, Bhattacharyya and the class separation distance.

In the following two methods will be used for model reduction. The first method is an open-loop feature selection algorithm that is based on Fisher's interclass separability criterion [65] calculated from the covariances of the clusters. See Section 3.3.4 for more details. The other method is the similarity-driven simplification proposed by Setnes et al. [240] (see also [230]). Differences in these reduction methods are: (i) Feature reduction based on the similarity analysis of fuzzy sets results in a closed-loop feature selection because it depends on the actual model while the open-loop feature selection can be used beforehand as it is independent from the model. (ii) In similarity analysis, a feature can be removed from individual rules. In the interclass separability method the feature is omitted in all the rules.

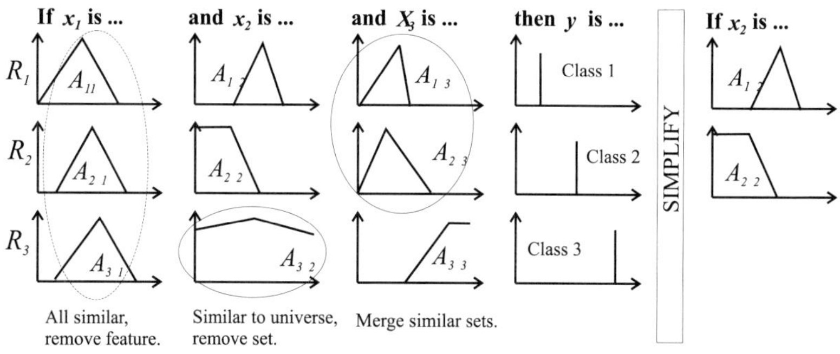

Figure 5.3: Similarity-driven simplification.

The similarity-driven rule base simplification method [240] uses a similarity measure to quantify the redundancy among the fuzzy sets in the rule base. A similarity measure based on the set-theoretic operations of intersection and union is applied:

$$S(A,B) = \frac{|A \cap B|}{|A \cup B|} \tag{5.25}$$

where $|.|$ denotes the cardinality of a set, and the $\cap$ and $\cup$ operators represent the intersection and union, respectively. If $S(A,B) = 1$, then the two membership functions A and B are equal. $S(A, B)$ becomes 0 when the membership functions are non-overlapping.

Similar fuzzy sets are merged when their similarity exceeds a user defined threshold $\theta \in [0, 1]$ (θ=0.5 is applied in the following). Merging reduces the number of different fuzzy sets (linguistic terms) used in the model and thereby increases the transparency. If all the fuzzy sets for a feature are similar to the universal set, or if merging led to only one membership function for a feature, then this feature is eliminated from the model. The method is illustrated in Figure 5.3

Example 5.1 (Classification of the Wine data based on iterative learning). *The Wine data contains the chemical analysis of* 178 *wines grown in the same region in Italy but derived from three different cultivars. The problem is to distinguish the three different types based on* 13 *continuous attributes derived from chemical analysis as can be seen in Figure* 5.4 *(see also Example* 2.1*). Corcoran and*

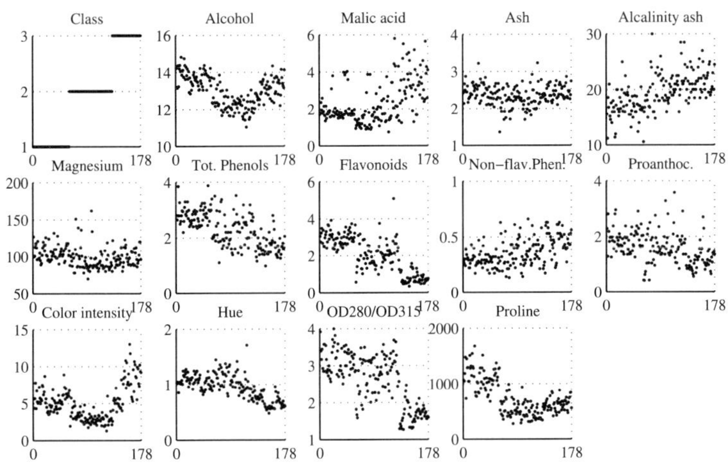

Figure 5.4: Wine data: 3 classes and 13 attributes.

Sen [66] *applied all the* 178 *samples for learning* 60 *non-fuzzy if-then rules in a real-coded genetic based-machine learning approach. They used a population of* 1500 *individuals and applied* 300 *generations, with full replacement, to come up with the following result for ten independent trials: best classification rate* 100%, *average classification rate* 99.5% *and worst classification rate* 98.3% *which is* 3 *misclassifications. Ishibuchi et al.* [125] *applied all the* 178 *samples designing a fuzzy classifier with* 60 *fuzzy rules by means of an integer-coded genetic algorithm and grid partitioning. Their population contained* 100 *individuals and they applied* 1000 *generations, with full replacement, to come up with the following result for ten independent trials: best classification rate* 99.4% (1 *misclassifications*), *average*

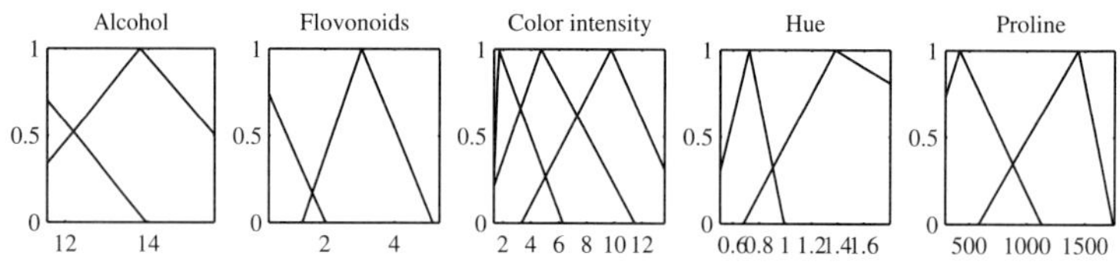

Figure 5.5: The fuzzy sets of the optimized three rule classifier for the Wine data.

classification rate 98.5% *and worst classification rate* 97.8% (4 *misclassifications*). *In both approaches the final rule base contains* 60 *rules. The main difference is the number of model evaluations that was necessary to come to the final result.*

An initial classifier with three rules was constructed with the presented covariance-based model initialization by using all samples resulting in 90.5% *correct,* 1.7% *undecided and* 7.9% *misclassifications with the following average certainty factors* CF, $[82.0, 99.6, 80.5]$. *Improved classifiers are developed based on the three schemes given in Figure* 5.2*:*

- *Scheme* 1*: The Fisher interclass separability criterion gives the following feature ranking* $\{8, 5, 11, 2, 3, 9, 10, 6, 7, 4, 1, 12, 13\}$. *Classifiers where made by adding one-by-one feature and* 400 *iterations with the GA-optimization. The two best classifiers were obtained by using the first* 5 *or* 7 *features* (15 *or* 21 *fuzzy sets*). *This results in* 98.9%, 99.4% *with* CF *for the three classes* $[0.95, 0.94, 0.84]$, $[0.94, 0.99, 0.97]$, *respectively.*

- *Scheme* 2*: The similarity driven simplification removed eight inputs in 3 steps: (i)* $\{3, 5\}$, (ii) $\{2, 4, 8, 9\}$, (iii) $\{6, 12\}$. *After each reduction,* 200 *GA-iterations were done and* 400 *after the last reduction. The final five-rule classifier (Figure* 5.5*) contains only* 11 *fuzzy sets. The result on learning data was* 99.4% *correct and* CF *for the three classes is* $[0.96, 0.94, 0.94]$.

- *Scheme* 3*:* 5 *features are selected based on the feature ranking initially resulting* 5% *misclassification,* 2 *sets and* 1 *feature are removed based on similarity,* 200 *GA iterations led to a reduction with* 1 *set, and the final result, after* 400 *GA-iterations is* 98.3% *with* CF *for the three classes is* $[0.93, 0.91, 0.91]$. *The final model contains features* $\{1, 7, 12, 13\}$. *The obtained sets for* $\{1, 7, 13\}$ *are similar as those shown in Figure* 5.5.

Concluding, the repetitive simplification and optimization left 6–9 *features without antecedent terms. Thus, feature reduction is obtained in the different schemes. Differences in the reduction methods are:* (i) *Similarity analysis results in a closed-loop feature selection because it depends on the actual model structure while the applied open-loop feature selection can be used beforehand as it is independent from the model.* (ii) *In similarity analysis, features can be removed from individual rules while in the interclass separability method, a feature is omitted in all the rules.*

Table 5.2: Three rule fuzzy classifier for the Wine data. The labels L, M, and H, corresponds to Low, Medium and High, respectively.

	1 Alc	2 Mal	3 Ash	4 aAsh	5 Mag	6 Tot	7 Fla	8 nFlav	9 Pro	10 Col	11 Hue	12 OD2	13 Pro	Class
R_1	H	-	-	-	-	-	H	-	-	M	L	-	L	1
R_2	L	-	-	-	-	-	-	-	-	L	L	-	H	2
R_3	H	-	-	-	-	-	L	-	-	H	H	-	H	3

Table 5.3: Classification rates on the Wine data for ten independent runs.

Method	Best result	Aver result	Worst result	Rules	Model eval
Corcoran and Sen [66]	100%	99.5%	98.3%	60	150000
Ishibuchi et al. [125]	99.4%	98.5%	97.8%	60	6000
Proposed method	99.4 %	varying schemes	98.3%	3	4000-8000

The obtained classifiers result in comparable results to those in [66] *and* [125], *but use far less rules* (3 *compared to* 60) *and less features. Comparing the fuzzy sets in Figure* 5.5 *with the data in Figure* 5.4 *shows that the obtained rules are highly interpretable.*

□

5.2.2 Conclusions

The design of fuzzy rule base classifiers is approached by combining separate tools for feature selection, model initialization, model reduction and model tuning. It is shown that these can be applied in an iterative way. A covariance-based model initialization method is applied to obtain an initial fuzzy classifier. Successive application of feature selection, rule base simplification and GA-based tuning resulted in compact and accurate classifiers. The presented approach was successfully applied to the Wine data.

5.3 Supervised Fuzzy Clustering for the Identification of Fuzzy Classifiers

The automatic determination of compact fuzzy classifiers rules from data has been approached by several different techniques. Generally, the bottleneck of the data-driven identification of fuzzy systems is the structure identification that requires nonlinear optimization as it was shown by Example 5.1 as well. Thus for high-dimensional problems, the initialization of the fuzzy model becomes very significant. Common initialization methods such as grid-type partitioning [125] and *rule generation on extrema* initialization, result in complex and non-interpretable initial models and the rule-base simplification and reduction steps become computationally demanding. To avoid these problems, fuzzy clustering algorithms [239] were put forward. However, the obtained membership values have to be projected onto the input variables and approximated by parameterized membership functions that deteriorates the performance of the classifier. This decomposition error can be reduced by using eigenvector projection [151], but the obtained linearly transformed input variables do not allow the interpretation of the model. To avoid the projection error and maintain the interpretability of the model, the presented approach is based on the Gath–Geva (GG) clustering algorithm [93] instead of the widely used Gustafson–Kessel (GK) algorithm [108], because the simplified version of GG clustering allows the direct identification of fuzzy models with exponential membership functions [119].

Neither the GG nor the GK algorithm utilizes the class labels. Hence, they give suboptimal results if the obtained clusters are directly used to formulate a classical fuzzy classifier. Hence, there is a need for fine-tuning of the model. This GA or gradient-based fine-tuning, however, can result in overfitting and thus poor generalization of the identified model. Unfortunately, the severe computational requirements of these approaches limit their applicability as a rapid model-development tool.

This section focuses on the design of interpretable fuzzy rule based classifiers from data with low-human intervention and low-computational complexity. Hence, a new modelling scheme is introduced based only on fuzzy clustering (see also in [17]). The presented algorithm uses the class label of each point to identify the optimal set of clusters that describe the data. The obtained clusters are then used to build a fuzzy classifier.

The contribution of this approach is twofold.

- The classical fuzzy classifier consists of rules each one describing one of the C classes. In this section a new fuzzy model structure is presented where the consequent part is defined as the probabilities that a given rule represents the $c_1, \ldots, c_C$ classes. The novelty of this new model is that one rule can represent more than one classes with different probabilities.
- Classical fuzzy clustering algorithms are used to estimate the distribution of the data. Hence, they do not utilize the class label of each data point available for the identification. Furthermore, the obtained clusters cannot be directly used to build the classifier. In this section a new cluster prototype and the related clustering algorithm have been introduced that allows the direct supervised identification of fuzzy classifiers.

The presented algorithm is similar to the Multi-Prototype Classifier technique [43, 227]. In this approach, each class is clustered independently from the other classes, and is modeled by few components (Gaussian in general). The main difference of this approach is that each cluster represents different classes, and the number of clusters used to approximate a given class have to be determined manually, while the presented approach does not suffer from these problems.

5.3.1 Supervised Fuzzy Clustering – the Algorithm

The objective of clustering is to partition the identification data $\mathbf{Z}$ into R clusters. This means, each observation consists of input and output variables, grouped into a row vector $\mathbf{z}_k = [\mathbf{x}_k^T \, y_k]$, where the k subscript denotes the $(k = 1, \ldots, N)$th row of the $\mathbf{Z} = [\mathbf{z}_k]_{N \times n+1}$ pattern matrix. The fuzzy partition, which can be found in the previous sections of this book, is represented by the $\mathbf{U} = [\mu_{i,k}]_{R \times N}$ matrix, where the $\mu_{i,k}$ element of the matrix represents the degree of membership, how the $\mathbf{z}_k$ observation is in the cluster $i = 1, \ldots, R$.

The clustering is based on the minimization of the sum of weighted $d^2(\mathbf{z}_k, r_i)$ squared distances between the data points and the η_i cluster prototypes that contain the parameters of the clusters.

$$J(\mathbf{Z}, \mathbf{U}, \eta) = \sum_{i=1}^{R} \sum_{k=1}^{N} (\mu_{i,k})^m \, d^2(\mathbf{z}_k, r_i). \tag{5.26}$$

Classical fuzzy clustering algorithms are used to estimate the distribution of the data. Hence, they do not utilize the class label of each data point available for the identification. Furthermore, the obtained clusters cannot be directly used to build the classifier. In the following a new cluster prototype and the related distance measure will be introduced that allows the direct supervised identification of fuzzy classifiers. As the clusters are used to obtain the parameters of the fuzzy classifier, the distance measure is defined similarly to the distance measure of the Bayes classifier (5.5):

$$\frac{1}{d^2(\mathbf{z}_k, r_i)} = \underbrace{P(r_i) \prod_{j=1}^{n} \exp\left(-\frac{1}{2} \frac{(x_{j,k} - v_{i,j})^2}{\sigma_{i,j}^2} \right)}_{\text{Gath–Geva clustering}} P(c_j = y_k | r_i). \tag{5.27}$$

This distance measure consists of two terms. The first term is based on the geometrical distance between the $\mathbf{v}_i$ cluster centers and the $\mathbf{x}_k$ observation vector, while the second is based on the probability that the r_ith cluster describes the density of the class of the kth data, $P(c_j = y_k | r_i)$. It is interesting to note that this distance measure only slightly differs from the unsupervised Gath–Geva clustering algorithm which can also be interpreted in a probabilistic framework [93]. However, the novelty of the presented approach is the second term, which allows the use of class labels.

Similarly to the update equations of Gath–Geva clustering algorithm, the following equations will result in a solution using Lagrange multipliers method.

Algorithm 5.3.1 (Supervised Fuzzy Clustering).

Initialization
Given a set of data $\mathbf{Z}$ *specify* R, *choose a termination tolerance* $\epsilon > 0$. *Initialize the* $\mathbf{U} = [\mu_{i,k}]_{R\times N}$ *partition matrix randomly, where* $\mu_{i,k}$ *denotes the membership that the* $\mathbf{z}_k$ *data is generated by the ith cluster.*

Repeat *for* $l = 1, 2, \ldots$

Step 1 *Calculate the parameters of the clusters*

- *Calculate the centers and standard deviation of the Gaussian membership functions (the diagonal elements of the* $\mathbf{F}_i$ *covariance matrices):*

$$\mathbf{v}_i^{(l)} = \frac{\sum_{k=1}^{N}\left(\mu_{i,k}^{(l-1)}\right)^m \mathbf{x}_k}{\sum_{k=1}^{N}\left(\mu_{i,k}^{(l-1)}\right)^m}, \quad \sigma_{i,j}^{2\,(l)} = \frac{\sum_{k=1}^{N}\left(\mu_{i,k}^{(l-1)}\right)^m (x_{j,k} - v_{j,k})^2}{\sum_{k=1}^{N}\left(\mu_{i,k}^{(l-1)}\right)^m}. \tag{5.28}$$

- *Estimate the consequent probability parameters,*

$$p(c_i|r_j) = \frac{\sum_{k|y_k=c_i}\left(\mu_{j,k}^{(l-1)}\right)^m}{\sum_{k=1}^{N}\left(\mu_{j,k}^{(l-1)}\right)^m}, 1 \le i \le C,\ 1 \le j \le R. \tag{5.29}$$

- A priori *probability of the cluster and the weight (impact) of the rules:*

$$P(r_i) = \frac{1}{N}\sum_{k=1}^{N}\left(\mu_{i,k}^{(l-1)}\right)^m, \quad w_i = P(r_i)\prod_{j=1}^{n}\frac{1}{\sqrt{2\pi\sigma_{i,j}^2}}. \tag{5.30}$$

Step 2 *Compute the distance measure* $d^2(\mathbf{z}_k, r_i)$ *by (5.27).*

Step 3 *Update the partition matrix*

$$\mu_{i,k}^{(l)} = \frac{1}{\sum_{j=1}^{R}\left(d(\mathbf{z}_k, r_i)/d(\mathbf{z}_k, r_j)\right)^{2/(m-1)}}, 1 \le i \le R,\ 1 \le k \le N. \tag{5.31}$$

until $||\mathbf{U}^{(l)} - \mathbf{U}^{(l-1)}|| < \epsilon$.

5.3.2 Performance Evaluation

In order to examine the performance of the presented identification method two well-known multidimensional classification benchmark problems are presented in this subsection. The studied Wisconsin Breast Cancer and Wine data come from the UCI Repository of Machine Learning Databases (*http://www.ics.uci.edu*).

The performance of the obtained classifiers was measured by ten-fold cross validation. The data divided into ten sub-sets of cases that have similar size and class distributions. Each sub-set is left out once, while the other nine are applied for the construction of the classifier which is subsequently validated for the unseen cases in the left-out sub-set.

Example 5.2 (The Wisconsin Breast Cancer classification based on supervised GG clustering). *The Wisconsin breast cancer data is widely used to test the effectiveness of classification and rule extraction algorithms. The aim of the classification is to distinguish between benign and malignant cancers based on the available nine measurements:* x_1 *clump thickness,* x_2 *uniformity of cell size,* x_3 *uniformity of cell shape,* x_4 *marginal adhesion,* x_5 *single epithelial cell size,* x_6 *bare nuclei,* x_7 *bland chromatin,* x_8 *normal nuclei, and* x_9 *mitosis (data shown in Figure* 5.6*). The attributes have integer value in the range* $[1, 10]$*. The original database contains* 699 *instances however* 16 *of these are omitted because these are incomplete, which is common with other studies. The class distribution is* 65.5% *benign and* 34.5% *malignant, respectively.*

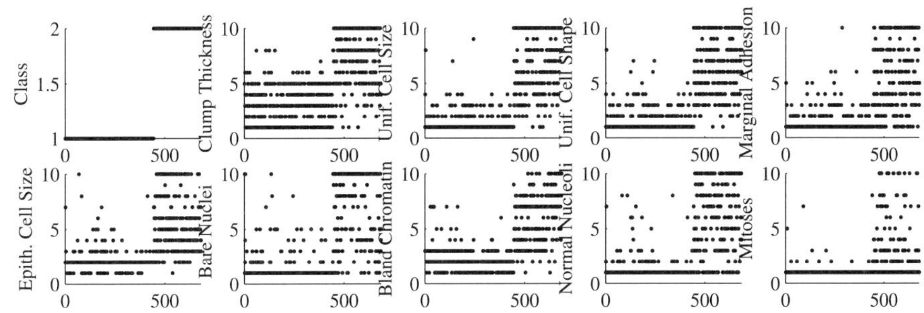

Figure 5.6: Wisconsin breast cancer data: 2 classes and 9 attributes (Class 1: 1-445, Class 2: 446-683).

The advanced version of C4.5 gives misclassification of 5.26% *on* 10*-fold cross validation* (94.74% *correct classification) with tree size* $25 \pm .5$ [225]*. Nauck and Kruse* [202] *combined neuro-fuzzy techniques with interactive strategies for rule pruning to obtain a fuzzy classifier. An initial rule-base was made by applying two sets for each input, resulting in* $2^9 = 512$ *rules which was reduced to* 135 *by deleting the non-firing rules. A heuristic data-driven learning method was applied instead of gradient descent learning, which is not applicable for triangular membership functions. Semantic properties were taken into account by constraining the search space. The final fuzzy classifier could be reduced to* 2 *rules with* 5*–*6 *features only, with a misclassification of* 4.94% *on* 10*-fold validation* (95.06% *classification accuracy). Rule-generating methods that combine GA and fuzzy logic were also applied to this problem* [199]*. In this method the number of rules to be generated needs to be determined* a priori*. This method constructs a fuzzy model that has four membership functions and one rule with an additional else part. Setiono*

Table 5.4: Classification rates and model complexity for classifiers constructed for the Wisconsin Breast Cancer problem. ♮ denotes results from averaging a ten-fold validation.

Author	Method	♯ Rules	♯ Conditions	Accuracy
Setiono [237]	NeuroRule 1f	4	4	97.36%
Setiono [237]	NeuroRule 2a	3	11	98.1%
Peña-Reyes and Sipper [199]	Fuzzy-GA1	1	4	97.07%
Peña-Reyes and Sipper [199]	Fuzzy-GA2	3	16	97.36%
Nauck and Kruse [202]	NEFCLASS	2	10-12	95.06% ♮

Table 5.5: Classification rates and model complexity for classifiers constructed for the Wisconsin Breast Cancer problem. Results from a ten-fold validation. GG: Gath–Geva clustering based classifier, Sup: presented method

Method	minAcc.	meanAcc.	maxAcc.	min ♯Feat.	mean♯Feat.	max ♯Feat.
GG: $R = 2$	84.28	90.99	95.71	8	8.9	9
Sup: $R = 2$	84.28	92.56	98.57	7	8	9
GG: $R = 4$	88.57	95.14	98.57	9	9	9
Sup: $R = 4$	90.00	95.57	98.57	8	8.7	9

[237] *has generated similar compact classifier by a two-step rule extraction from a feedforward neural network trained on preprocessed data.*

As Table 5.4 *shows, our fuzzy rule-based classifier is one of the most compact models in the literature with such high accuracy.*

In the current implementation of the algorithm after fuzzy clustering an initial fuzzy model is generated that utilizes all the 9 *information profile data about the patient. A step-wise feature reduction algorithm has been used where in every step one feature has been removed continuously checking the decrease of the performance of the classifier on the training data. To increase the classification performance, the classifier is re-identified in every step by re-clustering of reduced data which have smaller dimensionality because of the neglected input variables. As Table* 5.5 *shows, our supervised clustering approach gives better results than utilizing the Gath–Geva clustering algorithm in the same identification scheme.*

The ten-fold validation experiment with the presented approach showed 95.57% *average classification accuracy, with* 90.00% *as the worst and* 95.57% *as the best performance. This is really good for such a small classifier as compared with previously reported results (Table* 5.4*). As the error estimates are either obtained from* 10*-fold cross validation or from testing the solution once by using the* 50% *of the data as training set, the results given in Table* 5.4 *are only roughly comparable.*

□

Example 5.3 (The Wine data classification based on supervised GG clustering). *Firstly, for comparison purposes, a fuzzy classifier, that utilizes all the* 13 *information profile data about the wine has been identified by the presented clustering algorithm based on all the* 178 *samples (see Example* 2.1 *and Example* 5.1*). Fuzzy models with three and six rules were identified. The three rule-model gave only* 2 *misclassification (correct percentage* 98.9%*). When a cluster was added to improve the performance of this model, the obtained classifier gave only* 1 *misclassification (*99.4%*). The classification power of the identified models is then compared with fuzzy models with the same number of rules obtained by Gath–Geva clustering, as Gath–Geva clustering can be considered the unsupervised version of the presented clustering algorithm. The Gath–Geva identified fuzzy model achieves* 8 *misclassifications corresponding to a correct percentage of* 95.5%*, when three rules are used in the fuzzy model, and* 6 *misclassifications (correct percentage* 96.6%*) in the case of four rules. The results are summarized in Table* 5.6*. As it is shown, the performance of the obtained classifiers are comparable to those in* [66] *and* [125]*, but use far less rules (*3–5 *compared to* 60*) and less features.*

Table 5.6: Classification rates on the Wine data for ten independent runs.

Method	Best result	Aver result	Worst result	Rules	Model eval
Corcoran and Sen [66]	100%	99.5%	98.3%	60	150000
Ishibuchi et al. [125]	99.4%	98.5%	97.8%	60	6000
GG clustering	95.5 %	95.5 %	95.5 %	3	1
Sup (13 features)	98.9 %	98.9 %	98.9 %	3	1
Sup (13 features)	99.4 %	99.4 %	99.4 %	4	1

These results indicate that the presented clustering method effectively utilizes the class labels. As can be seen from Table 5.6*, because of the simplicity of the presented clustering algorithm, the presented approach is attractive in comparison with other iterative and optimization schemes that involves extensive intermediate optimization to generate fuzzy classifiers.*

The ten-fold validation is a rigorous test of the classifier identification algorithms. These experiments showed 97.77% *average classification accuracy, with* 88.88% *as the worst and* 100% *as the best performance (Table* 5.7*). The above presented automatic model reduction technique removed only one feature without the decrease of the classification performance on the training data. Hence, to avoid possible local minima, the feature selection algorithm is used to select only five features, and the presented scheme has been applied again to identify a model based on the selected five attributes. This compact model with average* 4.8 *rules showed* 97.15% *average classification accuracy, with* 88.23% *as the worst and* 100% *as the best performance. The resulted membership functions and the selected features*

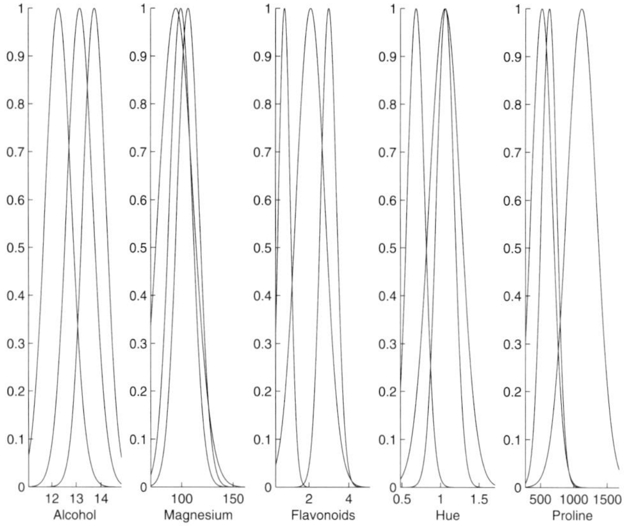

Figure 5.7: Membership functions obtained by fuzzy clustering.

Table 5.7: Classification rates and model complexity for classifiers constructed for the Wine classification problem. Results from averaging a ten-fold validation.

Method	minAcc.	meanAcc.	maxAcc.	min ♯Feat.	mean♯Feat.	max ♯Feat.
GG: $R = 3$	83.33	94.38	100	10	12.4	13
Sup: $R = 3$	88.88	97.77	100	12	12.6	13
GG: $R = 3$	88.23	95.49	100	4	4.8	5
Sup: $R = 3$	76.47	94.87	100	4	4.8	5
GG: $R = 6$	82.35	94.34	100	4	4.9	5
Sup: $R = 6$	88.23	97.15	100	4	4.8	5

are shown in Figure 5.7*. Comparing the fuzzy sets in Figure* 5.7 *with the data in Figure* 5.4 *shows that the obtained rules are highly interpretable. For example, the Flavonoids are divided in Low, Medium and High, which is clearly visible in the data.*

□

5.3.3 Conclusions

A new fuzzy classifier has been presented to represent Bayes classifiers defined by mixture of Gaussians density model. The novelty of this new model is that each rule can represent more than one classes with different probabilities. For the identification of the fuzzy classifier a supervised clustering method has been worked out that is the modification of the unsupervised Gath–Geva clustering algorithm. In addition, a method for the selection of the relevant input variables has been presented. The presented identification approach is demonstrated by the Wisconsin Breast Cancer and the Wine benchmark classification problems. The comparison to Gath–Geva clustering and GA-tuned fuzzy classifiers indicates that the presented supervised clustering method effectively utilizes the class labels and is able to identify compact and accurate fuzzy systems.

5.4 Supervised Clustering based Fuzzy Partitioning for Fuzzy Decision Tree Induction

This section will investigate how supervised clustering can be used for the effective partitioning of the input domains (see also in [208]). The application of fuzzy clustering for the quantization of the input variables is not a completely new idea. In [215] it has been shown that the results of the clustering coming in the form of a series of prototypes can be directly used to complete a quantization of the continuous attributes. In contrast with most discretization of continuous variables that deal with a single variable only, this approach concerns all the variables discussed at the same time. The discretization mechanism is straightforward: project the cluster prototypes on the respective variables (coordinate axes) and construct the discretization intervals (Figure 5.8).

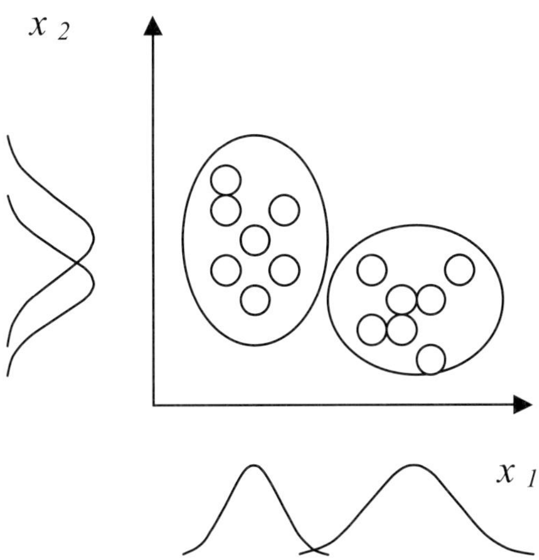

Figure 5.8: Clustering based partitioning of the input domains.

Our approach differs from the previously presented methods in the following main issues:

- It uses the *extended version of fuzzy classifiers* that is presented in Section 5.1.4.
- *Rule-based interpretation and rule-reduction.* After the induction of the tree the resulted decision tree can be transformed into a rule-based fuzzy system without any approximation error. Since FID is based on the ID3 algorithm (see Section 5.1.5), it generates trees where every branch of the tree contains all of the fuzzy tests defined on the domain of the selected variable. Hence, the generated fuzzy decision tree is often more complex than is needed, which often leads to unnecessarily complex classifiers due to the addition of meaningless rules. Hence, there is a place for rule-reduction tools as well.
- The main idea of the section is the application of the previously presented *supervised clustering algorithm* (see Section 5.3) to generate the fuzzy sets used by the fuzzy decision tree induction algorithms.
- *Similarity-driven simplification.* Since FID assumes fuzzy domains with small cardinalities, the performance and the size of the inducted trees are highly determined by the quality and the number of the membership functions extracted from the clustering. Hence, to obtain a parsimonious and interpretable fuzzy classifiers similarity-driven rule base simplification algorithm was applied (Section 5.2.1) to merge the similar fuzzy sets on each input domain.
- *Obtaining fuzzy data and fuzzy partitioning.* Contrary to other clustering based input partitioning approaches, the results of the applied supervised clustering algorithm can be directly used by fuzzy decision tree induction algorithms.

  Beside the effective utilization of the class label information, the main benefit of the applied clustering algorithm is that the clusters are represented by Gaussian membership functions, hence there is no need to project the resulted clusters into the input variables of the model, so there is no projection error that decreases the performance of the model building procedure (see also Algorithm 3.3.1, 3.4.2, 4.1.1, 4.2.1 and 5.3.1). According to the utilized results of the fuzzy clustering algorithm two approaches can be followed at the step of the induction of the decision tree.

  - SC-FID1: The obtained membership functions are only used for the quantization of the input variables.
  - SC-FID2: The fuzzy clustering is considered as a tool to obtain a compact representation of the class distribution in terms of a fuzzy model. In this approach, each rule of the fuzzy classifier identified by the clustering algorithm is considered as a fuzzy data, so the entire rule base is

a compact "fuzzy database". In this approach the FID algorithm does not utilize the original crisp data set, but it generates the fuzzy decision tree based on fuzzy data generated by the clustering algorithm.

This section deals with the SC-FID1 approach, since this approach can be used for the partitioning of the input variables.

The presented approach is applied to four well-known classification problems available from the Internet: to the Wisconsin Breast Cancer, the Iris, the Wine and the Thyroid classification problems in Example 5.4.

5.4.1 Fuzzy Decision Tree Induction

Decision trees can be represented in terms of logical rules, where each concept is represented by one disjunctive normal form, and where the antecedent consists of a sequence of attribute value tests. These attribute value tests partition the input domains of the classifier into intervals represented by the fuzzy sets, and the operating region of the rules is formulated by **and** connective of these domains. This interpretation of decision trees allows their transformation into rule-based (fuzzy) systems:

Algorithm 5.4.1 (Fuzzy Classification Tree Induction).

Step 0 *Index the leaves (the terminal nodes) of the tree by* l_i, *where* $i = 1, \ldots, R$, *where* R *is the number of leaves that will be identical to the number of the rules of the fuzzy classifier.*

Step 1 *Select a terminal node* l_i, $i \in \{1, \ldots, R\}$ *and collect all the attribute value tests* $A_{i,j}$ *related to the path of the selected terminal node to the root of the tree, where* $j \in \{1, \ldots, n\}$.

Step 2 *The* $A_{i,j}$ *attribute value tests (membership functions) define the antecedent part of the ith rule, and the conditional probabilities* $P(c_1|r_i), \ldots, P(c_C|r_i)$ *at the consequent part of the rule are given by the example counts computed by the FID algorithm* [133].

until $i = R$ *(all of the rules are recorded for all the leaves in the rule-based system).*

Since FID is based on the ID3 algorithm, it generates trees where every branches of the tree contain all of the fuzzy tests defined on the domain of the selected variable. Hence, the generated fuzzy decision tree is often more complex than is needed due to the addition of meaningless rules. Hence, the rules that are responsible for only a small number of data samples are erased form the rule-base, because they only cover exceptions or noise in the data.

In FID 3.3 trapezoidal membership functions are used to describe the fuzzy sets $A_{i,j}(x_j)$:

$$A_{i,j}(x_j; a_{i,j}, b_{i,j}, c_{i,j}, d_{i,j}) = \max\left(0, \min\left(\frac{x_j - a_{i,j}}{b_{i,j} - a_{i,j}}, 1, \frac{d_{i,j} - x_j}{d_{i,j} - c_{i,j}}\right)\right). \quad (5.32)$$

The problem that is addressed in this section is how the previously presented parameters of the fuzzy sets (5.32) can be obtained to provide an effective partition of the input variables to the fuzzy decision induction algorithm. The key idea of our solution is to use a supervised fuzzy clustering algorithm for this purpose.

5.4.2 Transformation and Merging of the Membership Functions

The supervised clustering algorithm presented in Section 5.3 obtains a rule-based fuzzy classifier defined with rules in the format given by (5.16) with Gaussian membership functions. Since the public implementation of the FID algorithm (FID 3.3) uses trapezoidal membership functions to describe the $A_{i,j}(x_j)$ fuzzy sets (5.32), there is a need for a method to transform the obtained Gaussian membership functions into trapezoidal ones. For this transformation the parameters of the Gaussian membership function can be used: $a_{i,j} = v_{i,j} - 3\sigma_{i,j}$, $b_{i,j} = v_{i,j} - 0.5\sigma_{i,j}$, $c_{i,j} = v_{i,j} + 0.5\sigma_{i,j}$ and $d_{i,j} = v_{i,j} + 3\sigma_{i,j}$. To obtain a parsimonious and interpretable fuzzy classifier a fuzzy set-merging algorithm can be applied which can be found in the introduction of this section.

The obtained membership functions are only used for the quantization of the input variables, so the FID algorithm generates a fuzzy decision tree based on the original training data and the obtained fuzzy partitioning. In the remaining part of the section this approach is referred as SC-FID1 algorithm, since **S**upervised **C**lustering is used to provide input data for the **F**uzzy **I**nduction of a **D**ecision tree.

For the induction of the decision trees the standard FID algorithm is used without any modification. To make the application of this program easy in the MATLAB® programming environment, we developed an interface which supports all the functions of FID 3.3 in the MATLAB® environment. This FID Toolbox is available from our website: *http://www.fmt.vein.hu/softcomp*. The main feature of this interface is that the generated tree can be transformed into a rule-based fuzzy model in the format of the Fuzzy Model Identification Toolbox (*http://www.dcsc.tudelft.nl/~babuska/*).

Example 5.4 (Comparative application study to demonstrate the effectiveness of the Fuzzy Classification Tree). *This subsection is intended to provide a comparative study based on a set of multivariate classification problems to present how the performance and the complexity of the classifier is changing trough the step-wise model building procedure of the previously presented SC-FID*1 *algorithm.*

The selected Iris, Wine, Wisc and Thyroid data sets, coming from the UCI Repository of Machine Learning Databases (http://www.ics.uci.edu), are examples of classification problems with different complexity, e.g., large and small numbers of features (see Table 5.8).

Table 5.8: Complexity of the classification problems.

Problem	♯Samples	♯Features	♯Classes
Iris	150	4	3
Wine	178	13	3
Wisc	699	9	2
Thyroid	215	5	3

During the experiments, the performances of the classifiers are measured by ten-fold cross validation. This means, that the data set is divided into ten sub-sets, and each sub-set is left out once, while the other nine are applied for the construction of the classifier which is subsequently validated for unseen cases in the left-out sub-set.

*To demonstrate the effectiveness of the presented SC-FID*1 *approach, the results obtained by the supervised (Supclust) and unsupervised Gath–Geva (GGclust) clustering based fuzzy classifiers* (5.16) *(see* [17] *for more details) are compared to the performances of fuzzy decision trees generated by the uniformly distributed (Ruspini-type) partitioning of the input variables. To illustrate how the supervision of the clustering improves the classification performance, not only partitioning based on the supervised clustering algorithm has been used for the induction of the trees (SC-FID*1*), but the results of the unsupervised Gath–Geva clustering (C-FID*1*). Table* 5.9 *shows the results of this comparison study, while the performances of the classifiers during the step-wise model building procedure are shown in Figure* 5.9*, Figure* 5.10*, Figure* 5.11 *and Figure* 5.12.

Table 5.9: Classification rates (acc.) achieved on the WINE classification problem. Average results of tenfold validation at the number of initial membership functions of the input variables are: 3–7.

♯Fuzzy Sets	Supclust	SC-FID1	GGclust	C-FID1	Ruspini
$c = 3$	96.63	96.08	93.85	94.90	96.60
$c = 4$	95.00	96.05	92.74	95.98	96.05
$c = 5$	94.41	96.60	90.46	95.46	93.79
$c = 6$	95.49	95.49	92.71	96.05	92.65
$c = 7$	97.22	95.46	94.44	96.60	89.77

In these figures the presented SC-FID1 is monitored by logging the number of rules, conditions and performances of the classifiers in the function of the number of clusters (i.e., the initial number of fuzzy sets on each input variables). As it appears from the figures, the best performances are usually obtained by the rule-based fuzzy classifiers by the supervised clustering algorithm (Supclust). The accuracy of these models decreases considerably after the transformation of the Gaussian membership function into trapezoidal ones. However, after the merging of membership functions and the induction of the decision tree accurate, yet compact classifiers can be obtained.

The fuzzy decision trees induced based on uniform (Ruspini-type) partition of the input variables gives worse classification performances and much complex classifiers compacted to the clustering based results, so the effectiveness of rule reduction method appears. Hence, these results confirm that the presented approach (SC-FID1) generates transparent and compact fuzzy classifiers, thanks to the effective input partitioning obtained by the supervised clustering.

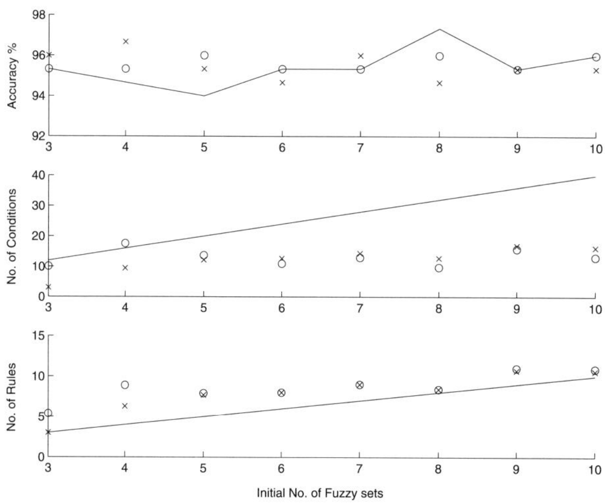

Figure 5.9: Classification performances and complexities of the classifiers on the IRIS data set in the function of the number of clusters. (0: Ruspini, X: SC-FID1, –: Supculst)

□

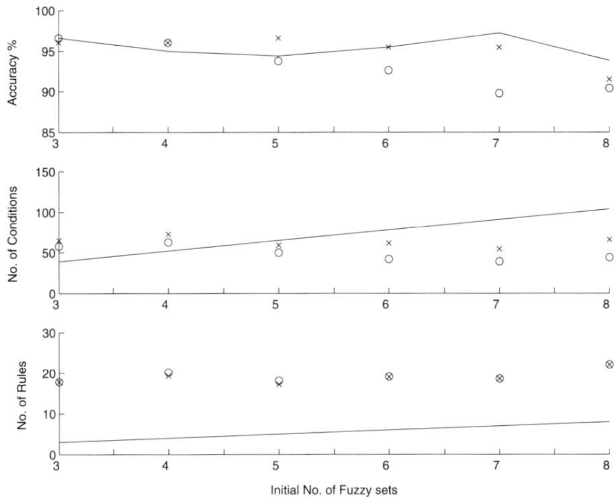

Figure 5.10: Classification performances on the WINE data set. (0: Ruspini, X: SC-FID1, –: Supculst)

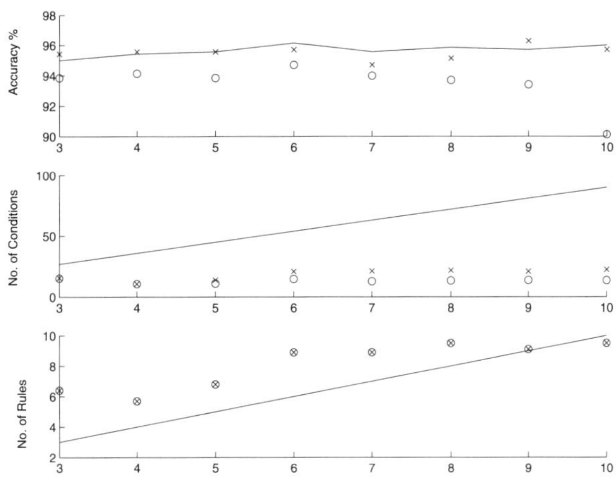

Figure 5.11: Classification performances on the WISC data set. (0: Ruspini, X: SC-FID1, –: Supculst)

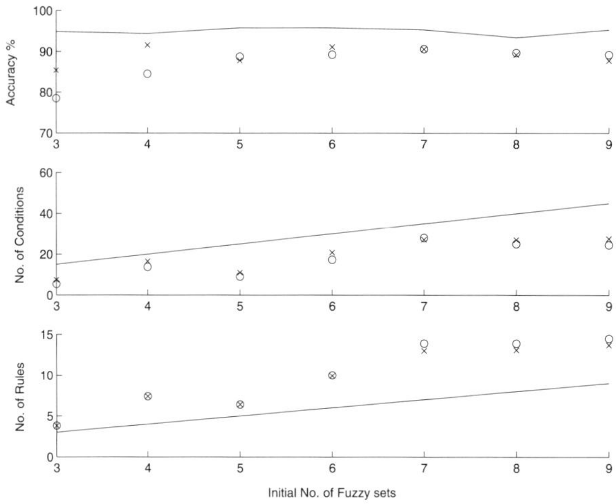

Figure 5.12: Classification performances on the THYROID data set. (0: Ruspini, X: SC-FID1, –: Supculst)

5.4.3 Conclusions

In this section a new approach to the identification of compact and accurate fuzzy classifiers has been presented. For the identification of the fuzzy classifier a supervised clustering method has been used to provide input partitioning to the fuzzy decision tree induction algorithm. The presented SC-FID1 approach is demonstrated by the Wisconsin Breast Cancer, the Iris, the Thyroid and the Wine benchmark classification problems. The comparison to the uniform partitioning based decision tree induction method indicates that the presented supervised clustering method effectively utilizes the class labels and is able to lead to compact and accurate fuzzy systems with the help of a standard decision tree induction algorithm.

Chapter 6

Segmentation of Multivariate Time-series

Partitioning a time-series into internally homogeneous segments is an important data mining problem. The changes of the variables of a multivariate time-series are usually vague and do not focus on any particular time point. Therefore, it is not practical to define crisp bounds of the segments. Although fuzzy clustering algorithms are widely used to group overlapping and vague objects, they cannot be directly applied to time-series segmentation, because the clusters need to be contiguous in time. This chapter proposes a clustering algorithm for the simultaneous identification of local Probabilistic Principal Component Analysis (PPCA) models used to measure the homogeneity of the segments and fuzzy sets used to represent the segments in time. The algorithm favors contiguous clusters in time and is able to detect changes in the hidden structure of multivariate time-series. A fuzzy decision making algorithm based on a compatibility criteria of the clusters have been worked out to determine the required number of segments, while the required number of principal components are determined by the screeplots of the eigenvalues of the fuzzy covariance matrices. The application example shows that this new technique is a useful tool for the analysis of historical process data.

6.1 Mining Time-series Data

A sequence of N observed data, $\mathbf{x}_1, \ldots, \mathbf{x}_N$, ordered in time, is called time-series. Sequence data are similar to time-series but without time label: they consist of sequences of ordered events (e.g., DNA-sequences). Mining of time-series data could be very important because

- time-series databases are very large, and long multivariate time-series cannot be handled well by human inspectors,

- they definitely can contain valuable information,
- real-life time-series can be taken from business, physical, social and behavioral science, economics, engineering, etc.

There are various tasks related to time-series, but they form basically two groups:

- to identify the nature of the phenomenon that generates the time-series,
- to predict the future values of the variables (see, e.g., [152]).

Related to that, particular tasks can be: trend analysis, seasonal pattern search, similarity search, segmentation etc. In the following sections the latter will be deeply described, but the others are briefly discussed here.

Time-series can be viewed (as in case of much other analysis as well) that the data contain a systematic pattern (whose components can be identified), and random noise (e.g., measuring error) that makes it difficult to recognize the patterns. The systematic pattern according to [110] can be decomposed into three main components or movements:

- trend movement (the main direct of the variable over a long period of time),
- cyclic movements (long period oscillations of the trend movement, may be periodic or not)
- seasonal movements (events that occur from time to time, e.g., in every year)

Time-series analysis can be viewed as time-series decomposition into the elements mentioned above.

Trend analysis. To determine the trend in a time-series one of the most often applied methods is the (weighted) moving average. It means that instead of the actual value of the variable the average of data in a 'window' is used. The length of the window has an effect on the range of the smoothing. There are other methods, e.g., free-hand method, where the user draws an approximation curve to the data. It cannot be applied in case of large datasets, but its 'automated version' can be, where a simple curve is fitted to the data based on the least squares criterion (e.g., spline fitting).

Seasonality analysis. It is formally defined as correlational dependency of order k between each ith element of the series and the $(i-k)$th element and measured by autocorrelation (i.e., a correlation between the two terms); k is usually called the lag. (See also Section 4.4.2.) If the measurement error is not too large, seasonality can be visually identified in the series as a pattern that repeats every k elements.

- Autocorrelation correlogram. Seasonal patterns of time series can be examined via correlograms. The correlogram (autocorrelogram) displays graphically and numerically the autocorrelation function, that is, serial correlation coefficients (and their standard errors) for consecutive lags in a specified range of lags (e.g., 1 through 30). Ranges of two standard errors for each lag are usually marked in correlograms but typically the size of auto correlation is of more interest than its reliability because we are usually interested only in very strong (and thus highly significant) autocorrelations.

- Examining correlograms. While examining correlograms one should keep in mind that autocorrelations for consecutive lags are formally dependent. Consider the following example. If the first element is closely related to the second, and the second to the third, then the first element must also be somewhat related to the third one, etc. This implies that the pattern of serial dependencies can change considerably after removing the first-order auto correlation (i.e., after differencing the series with a lag of 1).
- Partial autocorrelations. Another useful method to examine serial dependencies is to examine the partial autocorrelation function – an extension of autocorrelation, where the dependence on the intermediate elements (those within the lag) is removed. In other words the partial autocorrelation is similar to autocorrelation, except that when calculating it, the (auto) correlations with all the elements within the lag are partialled out.

One aim can be removing the serial dependency. Serial dependency for a particular lag of k can be removed by differencing the series, that is converting each ith element of the series into its difference from the $(i-k)$th element. There are two major reasons for such transformations.

First, one can identify the hidden nature of seasonal dependencies in the series. Remember that, as mentioned in the previous paragraph, autocorrelations for consecutive lags are interdependent. Therefore, removing some of the autocorrelations will change other auto correlations, that is, it may eliminate them or it may make some other seasonalities more apparent. The other reason for removing seasonal dependencies is to make the series stationary which is necessary for several techniques (e.g., ARIMA).

Similarity search. It finds data sequences that differ only slightly from the given query sequence. Given a set of time-series sequences, there are two types of similarity search. Subsequence matching finds all of the data sequences that are similar to the given sequence, while whole sequence matching finds those sequences that are similar to one other. As a similarity measure, Euclidean distance is typically used, but often useful to transform the data from time domain into frequency domain, discrete Fourier transform of discrete wavelet transform can be used.

6.2 Time-series Segmentation

In this chapter a fuzzy clustering based algorithm is presented which is useful for the fuzzy segmentation of multivariate temporal databases (these results were published in [10]). Time-series segmentation addresses the following data mining problem: given a time-series, T, find a partitioning of T into c segments that are internally homogeneous [272]. Depending on the application, the goal of the segmentation is to locate stable periods of time, to identify change points, or to simply compress the original time-series into a more compact representation [170]. Although in many real-life applications a lot of variables must be simultaneously tracked and monitored, most of the segmentation algorithms are used for the analysis of only one time-variant variable [154]. Hoverer, in some cases it is necessary to synchronously segment the time-series of the variables.

The segmentation of multivariate time-series is especially important in the data-based analysis and monitoring of modern production systems, where huge amount of historical process data are recorded with distributed control systems (DCS). These data definitely have the potential to provide information for product and process design, monitoring and control [290]. This is especially important in many practical applications where first-principles modelling of complex "data rich and knowledge poor" systems are not possible [143]. Therefore, KDD methods have been successfully applied to the analysis of process systems, and the results have been used in process design, process improvement, operator training, and so on [281]. Hence, the data mining algorithm presented in this chapter has been developed to the analysis of the historical process data of a medium and high-density polyethylene (MDPE, HDPE) plant. The operators of this polymerization process should simultaneously track many process variables. Of course, due to the hidden nature of the system the measured variables are correlated. Hence, it is useful to monitor only some principal components that is widely applied in advanced process monitoring. The main problem of this approach is the fact that in some cases the hidden process, which can be observed as the correlation among the variables, varies in time. In our example this phenomenon can occur when a different product is formed, and/or different catalyst is applied, or there are significant process faults, etc. The segmentation of only one measured variable is not able to detect such changes. Hence, the segmentation algorithm should be based on multivariate statistical tools.

To demonstrate this problem let us consider the synthetic dataset shown in Figure 6.1. The observed variables that can be seen in Figure 6.1(b) are not independent, they were generated by the latent variables shown in Figure 6.1(a). The correlation among the observed variables changes at the quarter of the time period, and the mean of the latent variables changes at the half of the time period. These changes are marked by vertical lines in Figure 6.1(a).

As it can be seen in Figure 6.1(b), such information can be detected neither by application of univariate segmentation algorithms, nor by the visual inspection of the observed variables. Hence, the aim of this chapter is to develop an algorithm that is able to handle time-varying characteristics of multivariate data: (i) changes in the mean; (ii) changes in the variance; and (iii) changes in the correlation structure among the variables.

To discover that type of changes of the hidden relationships of multivariate time-series, multivariate statistical tools should be applied by the segmentation algorithm. Among the wide range of possible tools, e.g., random projection, independent component analysis, the presented algorithm utilizes Principal Com-

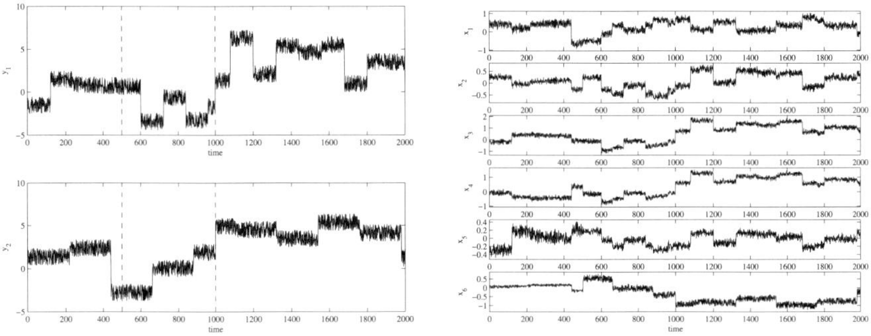

(a) Latent variables *(b) Observed variables*

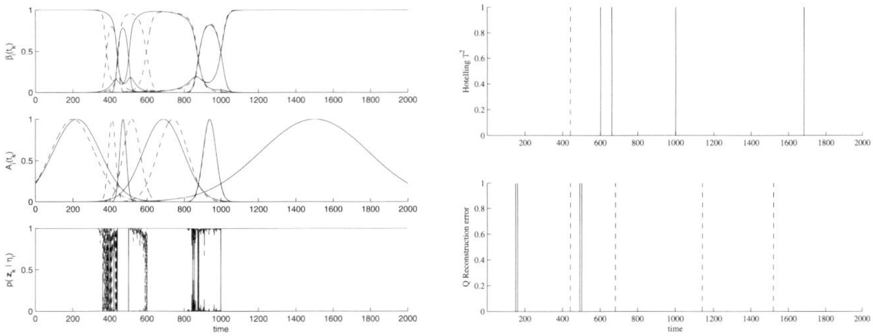

(c) Results obtained by fuzzy clustering, (–): $q = 2$,(- -): $q = 5$

(d) Results obtained by the bottom-up algorithm, (–): $q = 2$,(- -): $q = 5$

Figure 6.1: The synthetic dataset and its segmentation by different algorithms based on two and five principal components.

ponent Analysis (PCA) (for more details see Section 2.1.1). Linear PCA can give good prediction results for simple time-series, but can fail in the analysis of historical data having changes in regime or having nonlinear relations among the variables. The analysis of such data requires the detection of locally correlated clusters [61]. These algorithms do the clustering of the data to discover the local relationship among the variables similarly to mixture of Principal Component Models [265].

Mixtures have been extensively used as models where the data can be considered as originating from several populations mixed in varying proportions, and Expectation Maximization (EM) is widely used to estimate the parameters of the components in a mixture [44]. The clusters obtained by Gath–Geva (GG) clustering, also referred to Fuzzy Maximum Likelihood clustering, are multivariate Gaussian functions. The Alternating Optimization (AO) of these clusters is identical to the Expectation Maximization (EM) (maximum likelihood estimation) identification of the mixture of these Gaussian models with fuzzy weighting exponent $m = 2$ [40].

Time-series segmentation may be considered as clustering with a time-ordered structure. The contribution of this chapter is the introduction of a new fuzzy clustering algorithm which can be effectively used to segment large, multivariate time-series. Since the points in a cluster must come from successive time points, the time-coordinate of the data has to be also considered during the clustering. One possibility to deal with time is to define a new cluster prototype that uses time as an additional variable. Hence, the clustering is based on a distance measure which consists of two terms: the first distance term is based on how the data are in the given segment defined by the Gaussian fuzzy sets defined in the time domain, while the second term measures how far the data are from the hyperplane of the PCA model of the segments.

The fuzzy segmentation of time-series is an adequate idea. The changes of the variables of the time-series are usually vague and are not focused on any particular time point. Therefore, it is not practical to define crisp bounds of the segments. For example, if humans visually analyze historical process data, they use expressions like "this point belongs to this operating point less and belongs to the other more". A good example of this kind of fuzzy segmentation is how fuzzily the start and the end of *early morning* is defined. Fuzzy logic is widely used in various applications where the grouping of overlapping and vague objects is necessary [1], and there are many fruitful examples in the literature for the combination of fuzzy logic with time-series analysis tools [35, 96, 170, 284].

The key problem of the application of fuzzy clustering for time-series segmentation is the selection of the number of segments for the clustering process. Obviously, this is a standard problem also in the classical c-means clustering. In the context of time series, however, it appears to be even more severe. For this purpose a bottom-up algorithm has been worked out where the clusters are merged during a recursive process. The cluster merging is coordinated by a fuzzy decision making algorithm which utilizes a compatibility criterion of the clusters [146], where this criterion is calculated by the similarity of the Principal Component Models of the clusters [166].

Time-series Segmentation Problem Formulation

A time-series $T = \{\mathbf{x}_k | 1 \leq k \leq N\}$ is a finite set of N samples labeled by time points $t_1, \ldots, t_N$, where $\mathbf{x}_k = [x_{1,k}, x_{2,k}, \ldots, x_{n,k}]^T$. A segment of T is a set of consecutive time points $S(a,b) = \{a \leq k \leq b\}$, $\mathbf{x}_a, \mathbf{x}_{a+1}, \ldots, \mathbf{x}_b$. The c-segmentation of time-series T is a partition of T to c non-overlapping segments $S_T^c = \{S_i(a_i, b_i) | 1 \leq i \leq c\}$, such that $a_1 = 1, b_c = N$, and $a_i = b_{i-1} + 1$. In other words, a c-segmentation splits T to c disjoint time intervals by segment boundaries $s_1 < s_2 < \ldots < s_c$, where $S_i(s_{i-1} + 1, s_i)$.

Usually the goal is to find homogeneous segments from a given time-series. In such cases the segmentation problem can be defined as constrained clustering: data points should be grouped based on their similarity, but with the constraint that all points in a cluster must come from successive time points. (See [174] for the relationship of time series and clustering from another point of view.) In order to formalize this goal, a $\text{cost}(S(a,b))$ cost function with the internal homogeneity of individual segments should be defined. The cost function can be any arbitrary function. For example in [117, 272] the sum of variances of the variables in the segment was defined as $\text{cost}(S(a,b))$. Usually, the $cost(S(a,b))$ cost function is defined based on the distances between the actual values of the time-series and the values given by a simple function (constant or linear function, or a polynomial of a higher but limited degree) fitted to the data of each segment. Hence, the optimal c-segmentation simultaneously determines the a_i, b_i borders of the segments and the θ_i parameter vectors of the models of the segments by minimizing the cost of c-segmentation which is usually the sum of the costs of the individual segments:

$$\text{cost}(S_T^c) = \sum_{i=1}^{c} \text{cost}(S_i)\,. \tag{6.1}$$

This cost function can be minimized by dynamic programming, which is computationally intractable for many real datasets [117]. Consequently, heuristic optimization techniques such as greedy top-down or bottom-up techniques are frequently used to find good but suboptimal c-segmentations [149, 252]:

Sliding window: A segment is grown until it exceeds some error bound. The process repeats with the next data point not included in the newly approximated segment. For example a linear model is fitted on the observed period and the modelling error is analyzed.

Top-down method: The time-series is recursively partitioned until some stopping criterion is met.

Bottom-up method: Starting from the finest possible approximation, segments are merged until some stopping criterion is met.

Search for inflection points: Searching for primitive episodes located between two inflection points.

In the next section a different approach will be presented based on the clustering of the data.

Bottom-Up Segmentation Algorithm

In data mining, the bottom-up algorithm has been used extensively to support a variety of time-series data mining tasks [149]. The algorithm begins creating a fine approximation of the time-series, and iteratively merge the lowest cost pair of segments until a stopping criteria is met. When the S_i and S_{i+1} pair of adjacent segments are merged, the cost of merging the new segment with its right neighbor and the cost of merging the S_{i-1} segment with its new larger neighbor must be calculated. The pseudocode for algorithm is shown in Table 6.1.

Table 6.1: Bottom-up segmentation algorithm

- Create initial fine approximation.
- Find the cost of merging for each pair of segments:
  $mergecost(i) = cost(S(a_i, b_{i+1}))$
- while $min(mergecost) < maxerror$
  - Find the cheapest pair to merge: $i = argmin_i(mergecost(i))$
  - Merge the two segments, update the a_i, b_i boundary indices, and recalculate the merge costs.
    $mergecost(i) = cost(S(a_i, b_{i+1}))$
    $mergecost(i-1) = cost(S(a_{i-1}, b_i))$

  end

This algorithm is quite powerful since the merging cost evaluations requires simple identifications of PCA models which is easy to implement and computationally cheap to calculate. Because of this simplicities and because PCA defines linear hyperplane, the presented approach can be considered as the multivariate extension of the piecewise linear approximation (PLA) based time-series segmentation and analysis tools developed by Keogh [149, 150].

Principal Component Analysis (PCA) was described in details in Section 2.1.1. However, it is advantageous to overview this method just because of the notation as well. PCA is based on the projection of correlated high-dimensional data onto a hyperplane. This mapping uses only the first few q nonzero eigenvalues and the corresponding eigenvectors of the $\mathbf{F}_i = \mathbf{U}_i \Lambda_i \mathbf{U}_i^T$, covariance matrix, decomposed to the Λ_i matrix that includes the eigenvalues $\lambda_{i,j}$ of $\mathbf{F}_i$ in its diagonal in decreasing order, and to the $\mathbf{U}_i$ matrix that includes the eigenvectors corresponding to the eigenvalues in its columns. The vector $\mathbf{y}_{i,k} = \mathbf{W}_i^{-1}(\mathbf{x}_k) = \mathbf{W}_i^T(\mathbf{x}_k)$ is a q-dimensional reduced representation of the observed vector $\mathbf{x}_k$, where the $\mathbf{W}_i$ weight matrix contains the q principal orthonormal axes in its column $\mathbf{W}_i = \mathbf{U}_{i,q}\Lambda_{i,q}^{\frac{1}{2}}$.

Based on PCA the cost(S_i) can be calculated in two ways. This cost can be equal to the reconstruction error of this segment

$$\text{cost}(S_i) = \frac{1}{b_i - a_i + 1} \sum_{k=a_i}^{b_i} Q_{i,k} \tag{6.2}$$

where $Q_{i,k} = (\mathbf{x}_k - \hat{\mathbf{x}}_k)^T(\mathbf{x}_k - \hat{\mathbf{x}}_k) = \mathbf{x}_k^T(\mathbf{I} - \mathbf{U}_{i,p}\mathbf{U}_{i,p}^T)\mathbf{x}_k$. When the hyperplane of the PCA model has an adequate number of dimensions, the distance of the data from the hyperplane is resulted by measurement failures, disturbances and negligible information, so the projection of the data into this p-dimensional hyperplane does not cause significant reconstruction error.

Although the relationship among the variables can be effectively described by a linear model, in some cases it is possible that the data is distributed around some separated centers in this linear subspace. The Hotelling T^2 measure is often used to calculate the distance of the data point from the center in this linear subspace. This can be also used to compute $\text{cost}(S_i)$

$$\text{cost}(S_i) = \frac{1}{b_i - a_i + 1} \sum_{k=a_i}^{b_i} T_{i,k}^2 = \frac{1}{b_i - a_i + 1} \sum_{k=a_i}^{b_i} \mathbf{y}_{i,k}^T \mathbf{y}_{i,k}. \tag{6.3}$$

6.3 Fuzzy Cluster based Fuzzy Segmentation

When the variance of the segments are minimized during the segmentation, equation (6.1) results in the following equation:

$$\text{cost}(S_T^c) = \sum_{i=1}^{c} \sum_{k=s_{i-1}+1}^{s_i} \parallel \mathbf{x}_k - \mathbf{v}_i^x \parallel^2 = \sum_{i=1}^{c} \sum_{k=1}^{N} \beta_i(t_k) d^2(\mathbf{x}_k, \mathbf{v}_i^x), \tag{6.4}$$

where $d^2(\mathbf{x}_k, \mathbf{v}_i^x)$ represents the distance between the $\mathbf{v}_i^x$ mean of the variables in the ith segment (center of the ith cluster) and the $\mathbf{x}_k$ data point; and $\beta_i(t_k) \in \{0, 1\}$ stands for the crisp membership of the kth data point in the ith segment, and:

$$\beta_i(t_k) = \begin{cases} 1 & \text{if } s_{i-1} < k \leq s_i \\ 0, & \text{otherwise.} \end{cases} \tag{6.5}$$

This equation is well comparable to the typical error measure of standard k-means clustering but in this case the clusters are limited to contiguous segments of the time-series instead of the Voronoi regions in R^n.

The changes of the variables of the time-series are usually vague and are not focused on any particular time point. As it is not practical to define crisp bounds of the segments, in this chapter Gaussian membership functions, $A_i(t_k)$, are used to represent the $\beta_i(t_k) \in [0, 1]$ fuzzy segments of a time-series:

$$A_i(t_k) = \exp\left(-\frac{1}{2} \frac{(t_k - v_i^t)^2}{\sigma_{i,t}^2} \right), \; \beta_i(t_k) = \frac{A_i(t_k)}{\sum_{j=1}^{c} A_j(t_k)}. \tag{6.6}$$

(These terms are analogous to the Gaussian membership function and the degree of activation of the ith rule in classical fuzzy classifier as can be seen in Section 5.1.) For the identification of the v_i^t centers and $\sigma_{i,t}^2$ variances of the membership functions, a fuzzy clustering algorithm is introduced. The algorithm, which is similar to the modified Gath–Geva clustering [18], assumes that the data can be effectively modelled as a mixture of multivariate (including time as a variable) Gaussian distribution, so it minimizes the sum of the weighted squared distances between the $\mathbf{z}_k = [t_k, \mathbf{x}_k^T]^T$ data points and the η_i cluster prototypes

$$J = \sum_{i=1}^{c} \sum_{k=1}^{N} (\mu_{i,k})^m \, d^2(\mathbf{z}_k, \eta_i) \tag{6.7}$$

where $\mu_{i,k}$ represents the degree of membership of the observation $\mathbf{z}_k = [t_k, \mathbf{x}_k^T]^T$ is in the ith cluster ($i = 1, \dots, c$) and $m \in [1, \infty)$ is a weighting exponent that determines the fuzziness of the resulting clusters (usually chosen as $m = 2$).

The Gath–Geva clustering algorithm can be interpreted in a probabilistic framework, since the $d^2(\mathbf{z}_k, \eta_i)$ distance is inversely proportional to the probability that the $\mathbf{z}_k$ data point belongs to the ith cluster, $p(\mathbf{z}_k|\eta_i)$. The data are assumed to be normally distributed random variables with expected value $\mathbf{v}_i$ and covariance matrix $\mathbf{F}_i$. The Gath–Geva clustering algorithm is equivalent to the identification of a mixture of Gaussians that represents the $p(\mathbf{z}_k|\eta)$ probability density function expanded in a sum over the c clusters

$$p(\mathbf{z}_k|\eta) = \sum_{i=1}^{c} p(\mathbf{z}_k|\eta_i) p(\eta_i) \tag{6.8}$$

where the $p(\mathbf{z}_k|\eta_i)$ distribution generated by the ith cluster is represented by the Gaussian function

$$p(\mathbf{z}_k|\eta_i) = \frac{1}{(2\pi)^{\frac{n+1}{2}} \sqrt{\det(\mathbf{F}_i)}} \exp\left(-\frac{1}{2} (\mathbf{z}_k - \mathbf{v}_i)^T \mathbf{F}_i^{-1} (\mathbf{z}_k - \mathbf{v}_i) \right) \tag{6.9}$$

and

$$p(\eta_i) \text{ is the unconditional cluster probability}$$

(normalized such that $\sum_{i=1}^{c} p(\eta_i) = 1$ holds), where η_i represents the parameters of the ith cluster, $\eta_i = \{p(\eta_i), \mathbf{v}_i, \mathbf{F}_i | i = 1, \dots, c\}$.

Since the time variable is independent from the $\mathbf{x}_k$ variables, the presented clustering algorithm is based on the following $d^2(\mathbf{z}_k, \eta_i)$ distance measure

$$\begin{aligned} p(\mathbf{z}_k|\eta_i) = \frac{1}{d^2(\mathbf{z}_k, \eta_i)} &= \underbrace{\alpha_i}_{p(\eta_i)} \underbrace{\frac{1}{\sqrt{2\pi\sigma_{i,t}^2}} \exp\left(-\frac{1}{2} \frac{(t_k - v_i^t)^2}{\sigma_{i,t}^2} \right)}_{p(t_k|\eta_i)} \\ &\times \underbrace{\frac{1}{(2\pi)^{\frac{r}{2}} \sqrt{\det(\mathbf{A}_i)}} \exp\left(-\frac{1}{2} (\mathbf{x}_k - \mathbf{v}_i^x)^T \mathbf{A}_i^{-1} (\mathbf{x}_k - \mathbf{v}_i^x) \right)}_{p(\mathbf{x}_k|\eta_i)} \end{aligned} \tag{6.10}$$

which consists of three terms. The first α_i term represents the *a priori* probability of the cluster, while the second represents the distance between the kth data point and the v_i^t center of the ith segment in time. The third term represents the distance between the cluster prototype and the data in the feature space where $\mathbf{v}_i^x$ means the coordinate of the ith cluster center in the feature space and r is the rank of $\mathbf{A}_i$ distance norm corresponding to the ith cluster.

The presented cluster prototype formulated by (6.10) is similar to that used by the Gath–Geva clustering algorithm. However, it utilizes a different distance norm, $\mathbf{A}_i$. In the following section, it will be demonstrated how this norm can be based on the principal component analysis of the cluster.

6.3.1 PCA based Distance Measure

The $\mathbf{A}_i$ distance norm can be defined in many ways. It is wise to select this norm to scale the variables so that those with greater variability do not dominate the clustering. One can scale by dividing by standard deviations, but a better procedure is to use statistical (Mahalanobis) distance, which also adjusts for the correlations among the variables.

In this case $\mathbf{A}_i$ is the fuzzy covariance matrix $\mathbf{A}_i = \mathbf{F}_i$, where

$$\mathbf{F}_i = \frac{\sum_{k=1}^{N} (\mu_{i,k})^m (\mathbf{x}_k - \mathbf{v}_i^x)(\mathbf{x}_k - \mathbf{v}_i^x)^T}{\sum_{k=1}^{N} (\mu_{i,k})^m}. \tag{6.11}$$

When the variables are highly correlated, the $\mathbf{F}_i$ covariance matrix can be ill-conditioned and cannot be inverted. Recently two methods have been worked out to handle this problem [27].

The first method is based on fixing the ratio between the maximal and minimal eigenvalues of the covariance matrix. The second method is based on adding a scaled unity matrix to the calculated covariance matrix.

Both methods result in invertible matrices, but neither of them extracts the potential information about the hidden structure of the data.

One limiting disadvantage of PCA is the absence of an associated probability density or generative model which is required to compute $p(\mathbf{x}_k|\eta_i)$. Tipping and Bishop [265] developed a method called Probabilistic Principal Component Analysis (PPCA). In the PPCA the log-likelihood of observing the data under this model is

$$\mathcal{L} = \sum_{k=1}^{N} \ln(p(\mathbf{x}_k|\eta_i)) = -\frac{N}{2}\left\{n\ln(2\pi) + \ln\left(\det(\mathbf{A}_i)\right) + \operatorname{trace}(\mathbf{A}_i^{-1}\mathbf{F}_i)\right\} \tag{6.12}$$

where $\mathbf{A}_i = \sigma_{i,x}^2\mathbf{I} + \mathbf{W}_i\mathbf{W}_i^T$ is the modified covariance matrix of the ith cluster which can be used to compute the $p(\mathbf{x}_k|\eta_i)$ probability. The log-likelihood is maximized when the columns of $\mathbf{W}_i$ span the principal subspace of the data. Tipping and Bishop proofed that the only nonzero stationary points of the derivative of (6.12) with respect to $\mathbf{W}_i$ occur for

$$\mathbf{W}_i = \mathbf{U}_{i,q}\left(\Lambda_{i,q} - \sigma_{i,x}^2\mathbf{I}\right)^{1/2}\mathbf{R}_i \tag{6.13}$$

where $\mathbf{R}_i$ is an arbitrary $q \times q$ orthogonal rotation matrix and $\sigma_{i,x}^2$ is given by

$$\sigma_{i,x}^2 = \frac{1}{n-q}\sum_{j=q+1}^{n}\lambda_{i,j}. \tag{6.14}$$

The algorithmic description of the Expectation Maximization (EM) approach to PPCA model is given in [265] but it can also be found in the following section, where the estimation of this model is incorporated into the clustering procedure.

6.3.2 Modified Gath–Geva Clustering for Time-series Segmentation

One of the most important advantages of PPCA models is that it allows their combination into mixture of models. Mixtures have been extensively used as models where data can be viewed as arising from several populations mixed in varying proportions, and Expectation Maximization (EM) is widely used to estimate the parameters of the components in a mixture [44]. The clusters obtained by Gath–Geva (GG) clustering, also referred to Fuzzy Maximum Likelihood clustering, are multivariate Gaussian functions. The Alternating Optimization (AO) of these clusters is identical to the Expectation Maximization (EM) (maximum likelihood estimation) identification of the mixture of these Gaussian models when the fuzzy weighting exponent is $m = 2$ [40].

Similarly to GG clustering, in the presented algorithm the optimal parameters of the $\eta_i = \{\mathbf{v}_i^x, \mathbf{A}_i, v_i^t, \sigma_{i,x}^2, \alpha_i\}$ cluster prototypes are determined by the minimization of the (6.7) functional subjected to the classical clustering constraints (1.12), (1.13) and (1.14). The Alternating Optimization (see Section 1.5.3 and Section 1.5.4) results in the easily implementable algorithm described in Algorithm 6.3.1.

The usefulness and accuracy of the algorithm depends on the right choice of the q number of principal components (PCs) and the c number of the segments. Hence, the crucial question of the usefulness of the presented cluster algorithm is how these parameters can be determined in an automatic manner. This will be presented in the following two subsections.

Algorithm 6.3.1 (Clustering for Time-Series Segmentation).

Initialization

Given a time-series T *specify* c *and* q, *choose a termination tolerance* $\epsilon > 0$, *and initialize the values of* $\mathbf{W}_i, \mathbf{v}_i^x, \sigma_{i,x}^2, \mu_{i,k}$.

Repeat *for* $l = 1, 2, \ldots$

Step 1 *Calculate the* η_i *parameters of the clusters*

- a priori *probability of the cluster*

$$\alpha_i = \frac{1}{N}\sum_{k=1}^{N}\mu_{i,k}. \tag{6.15}$$

- *the cluster centers*

$$\mathbf{v}_i^x = \frac{\sum_{k=1}^{N}(\mu_{i,k})^m\left(\mathbf{x}_k - \mathbf{W}_i\langle\mathbf{y}_{i,k}\rangle\right)}{\sum_{k=1}^{N}(\mu_{i,k})^m} \tag{6.16}$$

where the expectation of the latent variables is $\langle\mathbf{y}_{i,k}\rangle = \mathbf{M}_i^{-1}\mathbf{W}_i^T\left(\mathbf{x}_k - \mathbf{v}_i^x\right)$ *and the* $q \times q$ *matrix* $\mathbf{M}_i = \sigma_{i,x}^2\mathbf{I} + \mathbf{W}_i^T\mathbf{W}_i$.

- *the new values of* $\mathbf{W}_i$

$$\widetilde{\mathbf{W}}_i = \mathbf{F}_i\mathbf{W}_i\left(\sigma_{i,x}^2\mathbf{I} + \mathbf{M}_i^{-1}\mathbf{W}_i^T\mathbf{F}_i\mathbf{W}_i\right)^{-1} \tag{6.17}$$

where $\mathbf{F}_i$ *computed by* (6.11).

- *the new value of* $\sigma_{i,x}^2$

$$\sigma_{i,x}^2 = \frac{1}{q}\mathit{trace}(\mathbf{F}_i - \mathbf{F}_i\mathbf{W}_i\mathbf{M}_i^{-1}\widetilde{\mathbf{W}}_i^T). \tag{6.18}$$

- *the distance norm* ($n \times n$ *matrix*)

$$\mathbf{A}_i = \sigma_{i,x}^2\mathbf{I} + \widetilde{\mathbf{W}}_i\widetilde{\mathbf{W}}_i^T. \tag{6.19}$$

- *the model parameters in time: the center and the standard deviation*

$$v_i^t = \frac{\sum\limits_{k=1}^{N}(\mu_{i,k})^m t_k}{\sum\limits_{k=1}^{N}(\mu_{i,k})^m}, \; \sigma_{i,t}^2 = \frac{\sum\limits_{k=1}^{N}(\mu_{i,k})^m\left(t_k - v_i^t\right)^2}{\sum\limits_{k=1}^{N}(\mu_{i,k})^m}. \tag{6.20}$$

Step 2 *Compute the* $d^2(\mathbf{z}_k, \eta_i)$ *distance measures by* (6.10).

Step 3 *Update the partition matrix*

$$\mu_{i,k}^{(l)} = \frac{1}{\sum_{j=1}^{c}\left(d(\mathbf{z}_k,\eta_i)/d(\mathbf{z}_k,\eta_j)\right)^{2/(m-1)}}, 1 \le i \le c, \; 1 \le k \le N. \tag{6.21}$$

until $||\mathbf{U}^{(l)} - \mathbf{U}^{(l-1)}|| < \epsilon$.

6.3.3 Automatic Determination of the Number of Segments

In data mining, the bottom-up segmentation algorithm has been extensively used to support a variety of time series data mining tasks [149]. The algorithm starts by creating a fine approximation of the time series, and iteratively merges the lowest cost pair of segments until a stopping criteria is met. For the automatic selection of the number of segments, a similar approach is presented in this section. The presented recursive cluster merging technique evaluates the adjacent clusters for their compatibility (similarity) and merges the clusters that are found to be compatible. Then, after the proper initialization of the parameters of the new cluster the clustering is performed again. During this merging and re-clustering procedure the number of clusters is gradually reduced, until an appropriate number of clusters is found. This procedure is controlled by a fuzzy decision making algorithm based on the similarity between the PCA models.

Similarity of PCA Models

The similarity of two PCA models (i.e., hyperplanes) can be calculated by the PCA similarity factor, S_{PCA}, developed by Krzanowski [166, 246]. Consider two segments, S_i and S_j, of a dataset having the same n variables. Let the PCA models for S_i and S_j consist of q PC's each. The similarity between these subspaces is defined based on the sum of the squares of the cosines of the angles between each principal component of $\mathbf{U}_{i,q}$ and $\mathbf{U}_{j,q}$:

$$S_{PCA}^{i,j} = \frac{1}{q}\sum_{i=1}^{q}\sum_{j=1}^{q}\cos^2\theta_{i,j} = \frac{1}{q}\text{trace}\left(\mathbf{U}_{i,q}^T\mathbf{U}_{j,q}\mathbf{U}_{j,q}^T\mathbf{U}_{i,q}\right). \tag{6.22}$$

Because the $\mathbf{U}_{i,q}$ and $\mathbf{U}_{j,q}$ subspaces contain the q most important principal components that account for the most of the variance in their corresponding datasets, $S_{PCA}^{i,j}$ is also a measure of similarity between the segments S_i and S_j.

Since the purpose of the segmentation is also to detect changes in the mean of the variables, it is not sufficient to compute only the $S_{PCA}^{i,j}$ similarity factor but the distance among the cluster centers also has to be taken into account

$$d(\mathbf{v}_i^x, \mathbf{v}_j^x) = \|\mathbf{v}_i^x - \mathbf{v}_j^x\|. \tag{6.23}$$

Hence, the compatibility criterion has to consider the $c_{i,j}^1 = S_{PCA}^{i,j}$ and $c_{i,j}^2 = d(\mathbf{v}_i^x, \mathbf{v}_j^x)$ factors.

The Decision Making Algorithm

Because the compatibility criterion quantifies various aspects of the similarity of the clusters, the overall cluster compatibility should be obtained through an aggregation procedure. A fuzzy decision making algorithm can be used for this purpose [146].

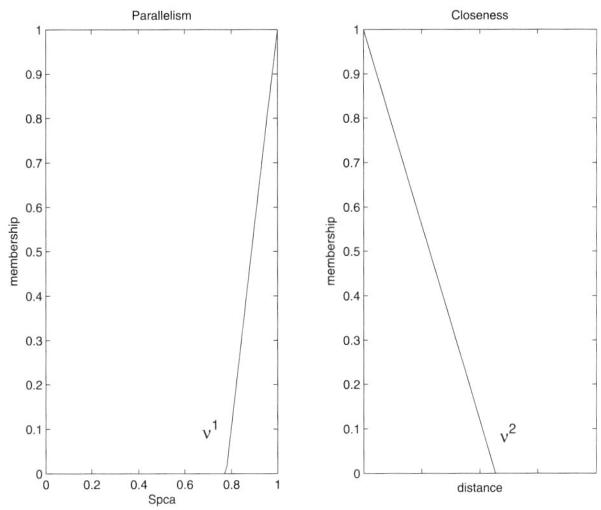

Figure 6.2: Membership functions for parallelism and closeness of clusters.

Compatibility criteria (6.22) and (6.23) are evaluated for each pair of clusters. The resulted compatibility matrix is descriptive concerning the structure of the whole time-series, e.g., repetitive motifs can be also detected from the analysis of the compatibility of the non-adjacent clusters. Following a fuzzy decision making approach, the decision goals for each criterion have to be defined using a fuzzy set. Figure 6.2 shows the triangular membership functions defined for the two criteria. The important parameters of the membership functions are the limits of their support, characterized by the ν^1 knot point for parallelism and ν^2 for closeness. The values of ν^1 and ν^2 are given by averaging compatibilities according to

$$\nu^1 = \frac{1}{c(c-1)} \sum_{i=1}^{c} \sum_{\substack{j=1 \\ j \neq i}}^{c} c_{i,j}^1, \tag{6.24}$$

$$\nu^2 = \frac{1}{c(c-1)} \sum_{i=1}^{c} \sum_{\substack{j=1 \\ j \neq i}}^{c} c_{i,j}^2. \tag{6.25}$$

Evaluating the membership functions with the values $c_{i,j}^1$ and $c_{i,j}^2$, one obtains the $\mu_{i,j}^1$ degree of parallelism and $\mu_{i,j}^2$ of closeness. The overall cluster compatibility is determined by the aggregation of the two criteria. A fuzzy aggregation operator is used for this purpose. The outcome of the decision procedure is the $\mathbf{O}$ overall compatibility matrix whose $O_{i,j}$ elements are given by

$$O_{i,j} = \left[\frac{(\mu_{i,j}^1)^2 + (\mu_{i,j}^2)^2}{2} \right]^{1/2}. \tag{6.26}$$

Given the **O** compatibility matrix, the clusters that will be merged must be identified and combined. Clusters can be merged in several ways. Our method merges the most similar pair of adjacent clusters as long as the value of the corresponding $O_{i,i+1}$ is above a threshold γ.

Cluster Merging

The applied bottom-up strategy merges two adjacent clusters in each iteration. To preserve the information represented by the clusters there is a need for a merging method that can directly compute the new initial parameters of the cluster from the merged clusters. This can be done by several ways (e.g., by Stroupe and Durrant-Whyte methods [188]). In this chapter the method developed by P. M. Kelly is applied [147].

The $\mathbf{v}_{i*}^x$ mean of the resulting cluster is computed from the individual cluster means

$$\mathbf{v}_{i*}^x = \frac{N_i}{N_{i*}}\mathbf{v}_i^x + \frac{N_j}{N_{i*}}\mathbf{v}_{i+1}^x \tag{6.27}$$

where $N_i = \sum_{k=1}^N \mu_{i,k}$, $N_{i+1} = \sum_{k=1}^N \mu_{i+1,k}$ and $N_{i*} = N_i + N_{i+1}$, while the $\mathbf{F}_{i*}$ new covariance matrix is calculated from the $\mathbf{F}_i$ and $\mathbf{F}_{i+1}$ old covariance matrices as

$$\mathbf{F}_{i*} = \frac{N_i - 1}{N_{i*} - 1}\mathbf{F}_i + \frac{N_{i+1} - 1}{N_{i*} - 1}\mathbf{F}_{i+1} + \frac{N_i N_{i+1}}{N_{i*}(N_{i*} - 1)}[(\mathbf{v}_i^x - \mathbf{v}_{i+1}^x)(\mathbf{v}_i^x - \mathbf{v}_{i+1}^x)^T]. \tag{6.28}$$

The new values of $\mathbf{W}_{i*}$ and $\sigma_{i*,x}^2$ can be computed by (6.13) and (6.14) based on the eigenvector-eigenvalue decomposition of $\mathbf{F}_{i*}$. Based on the obtained results the $p(\mathbf{x}_k|\eta_{i*})$ probabilities can be easily computed (see the algorithm in Section 6.3.2).

This method can also be applied to merge the membership functions characterized by the parameters v_i^t, $\sigma_{i,t}^2$ and v_{i+1}^t, $\sigma_{i+1,t}^2$, and the $p(t_k|\eta_{i*})$ probabilities can be determined from the new values of v_{i*}^t and $\sigma_{i*,t}^2$ by (6.10). The unconditional probability of the $i*$th cluster is equal to the sum of the probabilities of the previous clusters $p(\eta_{i*}) = p(\eta_i) + p(\eta_{i+1})$. After the previously presented initialization of the new cluster prototype the new $\mu_{i*,k}$ membership values can be computed by (6.10) and (6.21).

6.3.4 Number of Principal Components

Beside the selection of the right number of clusters, the second bottleneck of the successful application of the presented algorithm is the selection of the right number of principal components. This can be done by the analysis of the eigenvalues of the covariance matrices of the initial segments. For this purpose a so-called screenplot can be drawn that plots the ordered eigenvalues according to their contribution to the variance of data. Another possibility is to define q based on the desired accuracy (loss of variance) of the PPCA models:

$$\sum_{j=1}^{q-1} \lambda_{i,j} / \sum_{j=1}^{n} \lambda_{i,j} < \text{ accuracy } \leq \sum_{j=1}^{q} \lambda_{i,j} / \sum_{j=1}^{n} \lambda_{i,j}. \tag{6.29}$$

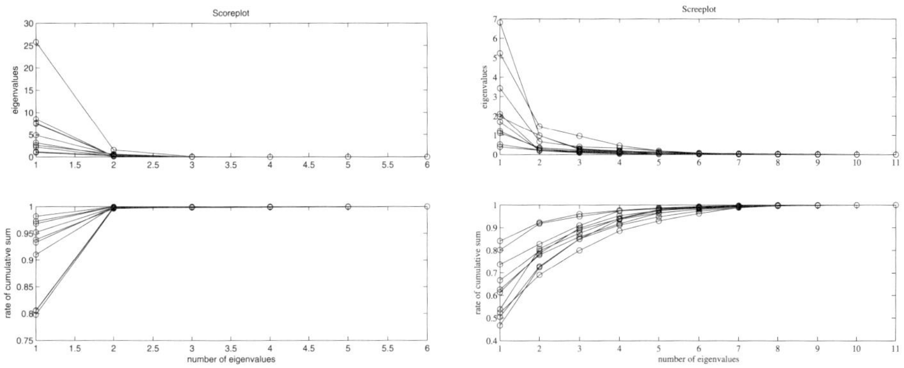

(a) *Synthetic dataset.* (b) *Industrial dataset.*

Figure 6.3: Screeplots of the synthetic and the industrial data shown in Figure 6.1 and Figure 6.4, respectively.

The synthetic dataset shown in Figure 6.1 was initially partitioned into ten segments. As Figure 6.3(a) illustrates, the magnitude of the eigenvalues and their cumulative rate to their sum show that two PCs are sufficient to approximate the distribution of the data with 98% accuracy. Obviously, this analysis can be fully automatized.

It is interesting to note that the analysis of the fuzzy hypervolume cluster validity measure is similar to the approach of the analysis of the eigenvalues, because the $V_i = (\det(\mathbf{F}_i))^{1/2}$ hypervolume of a cluster is proportional to the product of the eigenvalues. Hence, the right number of principal components can be also determined based on the product of the q largest eigenvalues.

6.3.5 The Segmentation Algorithm

Based on the previously presented building blocks the presented clustering based segmentation algorithm is formulated as shown in Algorithm 6.3.2.

This formulation of the algorithm shows that although there is a need to define some parameters of the algorithm before its application (γ, ϵ, and the initial number of clusters), it is possible to apply the method for high-dimensional time-series even if almost nothing is known about the structure of these series in advance.

Algorithm 6.3.2 (Time-Series Segmentation Algorithm).

Step 1 *Uniformly segment the data by a large number of segments (in the examples given in this chapter ten segments were used as starting point). Determine the q number of the principal components based on the analysis of the eigenvalues of these segments. For this purpose screeplot or cluster validity measure can be used (see Section 6.3.4 for more details).*

Step 2 *The values of m fuzziness parameter, the γ threshold for the $\mathbf{O}$ compatibility matrix and the ϵ termination tolerance must be chosen. In the case studies $m = 2, \epsilon = 10^{-4}$, the value of γ is usually between 0.3–0.75 depending of the homogeneity of the time-series.*

Step 3 *Execute the clustering algorithm (see Section 6.3.2). The cluster merging must be evaluated after a predefined number of iteration steps (see Section 6.3.3). In all of our applications this number was 100. The algorithm stops if the termination tolerance is reached and cluster merging is not necessary.*

Hence, the method is useful for knowledge discovery. Of course, data mining is an iterative procedure. The results of the segmentation should be evaluated by human experts or by the performances of other modelling and data mining tools based on the segmented data, and if it is needed, the "knowledge worker" should return to the segmentation task with a new set of these parameters.

6.3.6 Case Studies

In this section the effectiveness of the developed algorithm will be illustrated by two examples: the synthetic dataset introduced in Section 6.2 and an industrial dataset taken from an industrial polymerization reactor, and the obtained results will be compared to the results given by the multivariate extension of the bottom-up segmentation algorithm of Keogh [149].

Example 6.1 (Synthetic time-series segmentation based on GG clustering). *The synthetic dataset given in Figure 6.1 is designed to illustrate how a multivariate segmentation algorithm should detect the changes of the latent process behind the high-dimensional data.*

In Figure 6.3 it has been shown that the screeplot of the eigenvalues suggests that the clustering algorithm should take into account two principal components. From Figure 6.1(c) – which shows the β_i normalized and the $A_i(t) = p(t_k|\eta_i)$ Gaussian membership functions, and the $p(\mathbf{z}_k|\eta_i)$ probabilities – it can be seen that with this parameter the presented method found five segments and it is able to detect *the changes in the correlation structure and in the mean of the data. The* $S_{PCA}^{i(i+1)}$ *similarity measures of the adjacent clusters are* 0.99, 0.17, 0.99, *and* 0.99812, *which suggest that the correlation among the variables has significantly changed between the first and the second segment, while the other segments differ mainly in their mean. These results agree with Figure* 6.1(a) *and justify the accuracy and the usefulness of the presented method.*

To illustrate the importance of the selection of the right number of PCs the same segmentation has been performed with only one PC. In this case, the algorithm found 10 *nearly symmetric segments, hence it was not able to explore the hidden information behind the data. As it is depicted in Figure* 6.1(c), *in case of five PCs the algorithm gave reasonable, but not so characteristic result.*

These results were compared to the results of the bottom-up method based on the Hotelling T^2 *(top) and the reconstruction error* Q *(bottom) shown in Figure* 6.1(d). *The bottom-up algorithm based on the reconstruction error* Q *is sensitive to the change in the correlation structure but it was not able to find the change in the mean. The method based on the Hotelling* T^2 *measure is on the contrary. The method based on the* Q *measure is very sensitive to the number of PCs. As can be seen in Figure* 6.1(d) *when* $q = 2$ *the result is very different from that obtained by* $q = 5$, *but in both cases the algorithm finds the change in the correlation structure.*

This comparison showed contrary to the multivariate extensions of the classical bottom-up segmentation algorithm, the developed cluster analysis based segmentation algorithm can simultaneously handle the problems of the detection of the change of the latent process and the change of the mean of the variables and it is more robust with respect to the number of principal components.

□

Example 6.2 (Application of clustering based time-series segmentation to process monitoring). *Manual process supervision relies heavily on visual monitoring of characteristic shapes of changes in process variables, especially their trends. Although humans are very good at visually detecting such patterns, it is a difficult problem for a control system software. Researchers with different background, for example from pattern recognition, digital signal processing and data mining, have contributed to the process trend analysis development* [154, 252, 284].

The aim of this example is to show how the presented algorithm is able to detect meaningful temporal shapes from multivariate historical process data. The monitoring of a medium and high-density polyethylene (MDPE, HDPE) plant is considered. The plant is operated by TVK Ltd. (www.tvk.hu), which is the largest Hungarian polymer production company and produces raw materials for versatile plastics used for household goods, packaging, car parts and pipe. An interesting problem with the process is that it requires the production about ten product grades according to the market demand. Hence, there is a clear need to minimize the time of changeover between the different products because off-specification product may be produced during the process transition. The difficulty of the analysis of the production and the transitions comes from the fact that there are more than ten process variables that need to be simultaneously monitored. Measurements are available in every 15 *seconds on process variables* $\mathbf{x}_k$*, which are the polymer production intensity* (PE)*, the inlet flowrates of hexene* ($C6_{in}$)*, ethlylene* ($C2_{in}$)*, hydrogen* ($H2_{in}$)*, the isobutane solvent* (IB_{in}) *and the catalyzator* (Kat)*, the concentrations of ethylene* (C_2)*, hexene* (C_6)*, and hydrogen* (H_2) *and the slurry in the reactor* ($slurry$)*, and the temperature of the reactor* (T)*.*

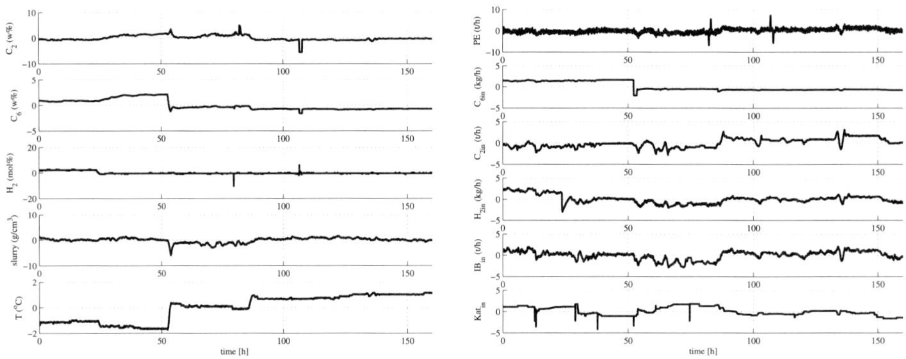

(a) States of the reactor.

(b) Input variables of the reactor.

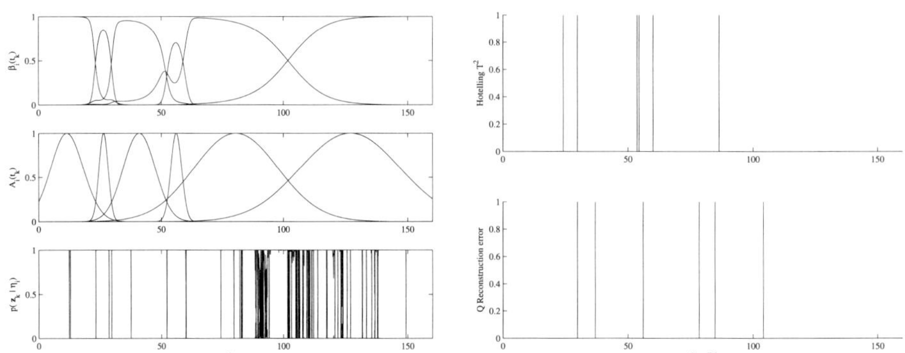

(c) Fuzzy segmentation results, $q = 5$, $\gamma = 0.4$.

(d) Bottom-up segmentation results.

Figure 6.4: Segmentation of the industrial dataset.

The dataset used in this example represents 160 *hours of operation and includes three product transitions around the* 24, 54, *and* 86*th hour. The initial number of the segments was ten and the* γ *threshold was chosen to* $\gamma = 0.4$*. In the Figure* 6.3 *it can be seen that* $q = 5$ *principal components must be considered for* 95% *accuracy.*

As Figure 6.4 shows, both the bottom-up and the clustering based algorithm are able to detect the product transitions, and all the three methods gave similar results. This reflects that the mean and the covariance of the data were not independently changed. This is also confirmed by the analysis of the compatibilities of the adjacent clusters. As it can be seen, the product transitions are represented by two independent clusters, while the third transition was not so characteristic that it would require an independent segment. This smooth transition between the third and the fourth product is also reflected by how the $p(\mathbf{z}_k|\eta_i)$ *probabilities overlap between the* 75–125*th hour of operation. The changes of the* $p(\mathbf{z}_k|\eta_i)$ *probabilities around the* 135*th hour of operation are also informative, as the period of lower or drastically changing probabilities reflect some erroneous operation of the process. The results are similar if more than* 5 *principal components are taken into account.*

This example illustrated that the presented tool can be applied for the segmentation of a historical database and with the application of this tool useful information can be extracted concerning the changes of the operation regimes of the process and process faults. In the current state of our project we use this tool to compare the production of different products and extract homogenous segments of operation that can be used by a Kalman-filter based state estimation algorithm for the identification of useful kinetic parameters and models which are able to predict the quality of the products [4].

□

6.4 Conclusions

This chapter presented a new clustering algorithm for the fuzzy segmentation of large multivariate time-series. The algorithm is based on the simultaneous identification of fuzzy sets which represent the segments in time and the hyperplanes of local PCA models used to measure the homogeneity of the segments. The algorithm favors contiguous clusters in time and is able to detect changes in the hidden structure of multivariate time-series. A fuzzy decision making algorithm based on a compatibility criterion of the clusters has been worked out to determine the required number of segments, while the required number of principal components are determined by the screeplots of the eigenvalues of the fuzzy covariance matrices.

The results suggest that the presented tool can be applied to extract useful information from temporal databases, e.g., the detected segments can be used to classify typical operational conditions and analyze product grade transitions.

Beside the industrial application example a synthetic dataset was analyzed to convince the readers about the usefulness of the method. Furthermore, the MATLAB® code of the algorithm is available from our website

(www.fmt.vein.hu/softcomp/segment),

so the readers can easily test the presented method on their own datasets.

The application of the identified fuzzy segments in intelligent query systems designed for multivariate historical process databases is an interesting and useful idea for future research.

Appendix

Hermite Spline Interpolation

Lets define a cubic spline for a knot sequence $\Xi, a = \xi_0 < \xi_1 < \ldots < \xi_N = b$. The cubic spline is then based on cubic polynomes of the form $S(x)$, given for each subinterval $[\xi_0, \xi_1]$, $[\xi_1, \xi_2], \ldots, [\xi_{N-1}]$. These polynomes are connected at the interior knots Ξ in such a way that $S(x)$ has *two* continuous derivatives on $[a, b]$. Every cubic splines can be written in the form:

$$S(x) = \alpha x^3 + \beta x^2 + \gamma x + \delta + \sum_{i=1}^{B-1} a_i |x - \xi_i|^3 . \tag{A.1}$$

Thus, $S(x)$ is simply a linear combination of the functions

$$x^3, x^2, x, 1, |x - \xi_1|^3, \ldots, |x - \xi_{N-1}|^3 . \tag{A.2}$$

An efficient notation for identification is given by Horiuchi [120], who defined the piecewise polynoms by a combination of the function-value and its first derivative at the knots:

$$\begin{aligned} S_i(x) \quad = \quad & \frac{dS(\xi_i)}{dx} \cdot \frac{(\xi_{i+1} - x)^2 (x - \xi_i)}{h_i^2} \\ & - \frac{dS(\xi_{i+1})}{dx} \cdot \frac{(x - \xi_i)^2 (\xi_{i+1} - x)}{h_i^2} \\ & + S(\xi_i) \cdot \frac{(\xi_{i+1} - x)^2 (2(x - \xi_i) + h_i)}{h_i^3} \\ & + S(\xi_{i+1}) \cdot \frac{(x - \xi_i)^2 (2(\xi_{i+1} - x) + h_i)}{h_i^3} , \end{aligned} \tag{A.3}$$

where $h_i = (\xi_{i+1} - \xi_i)$.

Denote:

$$a_i(x) = \frac{(\xi_{i+1}-x)^2(x-\xi_i)}{h_i^2}, \tag{A.4}$$

$$b_i(x) = -\frac{(x-\xi_i)^2(\xi_{i+1}-x)}{h_i^2}, \tag{A.5}$$

$$c_i(x) = \frac{(\xi_{i+1}-x)^2(2(x-\xi_i)+h_i)}{h_i^3}, \tag{A.6}$$

$$d_i(x) = \frac{(x-\xi_i)^2(2(\xi_{i+1}-x)+h_i)}{h_i^3}, \tag{A.7}$$

and

$$y_i = S(\xi_i), \tag{A.8}$$

$$y_i' = \frac{dS(\xi_i)}{dx}, \tag{A.9}$$

then (A.3) can be given as:

$$S_i(x) = y_i a_i(x) + y_{i+1} b_i(x) + y_i' c_i(x) + y_{i+1}' d_i(x). \tag{A.10}$$

Now, the unknown variables y_i and y_i' can be determined by a least square method. Define a quadratic objective function Q by:

$$Q(y_1, y_1', \ldots, y_N, y_{N+1}') = \sum_{k=1}^{N} \left(S(x_k) - y_k\right)^2. \tag{A.11}$$

The function Q can be minimized by solving the normal equations derived by partial differentiation of (A.11) with $\theta = [y_1, y_1', \ldots, y_N, y_N']$:

Define for $\nu \in \{a, b, c, d\}$:

$$\nu_i = \sum_{k=p_i}^{q_i} \nu_1(x_k). \tag{A.12}$$

$$\frac{dQ}{d\theta} = 2 \left(\begin{bmatrix} \cdot & \cdot & \cdot & \cdot & \cdot & \cdot & \cdot & \cdot & \cdot \\ \cdot & \cdot & \cdot & \cdot & \cdot & \cdot & \cdot & \cdot & \cdot \\ c_1 d_1 & a_1 d_1 & d_1^2 & b_1 d_1 & \cdot & \cdot & \cdot & \cdot & -d_1 y_k \\ c_1 b_1 & a_1 b_1 & b_1 d_1^2 & b_1^2 & \cdot & \cdot & \cdot & \cdot & -b_1 y_k \\ \cdot & \cdot & \cdot & \cdot & \cdot & \cdot & \cdot & \cdot & \cdot \\ \cdot & \cdot & \cdot & \cdot & c_N d_N & a_N d_N & d_N^2 & b_N d_N & -d_N y_k \\ \cdot & \cdot & \cdot & \cdot & c_N b_N & a_N b_N & b_N d_N^2 & b_N^2 & -b_N y_k \end{bmatrix} + \begin{bmatrix} c_1^2 & a_1 c_1 & c_1 d_1 & b_1 c_1 & \cdot & \cdot & \cdot\ \cdot & -c_1 y_k \\ a_1 c_1 & a_1^2 & a_1 d_1 & a_1 b_1 & \cdot & \cdot & \cdot\ \cdot & -a_1 y_k \\ \cdot & \cdot & c_2^2 & a_2 c_2 & c_2 d_2 & b_2 c_2 & \cdot\ \cdot & -c_2 y_k \\ \cdot & \cdot & a_2 c_2 & a_2^2 & a_2 d_2 & a_2 b_2 & \cdot\ \cdot & -a_2 y_k \\ \cdot & \cdot & \cdot & \cdot & \cdot & \cdot & \cdot\ \cdot & \cdot \\ \cdot & \cdot & \cdot & \cdot & \cdot & \cdot & \cdot\ \cdot & \cdot \\ \cdot & \cdot & \cdot & \cdot & \cdot & \cdot & \cdot\ \cdot & \cdot \end{bmatrix} \right) \begin{bmatrix} \theta^T \\ 1 \end{bmatrix} \tag{A.13}$$

Bibliography

[1] J. Abonyi. *Fuzzy Model Identification for Control*. Birkhäuser, Cambridge, MA, 2003.

[2] J. Abonyi. *Instrument Engineers' Handbook – ed: B.G. Lipták*, volume 2, chapter Neural Networks for Process Modelling, pages 253–264. 4 edition, 2005.

[3] J. Abonyi. *Instrument Engineers' Handbook – ed: B.G. Lipták*, volume 2, chapter Software for Fuzzy Logic Control, pages 360–374. 4 edition, 2005.

[4] J. Abonyi, P. Arva, S. Nemeth, Cs. Vincze, B. Bodolai, Zs.Horvath, G. Nagy, and M. Nemeth. Operator support system for multi product processes – application to polyethylene production. In *European Symposium on Computer Aided Process Engineering*, pages 347–352. Lappeenranta, Finland, 2003.

[5] J. Abonyi, R. Babuska, H. Verbruggen, and F. Szeifert. Using a priori knowledge in fuzzy model identification. *International Journal of Systems Science*, 31:657–667, 2000.

[6] J. Abonyi and R. Babuška. Local and global identification and interpretation of parameters in Takagi–Sugeno fuzzy models. In *Proceedings IEEE International Conference on Fuzzy Systems*, San Antonio, USA, May 2000.

[7] J. Abonyi, R. Babuška, T. Chovan, and F. Szeifert. Incorporating prior knowledge in fuzzy c–regression models – application to system identification. In *Proceedings of the Intelligent Systems in Control and Measurement Symposium, INTCOM* 2000, pages 99–110, Veszprem, Hungary, 2000.

[8] J. Abonyi, A. Bodizs, L. Nagy, and F. Szeifert. Hybrid fuzzy convolution model and its application in predictive control. *Chemical Engineering Research and Design*, 78:597–604, 2000.

[9] J. Abonyi, B. Feil, and R. Babuska. State-space reconstruction and prediction of chaotic time series based on fuzzy clustering. International Conference on Systems, Man and Cybernetics, October 2004.

[10] J. Abonyi, B. Feil, S. Nemeth, and P. Arva. Modified Gath–Geva clustering for fuzzy segmentation of multivariate time-series. *Fuzzy Sets and Systems – Fuzzy Sets in Knowledge Discovery*, 149(1):39–56, 2005.

[11] J. Abonyi and F. Szeifert J.A. Roubos, M. Oosterom. Compact ts-fuzzy models through clustering and OLS plus FIS model reduction. In *Proc. of IEEE international conference on fuzzy systems*, Sydney, Australia, 2001.

[12] J. Abonyi, B.G. Lakatos, and Zs. Ulbert. Modelling and control of isothermal crystallizers by self-organising maps. *Chem. Eng. Trans.*, 1(3):1329–1334, 2002.

[13] J. Abonyi and J. Madar. *Instrument Engineers' Handbook – ed: B.G. Lipták*, volume 2, chapter Genetic and Other Evolutionary Algorithms, pages 181–192. 4 edition, 2005.

[14] J. Abonyi, L. Nagy, and F. Szeifert. Hybrid fuzzy convolution modelling and identification of chemical process systems. *International Journal of Systems Science*, 31:457–466, 2000.

[15] J. Abonyi, H. Roubos, R. Babuska, and F. Szeifert. *Identification of Semi-Mechanistic Models with Interpretable TS-fuzzy Submodels by Clustering, OLS and FIS Model Reduction*, chapter 10. Fuzzy modelling and the interpretability-accuracy trade-off. Part I, interpretability issues. Studies in Fuzziness and Soft Computing, Physica-Verlag. 2003.

[16] J. Abonyi, H. Roubos, and F. Szeifert. Data-driven generation of compact, accurate, and linguistically sound fuzzy classifiers based on a decision tree initialization. *International Journal of Approximate Reasoning*, pages 1–21, 2003.

[17] J. Abonyi and F. Szeifert. Supervised fuzzy clustering for the identification of fuzzy classifiers. *Pattern Recognition Letters*, 24(14):2195–2207, 2003.

[18] J. Abonyi, F. Szeifert, and R. Babuska. Modified Gath-Geva fuzzy clustering for identification of Takagi-Sugeno fuzzy models. *IEEE Systems, Man and Cybernetics, Part B.*, pages 612–621, 2002.

[19] A. Ahalt, A.K. Khrisnamurthy, P. Chen, and D.E. Melton. Competitive learning algorithms for vector quantization. *Neural Networks*, 3:277–290, 1990.

[20] H. Akaike. A new look at the statistical model identification. *IEEE Trans. on Automatic Control*, 19:716–723, 1974.

[21] M.R. Anderberg. *Cluster Analysis for Applications.* Academic Press, Inc., 1973.

[22] H.C. Andersen, A. Lotfi, and L.C. Westphal. Comments on 'Functional Equivalence Between Radial Basis Function Networks and Fuzzy Inference

Systems'. pages 1529–1530, May 1998. IEEE Transactions on Neural Networks.

[23] Y. Ashkenazy. The use of generalized information dimension in measuring fractal dimension of time series. *Physica A*, 271:427–447, 1999.

[24] M. Ataei, B. Lohmann, A. Khaki-Sedigh, and C. Lucas. Model based method for estimating an attractor dimension from uni/multivariate chaotic time series with application to bremen climatic dynamics. *Chaos, Solitons and Fractals*, 19:1131–1139, 2004.

[25] J.G. Augustson and J. Minker. An analysis of some graph theoretical clustering techniques. *J. ACM*, 17(4):571–588, 1970.

[26] R. Babuška. *Fuzzy Modelling for Control*. Kluwer Academic Publishers, Boston, 1998.

[27] R. Babuška, P. J. van der Veen, and U. Kaymak. Improved covariance estimation for Gustafson-Kessel clustering. *IEEE International Conference on Fuzzy Systems*, pages 1081–1085, 2002.

[28] R. Babuška and H.B. Verbruggen. New approach to constructing fuzzy relational models from data. In *Proceedings Third European Congress on Intelligent Techniques and Soft Computing EUFIT'95*, pages 583–587, Aachen, Germany, August 1995.

[29] R. Babuška and H.B. Verbruggen. Constructing fuzzy models by product space clustering. In H. Hellendoorn and D. Driankov, editors, *Fuzzy Model Identification: Selected Approaches*, pages 53–90. Springer, Berlin, Germany, 1997.

[30] R. Babuška and H.B. Verbruggen. Fuzzy identification of Hammerstein systems. In *Proceedings Seventh IFSA World Congress*, volume II, pages 348–353, Prague, Czech Republic, June 1997.

[31] R. Babuška and H.B. Verbruggen. Fuzzy set methods for local modelling and identification. In R. Murray-Smith and T.A. Johansen, editors, *Multiple Model Approaches to Nonlinear Modelling and Control*, pages 75–100. Taylor & Francis, London, UK, 1997.

[32] R. Babuška, H.B. Verbruggen, and H.J.L. van Can. Fuzzy modelling of enzymatic penicillin–G conversion. *Engineering Applications of Artificial Intelligence*, 12(1):79–92, 1999.

[33] F.B. Backer and L.J. Hubert. A graph theoretic approach to goodness-of-fit in complete-link hierarchical clustering. *J. Am. Stat. Assoc.*, 71:870–878, 1976.

[34] R.A. Baeza-Yates. *Introduction to data structures and algorithms related to information retrieval*, pages 13–27. Information Retrieval: Data Structures and Algorithms. Prentice-Hall, Inc., Upper Saddle River, NJ, 1992.

[35] J.F. Baldwin, T.P. Martin, and J.M. Rossiter. Time series modelling and prediction using fuzzy trend information. *Proceedings of* 5th *International Conference on Soft Computing and Information Intelligent Systems*, pages 499–502, 1998.

[36] G.H. Ball and D.J. Hall. Isodata, a novel method of data analysis and classification. Technical report, Stanford University, Stanford, CA, 1965.

[37] A. Banerjee and Y. Arkun. Model predictive control of plant transitions using a new identification technique for interpolating nonlinear models. *Journal of Process Control*, 8:441–457, 1998.

[38] A.M. Bensaid, L.O. Hall, J.C. Bezdek, L.P. Clarke, M.L. Silbiger, J.A. Arrington, and R.F. Murtagh. Validity-guided (re)clustering with applications to image segmentation. *IEEE Transactions on Fuzzy Systems*, 4:112–123, 1996.

[39] J.C. Bezdek. *Pattern Recognition with Fuzzy Objective Function Algorithms.* Plenum Press, 1981.

[40] J.C. Bezdek and J.C. Dunn. Optimal fuzzy partitions: A heutristic for estimating the parameters in a mixture of normal distributions. *IEEE Transactions on Computers*, pages 835–838, 1975.

[41] J.C. Bezdek, R.J. Hathaway, R.E. Howard, C.A. Wilson, and M.P. Windham. Local convergence analysis of a grouped variable version of coordinate descent. *Journal of Optimization Theory and Applications*, 71:471–477, 1987.

[42] J.C. Bezdek and S.K. Pal. *Fuzzy Models for Pattern Recognition.* IEEE Press, New York, 1992.

[43] A. Biem, Katagiri S., McDermott E., and Juang BH. An application of discriminative feature extraction to filter-bank-based speech recognition. *IEEE Transactions On Speech And Audio Processing*, 9(2):96–110, 2001.

[44] C.M. Bishop. *Neural Networks for Pattern Recognition.* Oxford University Press, 1995.

[45] C.M. Bishop, M. Svensen, and C.K.I. Williams. Developments of the generative topographic mapping. *Neurocomputing*, 21:203–224, 1998.

[46] C.M. Bishop, M. Svensen, and C.K.I. Williams. Gtm: the generative topographic mapping. *Neural Comput.*, 10(1):215–234, 1998.

[47] C.M. Bishop and M.E. Tipping. A hierarchical latent variable model for data visualization. *IEEE Transactions on Pattern Analysis and Machine Intelligence*, 20(3):281–293, March 1998.

[48] T. Bohlin. A case study for grey box identification. *Automatica*, 30(2):307–318, 1994.

[49] J.D. Bomberger and D.E. Seborg. Determination of model order for NARX models directly from input–output data. *Journal of Process Control*, 8:459–468, Oct–Dec 1998.

[50] K.M. Bossley. *Neurofuzzy Modelling Approaches in System Identification.* PhD thesis, University of Southampton, 1997.

[51] R. Brachman and T. Anand. The process of knowledge discovery in databases. In *Advances in Knowledge Discovery and Data Mining*, pages 37–58. AAAI/MIT Press, 1994.

[52] V.L. Brailovsky. A probabilistic approach to clustering. *Pattern Recogn. Lett.*, 12(4):193–198, 1991.

[53] L. Breiman, J.H. Friedman, R.A. Olshen, and C.J. Stone. *Classification and Regression Trees.* Chapman and Hall (Wadsworth, Inc.), 1984.

[54] M. Leok B.T. Estimating the attractor dimension of the equatorial weather system. *Acta Physica Polonica A*, 85, 1994.

[55] Letellier C. and Aguirre L.A. Investigating nonlinear dynamics from time series: The influence of symmetries and the choice of observables. *Chaos*, 12(3):549–558, September 2002.

[56] J.A. Cadzow and O.M. Solomon. Algebraic approach to system identification. *IEEE Trans. on Acoust. Speach, Signal Processing*, 34(3):492–496, 1988.

[57] F. Camastra. Data dimensionality estimation methods: a survey. *Pattern Recognition*, 36:2945–2954, 2003.

[58] L. Cao. Practical method for determining the minimum embedding dimension of a scalar time series. *Physica D*, 110:43–50, 1997.

[59] G.A. Carpenter and S. Grossberg. Normal and amnesic learning, recognition, and memory by a neural model of cortico-hippocampal interactions. *Trends in Neuroscience*, 16(4):131–137, 1993.

[60] M. Casdagli. Nonlinear prediction of chaotic time series. *Physica D*, 35:335–356, 1989.

[61] K. Chakrabarti and S. Mehrotra. Local dimensionality reduction: A new approach to indexing high dimensional spaces. *Proceedings of the* 26*th VLDB Conference Cairo Egypt*, page P089, 2000.

[62] C. Chatterjee and V.P. Roychowdhury. On self-organizing algorithms and networks for class-separability features. *IEEE Transactions on Neural Networks*, 8(3):663–678, 1997.

[63] Y. Cheng. Mean shift, mode seeking, and clustering. *IEEE Trans. Pattern Anal. Mach. Intell.*, 17(7):790–799, 1995.

[64] S.L. Chiu. Selecting input variables for fuzzy models. *Journal of Intelligent and Fuzzy Systems*, 4:243–256, 1996.

[65] K.J. Cios, W. Pedrycz, and R.W. Swiniarski. *Data Mining Methods for Knowledge Discovery.* Kluwer Academic Press, Boston, 1998.

[66] A.L. Corcoran and S. Sen. Using real-valued genetic algorithms to evolve rule sets for classification. In *IEEE-CEC*, pages 120–124, Orlando, USA, 1994.

[67] D. Cubanski and D. Cyganski. Multivariable classification through adaptive Delaunay-based c^0 spline approximation. *IEEE Transactions on Pattern Analyzis and Machine Intelligence*, 17:403–417, 1995.

[68] Dick de Ridder and Robert P.W. Duin. Sammon's mapping using neural networks: A comparison. *Pattern Recognition Letters*, 18:1307–1316, 1997.

[69] A.P. Dempster, N.M. Laird, and D.B. Rubin. Maximum likelihood from incomplete data via the em algorithm. *J. Royal Stat. Soc. B.*, 39(1):1–38, 1977.

[70] E. Diday and J.C. Simon. *Clustering analysis.* Digital Pattern Recognition, 47–94. Springer-Verlag, Secaucus, NJ, 1976.

[71] A. Dobra and J. Gehrke. Secret: A scalable linear regression tree algorithm. In *Proceedings of the Eighth ACM SIGKDD International Conference on Knowledge Discovery and Data Mining*, July 2002. Edmonton, Alberta, Canada.

[72] D. Dong and T.J. McAvoy. Nonlinear principal component analysis – based on principal curves and neural networks. *Computers and Chemical Engineering*, 20(1):65–78, 1996.

[73] F.J. Doyle, B.A. Ogunnaike, and R.K. Pearson. Nonlinear model-based control using second-order volterra models. *Automatica*, 31:697, 1995.

[74] N.R. Draper and H. Smith. *Applied Regression Analysis, 3rd Edition.* John Wiley and Sons, Chichester, 1994.

[75] D. Driankov, H. Hellendoorn, and M. Reinfrank. *An Introduction to Fuzzy Control.* Springer-Verlag, Heidelberg, Germany, 1993.

[76] R.C. Dubes. How many clusters are best? – an experiment. *Pattern Recogn.*, 20(6):645–663, 1987.

[77] R.C. Dubes. Cluster analysis and related issues. In *Handbook of pattern recognition & computer vision*, pages 3–32, River Edge, NJ, USA, 1993. World Scientific Publishing Co., Inc.

[78] M.P. Dubuisson and A.K. Jain. A modified hausdorff distance for object matching. In *Proceedings of International Conference on Pattern Recognition* (*ICPR* 94), pages 566–568, 1994.

[79] J.C. Dunn. A fuzzy relative of the ISODATA process and its use in detecting compact well separated cluster. *Journal of Cybernetics*, 3:32–57, 1974.

[80] H.Al. Duwaish and N.M. Karim. A new method for identification of Hammerstein model. *Automatica*, 33(10):1871–1875, 1997.

[81] H.Al. Duwaish, N.M. Karim, and V. Chandrasekar. Use of multilayer feedforward neural networks in identification and control of Wiener model. *IEE Proceedings of Control Theory and Applications*, 143(3):255–258, 1996.

[82] Mendes EMAM and Billings S.A. An alternative solution to the model structure selection problem. *IEEE Transactions on Systems Man and Cybernetics part A-Systems and Humans*, 31(6):597–608, 2001.

[83] E. Eskinat, S.H. Johnson, and W. Luyben. Use of Hammerstein models in identification of nonlinear systems. *AIChE Journal*, 37(2):255–268, 1991.

[84] J.D. Farmer and J.J. Sidorowich. Predicting chaotic time series. *Physical Review Letters*, 59(8):845–848, 1987.

[85] B. Feil, J. Abonyi, and F. Szeifert. Model order selection of nonlinear input-output models – a clustering based approach. *Journal of Process Control*, 14(6):593–602, 2004.

[86] B.A. Foss, T.A. Johansen, and A.V. Sorensen. Nonlinear predictive control using local models – applied to a batch fermentation processes. *Control Engineering Practice*, 3:389–396, 1995.

[87] A.M. Fraser and H.L. Swinney. Independent coordinates for strange attractors from mutual information. *Physical Review A*, 33(2):1134–1140, 1986.

[88] J.H. Friedman. Multivariate adaptive regression splines. *The Annals of Statistics*, 19:1–141, 1991. with discussion.

[89] K.P. Fruzetti, A. Palazoglu, and K.A. McDonald. Nonlinear model predictive control using Hammerstein models. *Jounral of Process Control*, 7(1):31–41, 1997.

[90] K.S. Fu and S.Y. Lu. A clustering procedure for syntactic patterns. *IEEE Trans. Syst. Man Cybern.*, 7:734–742, 1977.

[91] K. Fukunaga and D.R. Olsen. An algorithm for finging intrinsic dimensionality of data. *IEEE Trans. Comput.*, 20(2):165–171, 1976.

[92] R. Gallion, D.C.St. Clair, C. Sabharwahl, and W.E. Bond. Dynamic id3: A symbolic learning algorithm for many-valued attribute domains. In *in Proc. 1993 Symp.Applied Computing.*, pages 14–20, New York, ACM Press, 1993.

[93] I. Gath and A.B. Geva. Unsupervised optimal fuzzy clustering. *IEEE Transactions on Pattern Analysis and Machine Intelligence*, 7:773–781, 1989.

[94] E.P. Gatzke and F. Doyle III. Multiple model approach for CSTR control. In *Proceedings of the 14th Triennieal IFAC World Congress*, pages N–7a–11–5, 1999.

[95] N. Gershenfeld, B. Schoner, and E. Metois. Cluster-weighted modelling for time-series analysis. *Nature*, 397:329–332, 1999.

[96] A.B. Geva. Hierarchical-fuzzy clustering of temporal-patterns and its application for time-series prediction. *Pattern Recognition Letters*, 20:1519–1532, 1999.

[97] D. Girimonte, R. Babuska, and J. Abonyi. Fuzzy clustering for selecting structure of nonlinear models with mixed discrete and continuous inputs. In *FUZZ-IEEE'04 Conference*, Budapest, Hungary, 2004.

[98] G.C. Gotlieb and S. Kumar. Semantic clustering of index terms. *J. ACM*, 15:493–513, 1968.

[99] K.C. Gowda and E. Diday. Symbolic clustering using a new dissimilarity measure. *IEEE Trans. Syst. Man Cybern.*, 22:368–378, 1992.

[100] K.C. Gowda and G. Krishna. Agglomerative clustering using the concept of mutual nearest neighborhood. *Pattern Recogn.*, 10:105–112, 1977.

[101] J.C. Gower and G.J.S. Ross. Minimum spanning trees and single-linkage cluster analysis. *Appl. Stat.*, 18:54–64, 1969.

[102] P. Grassberger and I. Procaccia. Measuring the strangeness of strange attractors. *Physica D*, 9:189–208, 1983.

[103] P. Grassberger and I. Procaccia. Measuring the strangeness of strange attractors. *Physica D*, 9:189–208, 1983.

[104] A. Grauel and H. Mackenberg. Mathematical analysis of the Sugeno controller leading to general design rules. *Fuzzy Sets and Systems*, 85:165–175, 1997.

[105] S. Grossberg. How does a brain build a cognitive code? *Psychological Review*, 87:1–51, 1980.

[106] S. Grossberg. The link between brain, learning, attention, and consciousness. *Consciousness and Cognition*, 8:1–44, 1999.

[107] J.K. Gugaliya, R.D. Gudi, and S. Lakshminarayanan. Multi-model decomposition of nonlinear dynamics using a fuzzy-cart approach. *Journal of Process Control*, 15:417–434, 2005.

[108] D.E. Gustafson and W.C. Kessel. Fuzzy clustering with fuzzy covariance matrix. In *Proceedings of the IEEE CDC, San Diego*, pages 761–766, 1979.

[109] I. Guyon and A. Elisseeff. An introduction to variable and feature selection. *Journal of Machine Learning Research*, 3:1157–1182, 2003.

[110] J. Han and M. Kamber. *Data Mining: Concepts and Techniques.* Morgan Kaufmann Publishers, 2001.

[111] M. Hanesch, R. Scholger, and M.J. Dekkers. The application of fuzzy c-means cluster analysis and non-linear mapping to a soil data set for the detection of polluted sites. *Phys. Chem. Earth*, 26:885–891, 2001.

[112] E.J. Hartman, J.D. Keeler, and J.M. Kowalski. Layered neural networks with gaussian hidden units as universal approximations. *Neural Computation*, 2:210–215, 1990.

[113] R.J. Hathaway and J.C. Bezdek. Switching regression models and fuzzy clustering. *IEEE Transactions on Fuzzy Systems*, 1:195–204, 1993.

[114] X. He and H. Asada. A new method for identifying orders of input-output models for nonlinear dynamic systems. *Proceedings of the ACC*, San Fransisco, CA:2520–2523, 1993.

[115] H. Hellendoorn and D. Driankov, editors. *Fuzzy Model Identification: Selected Approaches.* Springer, Berlin, Germany, 1997.

[116] E. Hernandez and Y. Arkun. Control of nonlinear systems using polynomial ARMA models. *AICHE Journal*, 39(3):446–460, 1993.

[117] J. Himberg, K. Korpiaho, H. Mannila, J. Tikanmaki, and H. T. Toivonen. Time-series segmentation for context recognition in mobile devices. *IEEE International Conference on Data Mining* (*ICDM* 01)*, San Jose, California*, pages 203–210, 2001.

[118] F. Hoffmann and O. Nelles. Genetic programming for model selection of TSK-fuzzy systems. *Information Sciences*, 136(1-4):7–28, August 2001.

[119] F. Hoppner, F. Klawonn, R. Kruse, and T. Runkler. *Fuzzy Cluster Analysis – Methods for Classification, Data Analysis and Image Recognition.* John Wiley and Sons, 1999.

[120] J. Horiuchi, M. Kishimoto, M. Kamasawa, and H. Miyakawa. Data base simulation of batch and fed-batch cultures for α-amylase production using a culture data base and a statistical procedure. *Journal of Fermentation and Bioengineering*, 76(4):326–332, 1993.

[121] K.J. Hunt, R. Haas, and R. Murray-Smith. Extending the functional equivalence of radial basis function networks and fuzzy inference systems. *IEEE Transactions on Neural Networks*, 7(3):776–781, 1996.

[122] K.J. Hunt, D. Sbarbaro, R. Zbikowki, and P.J. Gawthrop. Neural networks for control systems – A survey. *IEEE Transactions on Neural Networks*, 28:1083–1112, 1992.

[123] D.P. Huttenlocher, G.A. Klanderman, and W.J. Rucklidge. Comparing images using the hausdorff distance. *IEEE Trans. Pattern Anal. Mach. Intell.*, 15(9):850–863, 1993.

[124] M. Ichino and H. Yaguchi. Generalized minkowski metrics for mixed feature-type data analysis. *IEEE Trans. Syst. Man Cybern.*, 24:698–708, 1994.

[125] H. Ishibuchi, T. Nakashima, and T. Murata. Performance evaluation of fuzzy classifier systems for multidimensional pattern classification problems. *IEEE Transaction on Systems, Man, and Cybernetics: Part B*, 29:601–618, 1999.

[126] I. Ivanova and M. Kubat. Initialization of neural networks by means of decision trees. *Knowledge-Based Systems*, 8:333–344, 1995.

[127] R.A. Jacobs and M.I. Jordan. Learning piecewise control strategies in a modular neural network architecture. *IEEE Transactions on Systems, Man, and Cybernetics*, 23(2):337–345, 1993.

[128] A.K. Jain and R.C. Dubes. *Algorithms for Clustering Data.* Prentice-Hall advanced reference series. Prentice-Hall, Inc., 1988.

[129] J.-S. R. Jang and C.-T. Sun. Functional Equivalence Between Radial Basis Function Networks and Fuzzy Inference Systems. *IEEE Transactions on Neural Networks*, 4(1):156–159, Jan. 1993.

[130] J.-S.R. Jang. Input selection for ANFIS learning. In *Proceedings of the IEEE International Conference on Fuzzy Systems*, volume 2, pages 1493–1499, New York, USA, 1996.

[131] J.-S.R. Jang, C.-T. Sun, and E. Mizutani. *Neuro–Fuzzy and Soft Computing; a Computational Approach to Learning and Machine Intelligence.* Prentice–Hall, Upper Sadle River, 1997.

[132] J.S.R. Jang and C.T. Sun. Neuro-fuzzy modelling and control. *Proceedings of the IEEE*, 83:378–406, 1995.

[133] C.Z. Janikow. Fuzzy decision trees: Issues and methods. *IEEE Trans. SMC-B*, 28:1–14, 1998.

[134] R.A. Jarvis and E.A. Patrick. Clustering using a similarity method based on shared near neighbors. *IEEE Trans. Comput. C*, 22(8):1025–1034, 1973.

[135] X. Jiang and H. Adeli. Fuzzy clustering approach for accurate embedding dimension identification in chaotic time series. *Integrated Computer-Aided Engineering*, 10:287–302, 2003.

[136] Y. Jin. Fuzzy modelling of high–dimensional systems. *IEEE Transactions on Fuzzy Systems*, 8:212–221, 2000.

[137] T.A. Johansen. *Operating regime based process modelling and identification.* PhD thesis, Department of Engineering Cybernetics, Norwegian Institute of Technolgy, University of Trondheim, Norway, 1994.

[138] T.A. Johansen. Identification of non–linear systems using empirical data and a priori knowledge – an optimisation approach. *Automatica*, 32:337–356, 1996.

[139] T.A. Johansen and R. Babuska. On multi-objective identification of takagi-sugeno fuzzy model parameters. In *Preprints* 15*th IFAC Word Congress*, Barcelona, Spain, 2002.

[140] T.A. Johansen, R. Shorten, and R. Murray-Smith. On the interpretation and identification of Takagi–Sugeno fuzzy models. *IEEE Transactions on Fuzzy Systems*, 8:297–313, 2000.

[141] J.H. Ward Jr. Hierarchical grouping to optimize an objective function. *J. Am. Stat. Assoc.*, 58:236–244, 1963.

[142] Norman F. Jr. Hunter. Nonlinear prediction of speach signals. *In M. Casdagli and S. Eubanks, editors, Nonlinear Modelling and Forecasting, Addison-Wesley*, pages 467–492, 1992.

[143] J.Zhang, E.B. Martin, and A.J. Morris. Process monitoring using non-linear statistical techniques. *Chemical Engineering Journal*, 67:181–189, 1997.

[144] N. Kambhatala. *Local Models and Gaussian Mixture Models for Statistical Data Processing.* Ph.D. Thesis, Oregon Gradual Institute of Science and Technology, 1996.

[145] T. Kavli. ASMOD – an algorithm for adaptive spline modelling of observation data. *International Journal of Control*, 58(4):947–967, 1993.

[146] U. Kaymak and R. Babuska. Compatible cluster merging for fuzzy modelling. In *Proceedings of the IEEE International Conference on Fuzzy Systems*, pages 897–904. Yokohama, Japan, 1995.

[147] P.M. Kelly. An algorithm for merging hyperellipsoidal clusters. *Technical Report LA-UR*-94-3306*, Los Alamos National Laboratory, Los Alamos, NM*, 1994.

[148] M.B. Kennen, R. Brown, and H.D.I. Abarbanel. Determining embedding dimension for phase-space reconstruction using a geometrical construction. *Physical Review*, A:3403–3411, 1992.

[149] E. Keogh, S. Chu, D. Hart, and M. Pazzani. An online algorithm for segmenting time series. *IEEE International Conference on Data Mining*, page http://citeseer.nj.nec.com/keogh01online.html, 2001.

[150] E. Keogh and M. Pazzani. An enhanced representation of time series which allows fast and accurate classification, clustering and relevance feedback. 4*th Int. Conf. on KDD.*, pages 239–243, 1998.

[151] E. Kim, S. Kim, and M. Park. A transformed input-domain approach to fuzzy modelling. *IEEE Transactions on Fuzzy Systems*, 6:596–604, 1998.

[152] L. Kindermann and A. Lewandowski. Natural interpolation of time series. BSIS Technical Reports 03-3, RIKEN Brain Science Institute, Lab for Mathematical Neuroscience, Japan ÖFAI, 2004.

[153] B. King. Step-wise clustering procedures. *J. Am. Stat. Assoc.*, 69:86–101, 1967.

[154] S. Kivikunnas. Overview of process trend analysis methods and applications. *ERUDIT Workshop on Applications in Pulp and Paper Industry*, page CD ROM, 1998.

[155] D. Knuth. *The Art of Computer Programming.* Addison-Wesley, Reading, MA, 1973.

[156] T. Kohonen. *Self-organization and associative memory.* Springer, Berlin, 2nd edition, 1984.

[157] T. Kohonen. The self-organizing map. *Proceedings of the IEEE*, 78(9):1464–1480, 1990.

[158] T. Kohonen. The self-organizing map. *Neurocomputing*, 21:1–6, 1998.

[159] I. Konenko, I. Bratko, and E. Roskar. *Experiments in automatic learning of medical diagnostic rules.* Tech. Rep., J. Stefan Inst. Yugoslavia, 1994.

[160] A. Kovacs and J. Abonyi. Vizualization of fuzzy clustering results by modified Sammon mapping. In *Proceedings of the 3rd International Symposium of Hungarian Researchers on Computational Intelligence*, pages 177–188, 2004.

[161] M.A. Kramer. Nonlinear principal component analysis using autoassociative neural networks. *Neural Computation*, 9(7):1493–1516, 1991.

[162] J.B. Kruskal. On the shortest spanning subtree of a graph and the traveling salesman problem. *American Mathematical Society*, 7:48–50, 1956.

[163] J.B. Kruskal. Multidimensional scaling by optimizing goodness of fit to a nonmetric hypothesis. *Psychometrika*, 29:1–27, 1964.

[164] J.B. Kruskal. Nonmetric multidimensional scaling: a numerical method. *Psychometrika*, 29:115–130, 1964.

[165] J.B Kruskal and M. Wish. Multidimensional scaling. *Sage University Papers on Quantitative Applications in the Social Sciences*, 07(011), 1978. Newbury Park, CA.

[166] W.J. Krzanowsky. Between group comparison of principal components. *J. Amer. Stat. Assoc.*, pages 703–707, 1979.

[167] M. Kubat. Decision trees can initialize radial-basis-function networks. *IEEE Trans. NN*, 9:813–821, 1998.

[168] Aguirre L.A. and Mendes E.M.A.M. Global nonlinear polynomial models: Structure, term clusters and fixed points. *International Journal of Bifurcation and Chaos*, 6(2):279–294, Februar 1996.

[169] Aguirre L.A. and Billings S.A. Improved structure selection for nonlinear models based on term clustering. *International Journal of Control*, 62(3):569–587, September 1995.

[170] M. Last, Y. Klein, and A. Kandel. Knowledge discovery in time series databases. *IEEE Transactions on Systems, Man, and Cybernetics*, 31(1):160–169, 2000.

[171] M. Lebowitz. Categorizing numeric information for generalization. *Cognitive Science*, 9:285–308, 1985.

[172] I.J. Leontaritis and S.A. Billings. Experimental design and identifiably for nonlinear systems. *International Journal of Systems Science*, 18:189–202, 1987.

[173] G. Liang, D.M. Wilkes, and J.A. Cadzow. Arma model order estimation based on the eigenvalues of the covariance matrix. *IEEE Trans. on Signal Processing*, 41(10):3003–3009, 1993.

[174] T.W. Liao. Clustering of time series data – a survey. *Pattern Recognition*, 2005. In Press.

[175] G. Lightbody, P. O'Reilly, K. Kelly, and J. McCormick. Neural modelling of chemical plant using MLP and B-spline networks. *Control Engineering Practice*, 5(11):150–1515, 1997.

[176] B. Lillekjendlie, D. Kugiumtzis, and N. Christophersen. Chaotic time series – part II: System identification and prediction. *Modelling, Identification and Control*, 15(4):225–243, 1994.

[177] P. Lindskog and L. Ljung. Tools for semi-physical modelling. In *Proceedings IFAC SYSID*, volume 3, pages 237–242, Kopenhagen, Danmark, 1994.

[178] D.A. Linkens and M.-Y. Chen. Input selection and partition validation for fuzzy modelling using neural network. *Fuzzy Sets and Systems*, pages 299–308, 1999.

[179] L. Ljung. *System Identification, Theory for the User.* Prentice–Hall, New Jersey, 1987.

[180] W.-Y. Loh. Regression trees with unbiased variable selection and interaction detection. *Statistica Sinica*, 12:361–386, 2002.

[181] S.Y. Lu and K.S. Fu. A sentence-tosentence clustering procedure for pattern analysis. *IEEE Trans. Syst. Man Cybern.*, 8:381–389, 1978.

[182] W. Luo, M.N. Karim, A.J. Morris, and E.B. Martin. A control relevant identification of a pH waste water neutralization process using adaptive radial basis function networks. *Computers and Chemical Engineering*, 20/S:1017–1022, 1996.

[183] Korenberg M., Billings S.A., Liu Y.P., and McIlroy P.J. Orthogonal parameter-estimation algorithm for nonlinear stochastic-systems. *International Journal of Control*, 48(1):193–210, 1988.

[184] E.H. Mamdani. Advances in the linguistic synthesis of fuzzy controllers. *International Journal of Man-Machine Studies*, 8:669–678, 1976.

[185] E.H. Mamdani, T. Teraqno, K. Asai, and M. Sugeno. Fuzzy-systems theory and its applications. *Nature*, 359:788–788, 1992.

[186] J. Mao and A.K. Jain. A self-organizing network for hyperellipsoidal clustering (hec). *IEEE Trans. Neural Netw.*, 7:16–29, 1996.

[187] J. Mao and K. Jain. Artificial neural networks for feature extraction and multivariate data projection. *IEEE Trans. on Neural Networks*, 6(2):296–317, 1995.

[188] P. Marcelino, P. Nunes, P. Lima, and M.I. Ribeiro. Improving object localization through sensor fusion applied to soccer robots. *Actas do Encontro Cientifico do Robotica*, 2003.

[189] T. Martinetz and K. Schulter. Topology representing networks. *Neural Networks*, 3:507–522, 1994.

[190] J. McQueen. Some methods for classification and analysis of multivariate observations. In *Proceedings of Fifth Berkeley Symposium on Mathematical Statistics and Probability*, pages 281–297, 1967.

[191] C. Merkwirth, U. Parlitz, I. Wedekind, and W. Lauterborn. *TSTOOL User Manual, Version* 1.11. http://www.physik3.gwdg.de/ tstool/indexde.html, April 12. 2002.

[192] R. Michalski, R.E. Stepp, and E. Diday. Automated construction of classifications: conceptual clustering versus numerical taxonomy. *IEEE Trans. Pattern Anal. Mach. Intell. PAMI*-5, 5:396–409, 1983.

[193] S. Migaly, J. Abonyi, and F. Szeifert. Fuzzy self-organizing map based on regularized fuzzy c-means clustering. *Advances in Soft Computing, Engineering Design and Manufacturing, Springer Engineering Series*, pages 99–108, 2002.

[194] T. Mitchell. *Machine Learning.* McGraw-Hill, Inc., New York, NY, 1997.

[195] J. Moody and C.J. Darken. Fast Learning in Networks of Locally-Tuned Processing Units. *Neural Computation*, 1:281–294, 1989.

[196] R. Murray-Smith and T.A. Johansen. *Multiple Model Appraoches to Modelling and Control.* Taylor and Francis, London, 1997.

[197] R. Murray-Smith and T.A. Johansen, editors. *Multiple Model Approaches to Nonlinear Modelling and Control.* Taylor & Francis, London, UK, 1997.

[198] F. Murtagh. A survey of recent advances in hierarchical clustering algorithms which use cluster centers. *Comput. J.*, 26:354–359, 1984.

[199] C.A. Pe na Reyes and M. Sipper. A fuzzy genetic approach to breast cancer diagnosis. *Artificial Intelligence in Medicine*, 17:131–155, 2000.

[200] G. Nagy. State of the art in pattern recognition. *Proc. IEEE*, 56:836–862, 1968.

[201] K.S. Narendra and K. Parthasarathy. Identification and control of dynamical systems. *IEEE Transactions on Neural Networks*, 1:4–27, 1990.

[202] D. Nauck and R. Kruse. Obtaining interpretable fuzzy classification rules from medical data. *Artificial Intelligence in Medicine*, 16:149–169, 1999.

[203] A. Negiz and A. Cinar. Monitoring of multivariable dynamic processes and sensor auditing. *Journal of Process Control*, 8(5):375–380, 1998.

[204] O. Nelles. *Nonlinear System Identification.* Springer, Berlin, Germany, 2001.

[205] S.J. Norquay, A. Palazoglu, and J.A. Romagnoli. Application of Wiener model predictive control (WMPC) to a pH neutralization experiment. *IEEE Transactions on Control Systems Technology*, 7:437–445, 1999.

[206] S.J. Norquay, A. Palazoglu, and J.A. Romagnoli. Application of Wiener model predictive control (WMPC) to an industrial c2–splitter. *Journal of Process Control*, 9:461–473, 1999.

[207] K. Ozawa. A stratificational overlapping cluster scheme. *Pattern Recogn.*, 18:279–286, 1985.

[208] P.F. Pach, J. Abonyi, S. Nemeth, and P. Arva. Supervised clustering and fuzzy decision tree induction for the identification of compact classifiers. In *5th International Symposium of Hungarian Researchers on Computational Intelligence*, Budapest, Hungary, 2004.

[209] N.R. Pal, J.C. Bezdek, and E.C.K. Tsao. Generalized clustering networks and Kohonen's self-organization scheme. *IEEE Transactions on Neural Networks*, 4(4):549–557, 1993.

[210] N.R. Pal and V.K. Eluri. Two efficient connectionist schemes for structure preserving dimensionality reduction. *IEEE Transactions on Neural Networks*, 9:1143–1153, 1998.

[211] R.D. Pascal-Marqui, A.D. Pascual Montano, K. Kochi, and J.M. Carazo. Smoothly distributed fuzzy c-means: a new self organizing map. *Pattern Recognition*, 34:2395–2402, 2001.

[212] K.M. Passino and S. Yurkovic. *Fuzzy Control.* Addison-Wesley, New York, USA, 1998.

[213] R.K. Pearson. Selecting nonlinear model structures for computer control. *Journal of Process Control*, 13(1):1–26, 2003.

[214] R.K. Pearson and B.A. Ogunnaike. Nonlinear process identification. In M.A. Henson and D.E. Seborg, editors, *Nonlinear Process Control*, pages 11–109. Prentice–Hall, Englewood Cliffs, NJ, 1997.

[215] W. Pedrycz and A. Zenon Sosnowskic. The design of decision trees in the framework of granular data and their application to software quality models. *Fuzzy Sets and Systems*, 123:271–290, 2001.

[216] M.H. Petrick and B.Wigdorowitz. A priori nonlinear model structure selection for system identification. *Control Engineering Practise*, 5(8):1053–1062, 1997.

[217] K. Pettis, T. Bailey, T. Jain, and R. Dubes. An intrinsic dimensionality estimator from near-neighbor information. *IEEE Trans. Pattern Anal. Mach. Intell.*, 1(1):25–37, 1979.

[218] M. Pottman and R.K. Pearson. Block-oriented narmax models with output multiplicities. *AIChE Journal*, 44:131–140, 1998.

[219] M. Pottman, H. Unbehauen, and D.E. Seborg. Application of a general muti–model approach for identification of highly nonlinear systems. *International Journal of Control*, 57:97–120, 1993.

[220] M.J.D. Powell. Radial basis functions for multivariable interpolation – A review. In *Algorithms for Approximation*, pages 143–167. Clarendon Press, Oxford, 1987.

[221] R. Prim. Shortest connection networks and some generalizations. *Bell System Technical Journal*, 36:1389–1401, 1957.

[222] D.C. Psichogios and L.H. Ungar. A hybrid neural network – first principles approach to process modelling. *AIChE J.*, 38:1499–1511, 1992.

[223] S.J. Qin and T.J. McAvoy. Nonlinear PLS modelling using neural networks. *Computers and Chemical Engineering*, 12(4):379–391, 1992.

[224] J.R. Quinlan. Induction of decision trees. *Machine Learning*, 1(1):81–106, 1986.

[225] J.R. Quinlan. Improved use of continuous attributes in c4.5. *Journal of Artificial Intelligence Research*, 4:77–90, 1996.

[226] V.V. Raghavan and C.T. Yu. A comparison of the stability characteristics of some graph theoretic clustering methods. *IEEE Trans. Pattern Anal. Mach. Intell.*, 3:393–402, 1981.

[227] A.F.R. Rahman and M.C. Fairhurst. Multi-prototype classification: improved modelling of the variability of handwritten data using statistical clustering algorithms. *Electron. Lett.*, 33(14):1208–1209, 1997.

[228] H. Rainer. Secret: A scalable linear regression tree algorithm. In *Proc. of the annual conf. of the North America Fuzzy Information Processing*, 444–449 1997.

[229] C. Rhodes and M. Morari. Determining the model order of nonlinear input/output systems. *AIChE Journal*, 44:151–163, 1998.

[230] J.A. Roubos and M. Setnes. Compact fuzzy models through complexity reduction and evolutionary optimization. In *Proc. of IEEE international conference on fuzzy systems*, pages 762–767, San Antonio, USA, 2000.

[231] J.A. Roubos, M. Setnes, and J. Abonyi. Learning fuzzy classification rules from data. In R. John and R. Birkenhead, editors, *Developments in Soft Computing*. Springer, Physica Verlag, 2001.

[232] D. Saez and A. Cipriano. Fuzzy modelling for a combined cycle power plant. In *Proceedings IEEE international Fuzzy Systems Conference*, volume 2, pages 1186–1190, Seoul, Korea, 1999.

[233] G. Saltong. Developments in automatic text retrieval. *Science*, 253:974–980, 1991.

[234] T. Sauer, J.A. Yorke, and M. Casdagli. Embedology. *Journal of Statistical Physics*, 65(3/4):579–615, 1991.

[235] J. Schubert, R. Simutis, M. Dors, and I. Havlik ab A. Lubbert. Bioprocess optimization and control: Application of hybrid modelling. *Journal of Biotechnology*, 35:51–68, 1994.

[236] L.K. Sethi. Entropy nets: From decision trees to neural networks. *Proc. IEEE*, 78:1605–1613, 1990.

[237] R. Setiono. Generating concise and accurate classification rules for breast cancer diagnosis. *Artificial Intelligence in Medicine*, 18:205–219, 2000.

[238] R. Setiono and W.K. Leow. On mapping decision trees and neural networks. *Knowledge Based Systems*, 13:95–99, 1999.

[239] M. Setnes and R. Babuška. Fuzzy relational classifier trained by fuzzy clustering. *IEEE Trans. SMC–B*, 29:619–625, 1999.

[240] M. Setnes, R. Babuška, U. Kaymak, and H.R. van Nauta Lemke. Similarity measures in fuzzy rule base simplification. *IEEE Transactions on Systems, Man and Cybernetics – Part B: Cybernetics*, 28(3):376–386, 1998.

[241] M. Setnes, R. Babuška, and H.B. Verbruggen. Rule-based modelling: Precision and transparency. *IEEE Trans. SMC-C*, 28:165–169, 1998.

[242] M. Setnes, V. Lacrose, and A. Titli. Complexity reduction methods for fuzzy systems design. In H.B. Verbruggen and R. Babuška, editors, *Fuzzy Logic Control: Advances in Applications*, pages 185–218. World Scientific, Singapore, 1999.

[243] M. Setnes and J.A. Roubos. Transparent fuzzy modelling using fuzzy clustering and GA's. In *In NAFIPS*, pages 198–202, New York, USA, 1999.

[244] C.E. Shannon. A mathematical theory of communication. *Bell System Technical Journal*, 27:379–432, 1948.

[245] R. Simutis and A. Lübbert. Exploratory analysis of bioprocesses using artificial neural network-based methods. *Biotechnology Progress*, 13:479–487, 1997.

[246] A. Singhal and D.E. Seborg. Matching patterns from historical data using PCA and distance similarity factors. *Proceedings of the American Control Conference*, pages 1759–1764, 2001.

[247] J. Sjöberg, Q. Zhang, L. Ljung, A. Benveniste, B. Deylon, P.-Y. Glorennec, H. Hjalmarsson, and A. Juditsky. Nonlinear black-box modelling in system identification: a unified overview. *Automatica*, 31:1691–1724, Dec 1995.

[248] A. Skeppstedt, L. Ljung, and M. Millnert. Construction of composite models from observed data. *International Journal of Control*, 55:141–152, 1992.

[249] S. Skogestad. Dynamics and control of distilation columns. *Chem. Eng. Res. Des.* (*Trans IChemE*), 75:539–562, 1997.

[250] P.H.A. Sneath and R.R. Sokal. *Numerical Taxonomy.* Freeman, London, UK, 1973.

[251] R. Srinivasan, C. Wang, W.K. Ho, and K.W. Lim. Dynamic principal component analysis based methodology for clustering process states in agile chemical plants. *Ind. Eng. Chem. Res.*, 43:2123–2139, 2004.

[252] G. Stephanopoulos and C. Han. Intelligent systems in process engineering: A review. *Comput. Chem. Engng.*, 20:743–791, 1996.

[253] H.T. Su and T.J. McAvoy. Integration of multilayer percepton networks and linear dynamic models: A Hammerstein modelling approach. *Ind. Eng. Chem. Res.*, 32:1927–1936, 1993.

[254] M. Sugeno and G.T. Kang. Fuzzy modelling and control of multilayer incinerator. *Fuzzy Sets and Systems*, 18:329–346, 1986.

[255] M. Sugeno and T. Yasukawa. A fuzzy-logic-based approach to qualitative modelling. *IEEE Transactions on Fuzzy Systems*, 1(1):7–31, 1993.

[256] M.J. Symon. Clustering criterion and multi-variate normal mixture. *Biometrics*, 77:35–43, 1977.

[257] T. Takagi and M. Sugeno. Fuzzy identification of systems and its application to modelling and control. *IEEE Transactions on Systems, Man and Cybernetics*, 15(1):116–132, 1985.

[258] F. Takens. Detecting strange attractor in turbulence. *D.A. Rand, L.S. Young (Eds.), Dynamical Systems and Turbulence, Springer, Berlin*, pages 366–381, 1981.

[259] E. Tanaka. Theoretical aspects of syntactic pattern recognition. *Pattern Recogn.*, 28:1053–1061, 1995.

[260] H.A.B. te Braake. *Neural Control of Biotechnological Processes*. PhD thesis, Delft University of Technology, Department of Electrical Engineering, Delft, The Netherlands, 1997.

[261] H.A.B. te Braake, M.A Botto, H.J.L. van Can, J.S. Costa, and H.B. Verbruggen. Linear predictive control based on approximate input-output feedback linearization. *IEE Proceedings – Control Theory and Applications*, 146(4):295–300, 1999.

[262] A. Tholudur and W.F. Ramirez. Optimization of fed-batch bioreactors using neural network parameter function models. *Biotechnology Progress*, 12:302–309, 1996.

[263] M.L. Thompson and M.A. Kramer. Modelling chemical processes using prior knowledge and neural networks. *AIChE Journal*, 40:1328–1340, 1994.

[264] W.D. Timmons, H.J. Chizeck, and P.G. Katona. Parameter-constrained adaptive control. *Ind. Eng. Chem. Res.*, 36:4894–4905, 1997.

[265] M.E. Tipping and C.M. Bishop. Mixtures of probabilistic principal component analyzers. *Neural Computation*, 11(2):443–482, 1999.

[266] G.T. Toussaint. The relative neighborhood graph of a finite planar set. *Pattern Recogn.*, 12:261–268, 1980.

[267] E.C.K. Tsao, J.C. Bezdek, and N.R. Pal. Fuzzy kohonen clustering networks. *Pattern Recognition*, 27(5):757–764, 1994.

[268] H.J.A.F Tulleken. Gray-box modelling and identification using physical knowledge and Bayesian techniques. *Automatica*, 29:285–308, 1993.

[269] T. Ullrich, K. Hohm, and H. Tolle. On the integration of expert knowledge in interpolating controllers. In *Proceedings of International ICSC/IFAC Symposium on Neural Computation (NC* 98), Vienna, Austria, 1998.

[270] T. Ullrich and H. Tolle. Delaunay–based local model networks for nonlinear system identification. In *Proceedings of IASTED International Conference Applied Modelling and Simulation*, Banff, Canada, 1997.

[271] H.J.L. van Can, H.A.B. te Braake, C. Hellinga, K.Ch.A.M. Luyben, and J.J. Heijnen. Strategy for dynamic process modelling based on neural networks and macroscopic balances. *AIChE Journal*, 42:3403–3418, 1996.

[272] K. Vasko and H.T.T. Toivonen. Estimating the number of segments in time series data using permutation tests. *IEEE International Conference on Data Mining*, pages 466–473, 2002.

[273] A. Vathy-Fogarassy, B. Feil, and J. Abonyi. Minimal spanning tree based fuzzy clustering. In Cemal Ardil, editor, *Transactions on Enformatika, Systems Sciences and Engineering*, volume 8, pages 7–12, 2005.

[274] J. Vesanto. Neural network tool for data mining: Som toolbox. *Proceedings of Symposium on Tool Environments and Development Methods for Intelligent Systems* (*TOOLMET*2000), pages 184–196, 2000.

[275] P. Vuorimaa, T. Jukarainen, and E. Karpanoja. A neuro-fuzzy system for chemical agent detection. *IEEE Transactions On Fuzzy Systems*, 4:403–414, 1995.

[276] B. Wahlberg. System–identification using laguerre models. *IEEE Transactions on Automatic Control*, 36(5):551–562, 1991.

[277] B. Wahlberg. System-identification using kautz models. *IEEE Transactions on Automatic Control*, 39(6):1276–1282, 1994.

[278] D.M. Walker and N.B. Tufillaro. Phase space reconstruction using input-output time series data. *HP Labs Technical Reports HPL*-1999-24, 990223, 1999.

[279] L.X. Wang and J.M. Mendel. Fuzzy Basis Functions, Universal Approximators, and Orthogonal Least-Squares Learning. *IEEE Trans Neural Networks*, 3(5):807–814, Sept. 1992.

[280] L.X. Wang. *A course in Fuzzy Systems and Control.* Prentice Hall, New York, USA, 1997.

[281] X.Z. Wang. *Data Mining and Knowledge Discovery for Process Monitoring and Control.* Springer, 1999.

[282] S. Watanabe. *Pattern Recognition: Human and Mechanical.* John Wiley and Sons, Inc., New York, NY, 1985.

[283] D.R. Wilson and T.R. Martinez. Improved heterogeneous distance functions. *J. Artif. Intell. Res.*, 6:1–34, 1997.

[284] J.C. Wong, K. McDonald, and A. Palazoglu. Classification of process trends based on fuzzified symbolic representation and hidden markov models. *Journal of Process Control*, 8:395–408, 1998.

[285] X.L. Xie and G.A. Beni. Validity measure for fuzzy clustering. *IEEE Trans. PAMI*, 3(8):841–846, 1991.

[286] G. Xinbo and X. Weixin. Advances in theory and applications of fuzzy clustering. *Chinese Science Bulletin*, 45(11):961–970, 2000.

[287] L. Xu, A. Krzyzak, and E. Oja. Rival penalized competitive learning for clustering analysis, RBF net and curve detection. *IEEE Transactions on Neural Networks*, 4(4):636–649, 1993.

[288] R.R. Yager and D.P. Filev. *Essentials of Fuzzy Modelling and Control.* John Wiley, New York, 1994.

[289] Y. Yam. Fuzzy approximation via grid point sampling and singular value decomposition. *IEEE Transactions on Systems, Man, and Cybernetics – Part B.*, 27(6):933–951, 1997.

[290] Y. Yamashita. Supervised learning for the analysis of the process operational data. *Computers and Chemical Engineering*, 24:471–474, 2000.

[291] H. Yan. Fuzzy curve-tracing algorithm. *IEEE Transactions on Systems, Man, and Cybernetics, Part B*, 5:768–773, 2001.

[292] W. Yao. *Improving Security of Communication via Chaotic Synchronization.* Phd thesis, the University of Western Ontario, 2002.

[293] J. Yen and L. Wang. Application of statistical information criteria for optimal fuzzy model construction. *IEEE Transactions on Fuzzy Systems*, 6:362–372, 1998.

[294] J. Yen and L. Wang. Simplifying fuzzy rule-based models using orthogonal transformation methods. *IEEE Transaction on Systems, Man, and Cybernetics: Part B*, 29:13–24, 1999.

[295] J. Yen, L. Wang, and C.W. Gillespie. Improving the interpretability of TSK fuzzy models by combining global learning and local learning. *IEEE Transactions on Fuzzy Systems*, 6(4):531–537, 1998.

[296] I.S. Yenyukov. *Data analysis learning symbolic and numeric knowledge*, chapter Indices for projection pursuit. Nova Science Publishers, New York, 1989.

[297] S.Y. Yi and M.J. Chung. Identification of fuzzy relational model and its application to control. *Fuzzy Sets and Systems*, 59(1):25–33, 1993.

[298] L.A. Zadeh. Fuzzy Sets. *Information and Control*, 8:338–353, 1965.

[299] C.T. Zahn. Graph-theoretical methods for detecting and describing gestalt clusters. *IEEE Trans. Comput. C*, 20:68–86, 1971.

[300] D. Zhang, M. Kamel, and M.T. Elmasry. Fuzzy clustering neural network (FCNN): competitive learning and parallel architecture. *Journal of Intelligent and Fuzzy Systems*, 2(4):289–298, 1994.

[301] K. Zhang. Algorithms for the constrained editing distance between ordered labeled trees and related problems. *Pattern Recogn.*, 28:463–474, 1995.

[302] Y. Zhu. Parametric Wiener model identification for control. In *Proceedings IFAC Word Congress*, pages H–3a–02–1, Bejing, China, July 1999.

Index

19407.1R.9 12/8/09

IdNo - BoxNo - Qty
3545 - 3063 - 32
3546 - 3064 - 4

Total - 36